A Theory of Language and Information

A Theory of Language and Information

A Mathematical Approach

ZELLIG HARRIS

CLARENDON PRESS · OXFORD
1991

Oxford University Press, Walton Street, Oxford OX2 6DP
Oxford New York Toronto
Delhi Bombay Calcutta Madras Karachi
Petaling Jaya Singapore Hong Kong Tokyo
Nairobi Dar es Salaam Cape Town
Melbourne Auckland
and associated companies in
Berlin Ibadan

Oxford is a trade mark of Oxford University Press

Published in the United States
by Oxford University Press, New York

British Library Cataloguing in Publication Data
Harris, Zellig S. (Zellig Sabbettai) 1909–
A theory of language and information: a mathematical
approach.
1. Mathematical linguistics
I. Title
410
ISBN 0–19–824224–7

Library of Congress Cataloging in Publication Data
Harris, Zellig Sabbettai, 1909–
A theory of language and information: a mathematical approach/
Zellig Harris.
Includes index.
1. Linguistics.   2. Language and languages.   3. Mathematics.
4. Information theory.
P121.H354   1990   415'.01'51—dc20   90–39472
ISBN 0–19–824224–7

Set by Hope Services (Abingdon) Ltd.
Printed in Great Britain by
Biddles Ltd
Guildford & King's Lynn

Preface

This book presents several interrelated matters: a formal theory of syntax; an example of how something in the real world—the basic structure of talk and writing—can be represented as a mathematical object; the intrinsic relation between language and information, and the structure of information of individual sciences; and, lastly, indications of how language is a self-organizing and evolving system.

The theory of syntax is stated in terms related to mathematical Information Theory: as constraints on word combination, each later constraint being defined on the resultants of a prior one. This structure not only permits a finitary description of the unbounded set of sentences, but also admits comparison of language with other notational systems, such as mathematics on the one hand and music on the other.

Since frameworks arise not arbitrarily but from how one considers the data, this book assesses first what methods are relevant to the data of language (ch. 4), and then what kind of theory can be reached from those methods. For reasons given in Chapter 4, one cannot rely on episodic examples; it is necessary to survey the full range of data in a language, in a way that establishes their regularities. Hence the analysis presented here draws upon a companion volume (the writer's *Grammar of English on Mathematical Principles*, Wiley-Interscience, New York, 1982), which covers in some detail the data of one language in terms of the present theory; references to this grammar (indicated by G) are given throughout the present book.

The analysis in this book thus arises from the results obtained by particular methods of description and decomposition of the word combinations found in a language. The intent is not only to understand language and information, but also to see what the methods used here can show about how structure comes to be.

One explanatory note may be needed here. In characterizing ('explaining') the structure of word combinations, a method is used whereby one combination is derived from another (its source). In some cases the derivation for a given form involves reconstructing a source which is not actually said in the language. These sources, however, are never abstract models. They are word combinations made in accordance with the actual grammar of the language, and from which the given form would be derived by the existing

derivational relations of the language. That they are not actually said may be
due to their complexity or length; in many cases it would be impossible to
formulate a regular grammar of the language that would exclude them as
ungrammatical. Therefore the unsaid sources can be counted upon as
derivational sources for the given word combinations.

Since some of the chapters may be read independently of the others, data
relevant to several different issues have been repeated in the respective
different chapters. A reader who seeks only the more general formulations
can safely omit various sections, chiefly 1.1.1, 3.2.2–3.2.5, 3.3.2.1–3, 3.4.2–
3.4.3, 5.3.2–5.3.4, and Chapters 7 and 8.

It may be in order to state what language-acquaintance underlies the
results presented here. The statements in this book have been based on
detailed grammatical formulations of several European and Semitic languages,
and on outlinings of the structures in a variety of American Indian languages
(especially Algonquian, Athapaskan, Siouan, Iroquoian) and Eskimo, and
also in some African languages (especially Swahili) and some Dravidian.
The statements have been checked against existing grammatical descriptions of
a sampling of languages: Basque; languages of ancient Asia Minor (where I
had the advantage of information from Anna Morpurgo Davies); Sumerian;
some languages of the following groups: Australian, Austronesian, Papuan
in New Guinea, Micronesian; and finally, Chinese and Korean. The variety
of structures in the languages surveyed is of course vast; but it was possible to
see that the dependence relation of operator on argument, on which rests
the syntactic structure presented here, is a possible underlying description
for them. There was no possibility of determining whether some of the other
constraints, defined here upon the basic dependence, hold for all the
languages.

The work here starts off from the distributional (combinatorial) methods
of Edward Sapir and of Leonard Bloomfield, to both of whom I am glad to
restate my scientific and personal debt. I am also indebted as always to
many enlightening and felicitious conversations with Henry Hiż, Henry
Hoenigswald, M. P. Schutzenberger, and Maurice Gross, and to very
valuable critiques and suggestions by André Lentin and Danuta Hiż.

Contents

II. THEORY OF SYNTAX

III. GRAMMATICAL ANALYSIS

IV. SUBSETS OF SENTENCES

V. INTERPRETATION

List of Figures

All figures are from Z. Harris, *Mathematical Structures of Language*, Wiley-Interscience, New York, 1968.

I

Introduction

1

Overview

1.0. *Introduction*

This book presents in the first place a formal theory of syntax (chs. 3, 4), in which a single relation on word-sequence gives to the set of these sequences the structure of a mathematical object, and produces a base set of sentences which carry all the information that is carried in the language. Within this set of sentences there acts a set of reductions which change the shape of words but not the information in the sentences, thereby producing the remaining sentences of the language. Prior material to this theory is an analysis of how words are constructed from sounds (7.1–4), and preliminary considerations of how words combine into sentences (7.5–8, 8.1). Following upon the establishment of sentences, we can go on to an analysis of how sentence sets, especially sentence-sequences, are constrained in longer utterances and in discourses (ch. 9), and finally in sublanguages (ch. 10). The relation of syntactic structure to information and to the development of language is then discussed in Chapters 11, 12.

The general structure of language consists then of: (1) the construction of words from sounds, in a way that does not affect the structure of syntax; (2) the construction of sentences from words, by a single partial-ordering relation; (3) the transformation of sentence shapes, primarily by a process of reduction. The independence of word-sounds from sentence construction leaves room for various languages to have sound–syntax correlations: particular sound similarities among words having the same sentential status. This is the chief basis for additional grammatical constructions (beyond those listed above), such as conjugations and declensions.

In the chain of operations, from individual sounds to complete discourses, each operation is defined on the resultant of prior operations, and each word that enters into the making of a sentence enters contiguously to the construction created thus far. Because of this, every construction (aside from statable exceptions) can be analyzed in an orderly way in terms of these operations as first principles.

It is thus not the case that language has no precise description and no structural relation to meaning, but rather that these properties are reached only via particular methods. These methods arise from a crucial property, namely that language has no external metalanguage in which one could define its elements and their relations (2.1). Hence, the elements can be established only by their redundancies in the set of utterances, i.e. by the departures from equiprobability in the observable parts of speech and writing. From this it follows that no matter what else a theory of language does, it must account for the departures from equiprobability in language. Furthermore, this accounting must not confuse the issue by adding to these any additional redundancies arising from its own formulation. That is, we need the least grammar sufficient to produce precisely the utterances of a language. It then turns out—necessarily so, on information-theoretic grounds (1.4)—that the elements and relations which are reached in a least–redundant description have fixed informational values which together yield the information of the sentences in which they occur. In this approach, finding the structure simultaneously exhibits the information (11.4).

To find the departures from equiprobability, we first show that not all combinations of elements occur (3.0), or are equally probable, in the utterances of a language (e.g. *Go a the* is not an English sentence), and then show what characterizes the occurring combinations as against the non-occurring. Given the great number and the changeability of the sentences of a language, the word combinations cannot be listed. However, they can be characterized by listing constraints on combination, such as can be understood to preclude all the non-occurring combinations, leaving those which are indeed found (3.0). It has been possible to find constraints that are relatively stable, and are such that each contributes a fixed meaning to the sentences formed by it. The initial and basic constraint shows that sentences are produced by a partial order on words in respect to their ability to occur with other words in the set of sentences. When the constraints can be formulated in such a way that each one acts on the resultant of another, we obtain not only an orderly description of the structure of the utterances, but even a way of quantifying the effect of each constraint without counting anything twice (12.4.5). These constraints are presented in 1.10 and in Chapter 3.

This least set of constraints, which characterizes language, is found to have mathematical properties that are relevant to their informational effect. In particular, the initial constraint, in producing the set of possible sentences, necessarily creates a mathematical object (1.2, ch. 6).

The same least–constraint method shows that sentence-sequences within longer utterances can best be described as satisfying additional constraints, of word repetition (ch. 9). This produces finally a double array of the sentence-making partial order among the words (9.3).

It is further found that when the constraint-seeking methods used to obtain the structure of a whole language are applied to the notes and articles of a single sub-science, they yield not the grammar of the language, but a somewhat different kind of grammar which has some of the properties of mathematical notation, and which reflects the structure of information in the given science. These methods provide a procedure for organizing the specific information in each document (or conversation) of the sub-science (1.3, ch. 10).

Because they can be formulated as constraints on combination, the entities and processes of this syntactic theory have an intrinsic relation to information, and suggest a theory of the information carried by language (1.4, ch. 11). Furthermore, the constraints can be regarded as processes which in creating the structure of language limit what comes to be sayable in it. As a result, they give some indication of the development of the construction of sentences, and of language as a whole (1.5, ch. 12).

There remain other aspects of language, such as its detailed historical changes (and dialectal differences), its analogic processes, and its use as art material. Some of these (e.g. analogy and change) can affect the structure, though most of them are domesticated into the structure created by the constraints. Others (e.g. social dialects and literary arts) involve the study of language as institutionalizations of a set of human behaviors, which for systemic reasons have to stay within the framework of the set of constraints described here.

1.1. *A theory in terms of constraints*

1.1.0. *The major constraints*

The theory presented in Chapter 3 states that all occurrences of language are word-sequences which satisfy certain combinatory constraints; furthermore, for reasons related to mathematical Information Theory, these constraints express and transmit information (1.4). The first constraint (3.1) is what gives a word-sequence the capacity to express fixed semantic relations among its words: it consists in a partial ordering on words (or morphemes, 7.4), whereby a word (or morpheme) in a sentence serves as 'operator' on other words, called its 'argument', in that sentence. (In *Sheep eat grass*, we take *eat* as operator on the argument pair *sheep*, *grass*.) The essential generality of this partial-order relation is considered in 1.2. All the other constraints act on the resultants of the first.

The second constraint (3.2) reflects the specific meaning in a sentence, by recognizing that for each argument word certain words are more likely than others to appear as operator on it. (*Eat* is more likely than *drink* and than *breathe* as operator on the pair *sheep*, *grass*.) The meanings of words are distinguished, and in part determined, by what words are their more likely operators or arguments; and the meaning of a particular occurrence of a word is determined by the 'selection' of what words are its operator or arguments in that sentence (11.1).

The third constraint (3.3) on word-sequence makes sentences more compact: it results from reducing—e.g. to an affix or to zero—certain words which have highest likelihood, and thus contribute little or no information, in a particular operator–argument sequence. (Pronouning and zeroing of repeated words under *and* reduces *Sheep eat grass and sheep eat weeds* to *Sheep eat grass and they eat weeds*, and to *Sheep eat grass and weeds*.)

In addition, the partially ordered operator–argument words are put into one or more linear forms (3.4), needed since speech is an event in the single dimension of time. (Thus '*eat* operating on *sheep, grass*' becomes *Sheep eat grass*.) These four steps create, out of the set of all possible permutations of words in a language, precisely the set of possible (grammatical or potentially grammatical) sentences of that language: the term 'sentence' here applies to something which is not only roughly recognizable by speakers of a language, but is also (7.7) delimited, by a stochastic process, as a segmentation of utterances (i.e. of connected occurrences of speaking or writing).

In addition to these constraints that make a sentence, there are two constraints on connected sentences. One of these demands interrelated word repetition in conjoined sentences. The other constraint demands repetition of the operator–argument relation (above) in interrelated sets of sentences: in connected discourse it can create a double array of words, beyond the linear form seen in sentences (ch. 9). And in specialized subject matters such as sciences it creates distinguished sentence-type formulas out of distinguished word subclasses; here the sentences are found to satisfy special conditions which make them a formally definable sublanguage that represents the special information in that science (ch. 10).

The theory presented here is not a model devised out of various considerations or assumptions about language, but a direct generalization of results obtained from syntactic analysis of languages.[1] The central result at this point is that when the sentences of a language are reconstructed into their unreduced form (by reversing the reductions that carry out the third constraint), the unreduced form is found to satisfy the first (operator–argument) constraint.

The discovery of such reconstruction, and of the form of unreduced sentences, arose out of an attempt to find the least system (grammar) sufficient to characterize, or produce, the sentences of a language. In following this line of research, it was found that after sentences had been described compactly in terms of component word-sequences (e.g. subject–verb–object and various modifiers), the description of language could still be tightened by showing that certain sentences contained in a regular way the

---

[1] That the picture of language presented here is similar in some respects or others to various views of language (cf. n. 6) does not obviate the need for reaching the present theory independently by detailed analyses of language. This is so partly because a more precise and detailed theory is then obtained, but chiefly because we then have a responsible theory in direct correspondence with the data of language rather than a model which is speculative no matter how plausible.

same component word-sequences as certain other sentences which were paraphrastic to them (i.e. which preserved the meanings of the first). There are two cases: (a) Many sentences consisted simply of other sentences plus additional (higher) words, with the meaning of the included (lower) sentence being both preserved and added to: *I know sheep eat grass*; *For sheep to eat grass is quite probable*; *Sheep's eating grass occurs all the time*. (b) Many sentences consist of another (source) sentence with no addition but with a change, in most cases a reduction (or transformation, 8.1), that leaves the meaning of the source sentence unaltered: *He's at the grocer's* has to be from *He's at the grocer's place*; *He reads all day* is from *He reads things all day* (this derivation avoids having both a transitive and an intransitive *read*); similarly for the two component source sentences in *Sheep eat grass and weeds* above. In (a), the sentences are derived from elementary base sentences (such as *Sheep eat grass*) plus further operators. In (b), the sentences are derived ultimately from base (i.e. unreduced) sentences by means of reductions or other transformations.

In a varied sample of languages (cf. the Preface), it was found that in each language there is a particular set of reductions (with possibly a few non-reductional transformations) and particular conditions necessary for their being carried out. Each reduction leaves a trace; in some cases two or more different reductions on two or more different sentences produce by a degeneracy the same word-sequence, which thereupon becomes a pluri-source, ambiguous, sentence. Given a sentence, one can recognize from its structural traces whether or not it has undergone reduction, and if so which reduction, up to this ambiguity. One can then reconstruct the unreduced source sentence from which it was derived. This derivability of sentences, from intermediate source sentences and ultimately from unreduced source sentences, was the first step toward the syntactic theory.

The next step was the finding that the unreduced sentences satisfy the first constraint above (4.1). More precisely: for the great bulk of sentences, the unreduced form is seen immediately to satisfy the operator–argument constraint; for the remaining sentences it has proved possible to define variants of the existing types and conditions of reduction, which would produce these remaining sentences as reduced forms of unreduced (operator–argument) sentences.

The first constraint, which creates sentences and is thus the essential one, holds that it is possible to collect the words of a language into classes which serve as operator and argument relatively to each other, such that in the set of unreduced sentences no word of an operator class appears unless the sentence also contains one or another word from each of its argument classes. Thus the occurrence of a word in an operator class depends on an occurrence of a word from its argument class. For example, *sleep* is an operator on one 'noun'-type argument, *eat* an operator on two: this produces not only *Sheep sleep* but also *Grass sleeps*, and not only *Sheep eat grass* but also *Sheep eat sheep*, *Grass eats sheep*, etc.; but all of these are syntactically possible sentences. This constraint suffices to create all possible unreduced sentences.

The operator–argument constraint imposes a partial order on the words of a language with respect to their defined ability to co-occur in sentences or utterances: in this order *eat* is higher than *sheep* or *grass*, while *sheep* is neither higher nor lower than *grass*; *know*, *probable*, *occur* are higher than *eat* (in (a) above). Each sentence is a set of word-occurrences satisfying this partial order (i.e. of word-occurrences within which the operator–argument relation holds), possibly with (statable) reductions. The partial ordering itself has a meaning: the operator is said about (i.e. predicated of) the argument (3.14). Hence, given the meanings of the words (by real-world referents, or by selection of operators and arguments), finding the operator–argument relations among the words of a sentence yields its meaning directly: that meaning is the hierarchy of predicatings among the meanings of the words of the sentence. The operator hierarchy, i.e. the partial ordering, is overt in the unreduced sentences; it and its meaning—though not its overtness—is preserved under reduction.

The grammar of a language is then the specification of these word classes in their partial ordering, and a specification of the reductions and linearizations. This seems quite far from the traditional contents of grammar. But the usual grammatical classes, such as verb and adjective, are obtainable from the partial order. Some (e.g. adjectives) are by-products of reductions on operators; others (e.g. verbal nouns) are secondary statuses such as that of an operator (*eat*) appearing as argument (*eating*) of a further operator (e.g. under *occurs all the time*, in *Sheep's eating grass occurs all the time*). And such syntactically local constructions as plural, tenses, conjugations, and declensions are merely patternings of the form, likelihood, and syntactic status of certain reductions. Thus the usual complex grammatical entities and constructions, important as they may be for practical descriptions of a language, are only incidental to (and derivable from) the universal and simple form-and-meaning constraints of language structure. The traditional constructions mentioned above can be brought out by seeking the similarities and interrelations among the reductions in a language, and correlations between the shape of words and their syntactic status.

In general, the constraints—including the vocabulary and the co-occurrence likelihoods of the words—are discovered from the regularities seen in sentences. Since each constraint acts on the resultant of previous constraints, the successive constraints have monotonically descending domains; and more importantly, at each stage of analysis the remaining constraints do not alter what has been obtained thus far. Furthermore, each sentence is a partially ordered structure of words and reductions. In its unreduced form, the words are ordered by their operator–argument relations there—in effect, by their ordered entry into the making of the sentence. The reductions and transformations which create a reduced sentence are ordered by the property that if they are carried out it is done as soon as the conditions for them are satisfied in the making of the sentence. Given all the properties noted above, the constraints make it possible to devise a strategy for analyzing each sentence in a language, recognizing the reduction traces and the operators and arguments, from the latest entries back to the ultimate elements. Conversely, the constraints make it possible to synthesize every sentence. For, when the words of each class are specified, with their ranges of meaning or their likelihood in respect to their argument words, the syntactic theory presented here produces actual sentences of the language, with direct indication of their meaning. The syntax of a sentence indicates its semantics because each constraint has a fixed meaning, and any difference in meaning that a word may have in its various operator–argument statuses is specifiable a priori.

In addition, this theory has certain useful properties. One is that while consisting of computable formal relations it has a behaviorally functional explanation (12.3.5). Another is that the theory is explanatory for grammatical detail: a great many special and seemingly arbitrary or irregular word combinations are seen to result from a constellation of regular processes in a way that explains both their form and their meaning.

1.1.1. *Example: The set-vs.-member connectives*

To give some impression of how analysis in terms of the theory explains grammatical construction in detail, we consider two particularly complex illustrations. One is the case of words, such as *only*, *except*, *but*, that can occur in various unique environments all of which are created by zeroings associated with the words *everything*, *everyone*. *Only* and the others are a subset of the bi-sentential operators, which can occur as connectives between any two (component) sentences. In this subset there is considerable likelihood for one of the component sentences to contain the word *everything*, or *everyone*, or the like, while the other component sentence contains a word for an individual or subset of that *every*. When the component sentences have this high-expectancy set-vs.-member matching, the *every* is predictable and hence zeroable. Under certain of these operators the occurrence of *not* in one of the component sentences but not in both is also highly likely and hence zeroable. It is these special zeroings, together with zeroings of repeated material, that can make the environments of these operators unique (G401).[2]

First, *only* appears in two environments, with different meanings: (1) *The meeting was fine, only John didn't come* (similar to saying *except, but*), which

---

[2] For the reference G, see n. 4 below.

conjoins any arbitrary two sentences; and the apparently independent (2) *Only John didn't come* (with stressed *only*, similar to saying *John alone didn't come*). However, the *only* of (2) turns out to be a case of the *only* of (1), if we derive (2) from (3) *Everyone (else) came, only John didn't come* (which is structurally of type (1) above). In (3) the whole first component sentence is zeroable (reducible to zero) on grounds of low information in respect to the set-vs.-member selection of the bi-sentential operator (conjunction) *only*. Note that the first component of (3) is identical with the second component, except that *everyone (else)* with optional *else* replaces the argument (*John*), and that the two verbs differ in negation alone (*didn't come*). When the two sentences which are arguments of *only* differ in just this way the whole of the first argument (here, *Everyone (else) came*) is zeroable, because of the set-vs.-member selection of *only*. This reduces (3) to (2). In other cases, e.g. (1) above, no zeroing occurs and *only* remains visibly bi-sentential.

Although the meaning of *only* in (2) seems to differ from that in (1) and in (3), that difference is seen on closer inspection to be merely a trace of the zeroing of the first sentence under *only*. Indeed, (3), with the meaning of *everyone else not, except John*, and (2), with the meaning of *John alone*, carry the same information. It is instructive to see here how meaning-preserving reductions (here, zeroings) in the first argument of *only* in (3) yield such a difference in the meaning of post-reduction *only* such as we see in (2).

Next, *except*. Here too we can find a juxtaposition of structurally arbitrary component sentences, as in *The meeting was fine, except (that) John didn't come* with optional *that*, and also of structurally contrasted sentences as in (4) *Everyone came, except that John didn't come*. What is zeroable in the contrasted sentences around *except* is just the *not* (optionally) plus verb plus *that* (here, *that . . . didn't come*) of the second component sentence, more rarely the verb alone (here, *came*). Then (4) reduces to (5) *Everyone came except John* or (5a) *Everyone except John came* or the rarer (5b) *Everyone came, except not John*. (5b) seems to negate (5), but is actually a paraphrase of it, differing only in not zeroing the expected *not*. (The second *came* must be zeroed if *not* is.) Again, the meaning of (4) is preserved in (5).

Then, *but*. We have any arbitrary component sentences in e.g. *The meeting was fine but John didn't come*, and contrasted ones in e.g. (6) *Everyone (else) came, but John didn't come*. Under *but*, in structurally contrasted sentences such as (6) one can zero the second verb, and if so also the high-expectability *not*. Then (6) reduces to (7a) *Everyone came, but not John* or *Everyone, but not John, came* (separated by commas) and to (7b) *Everyone came but John* or *Everyone but John came* (without commas), all preserving the meaning of (6). The *not* and the zeroability of the repeated verb may be in the first component sentence. Here the optionality of zeroing the *not* (compare 7a and 7b) yields the peculiar case of having identical meanings for a sentence and for its apparent negation: in *There weren't but two people left* and *There were but two people left*. (The zeroings that produce the last form are not available under *except*.)

This analysis shows *only*, etc., to be a subset of the sentence-introducing 'particles' (*therefore*, *however*, *nevertheless*, *too*, *so*, *even*, etc.), which do not normally appear on the first sentence of a discourse—this last because they are all derived from conjunctions under which various zeroings can eliminate parts of the component sentences, including (as under *only*) the whole first sentence (G393). In addition to this general derivation, the *only*, etc., discussed above have a high likelihood for *every* matched against a subset of whatever is referred to in the *every*; and *only*, *except*, *but*, also have high likelihood for *not* on one sentence but not both.

This illustration does more than show that explanatory analyses can be obtained by the syntactic theory presented here. It shows that regularities among suitable words can be discovered by comparing their environments. It also shows that the regularities apply to processes, such as zeroings here, on those environments. Furthermore, it shows that the analysis of a given environment has to be made in respect to all the other environments of the word in question, so that the given one is obtained from the others by the regular processes. This means taking into account all the environments of a word rather than saying—unhelpfully—that in different situations it is a different word or a different use or meaning of that word. It also means trying to take into account the etymology of a word, i.e. its earlier uses: *only* is from *one* + *-ly*; *except* is from Latin *ex* 'out of' + *capere* 'take'; *but* is from early forms of *by* + *out* + adverbial suffix. The etymology is relevant because: either the current word in its environment is a descendant (i.e. a repetition or a continuation) of its earlier source, appearing in the source of its current environment; or else at some time the old word was extended into the current environment by some regular or analogical relation to the earlier environments of the source word. Such etymological comparison is therefore not a checking of the modern word against its earliest known cognate, but a tracking of the changing use of the word in various environments, to the extent possible. And this evidential 'use' is a matter of specific environing words in various sentences, not a matter of meaning, because we have little knowledge of past meanings except via past word-environments. Furthermore, appeal to meaning would not suffice to predict the increasing preference for certain environments, which underlies the zeroabilities within them: why did the above sources of *only*, *except*, *but* (but not, for example, *unless*) develop the likelihoods and zeroings of *every* and *not* which they now have (while *unless*, not far in meaning, did not)? Why did *even* not move into a more contrastive use, with special likelihood for *not*, as it does slightly in *even so* (compare French *quand même*)?

1.1.2. *Example: The passive construction*

The other illustration is the case of the passive construction in English (G16, 362): *The book was taken by John*, by the side of the 'active' *John took the book*. It is first of all clear that the passive is not an independent combination of words, but rather a transform of the active (i.e. derivable from it), since for every passive sentence (except for those in subsets explained below) there corresponds an active sentence, as above (cf. 2.3). It can further be seen that the passive is not an independent transformation, but rather a resultant of three elementary transformations. (1) One is the *by* which arises in English on the subject (first argument) of a verb when a further operator acts on that verb, e.g. above in *taken by John* as argument under *was*. Note that this use of *by* differs syntactically from the 'prepositional' operator *by* which appears equally on active and passive sentences, as in *John read the book by candlelight* and *The book was read by John by candlelight*. (2) A second component of the passive which is known elsewhere in English grammar is the stative *-en* (*-ed*) suffix. This suffix appears also in the 'perfect' tense (*John has taken the book*), in the 'passive adjective' (*a crooked smile, a learned man*), and in a suffix on nouns (*a monied family, a two-wheeled vehicle*). (3) The third component is the moving ('permutation') of the object (second argument) of a verb to subject position, more precisely to being the subject of a further verb (here, *is*) that is operating on the original verb. This is seen in *The Armenians suffered repeated attacks by* (or: *from*) *Azerbaijanis*, where the grammatical (and physical) object of the attack were Armenians; the form is comparable to the passive *The Armenians were repeatedly attacked by Azerbaijanis*, from *Azerbaijanis repeatedly attacked the Armenians*.

As to the elementary transformations:

(1) *By* on the subject of a 'nominalized' sentence (i.e. a sentence appearing as argument of a further operator) is found independently of the other components of the passive: *We opposed the bombing of Dresden by Churchill, He had a scolding by* (or: *from*) *the principal*. In contrast, *of* and *'s* (and also absence of any such argument-markers) occur on both subject and object, under various conditions: *We opposed Churchill's bombing of Dresden* (*'s* on subject, *of* on object), *We opposed Dresden's bombing by Churchill* (*'s* on object, *by* on subject), *We heard the singing of birds* (*of* on subject), *We opposed Churchill bombing Dresden* (no markers).

(2) The *-en* suffix occurs independently of the subject–object permutation. In the perfect tense the *-en* means 'state', and *has -en* is simply the present tense of the completed 'having a state' (and *had -en* is its past, i.e. attaining the state at some past time). Thus the present perfect cannot be used in *Goethe has visited Lago di Garda*, because the no longer living Goethe

cannot be said to have such a state in the present time. But the present perfect can be used on the passive of the above, in *Lago di Garda has been visited by Goethe and others of his generation*, because the lake still exists and can be said to have that state. (If the passive were merely a direct reshuffling of the active, it should have the same range of tenses.) In the active, the present-tense state (on *visit*) would have been asserted of Goethe, but in the passive it is asserted (on *be*) of the lake. There is also a limited perfect tense with *is*, seen for example in *The Sun is risen, We are agreed, We are resolved, She is grown quite tall, I am finished* (or: *done*) *with it, He is possessed of great wealth*; this is a stative or completive of *We agree*, etc. Further, there is the passive adjective as in *the curved horizon, a learned scholar, a professed* (as well as *professing) Christian*. Finally, there is the stative *-ed* on nouns, as in *a monied family* (which means not a family which happens at the moment to have money, but a family in the ongoing state of having money).

(3) Having the object of a lower verb (e.g. *eat, praise* below) appear as subject of a higher verb (e.g. *is ready, deserve, is* below) occurs independently of the *-en*. Among the many forms, consider the following examples: *It is ready to eat, It is ready for eating* (where *it* is the zeroed object of *eat*); *He is to blame* (the missing object of *blame* is *him*); *He deserves our praise* (where *of him* can be added, as object of *praise*); *It improves with (the) telling, It is in process of dismantling* (where *it* is object of *tell, dismantle*, and can be said, e.g. . . . *of dismantling it*); *This is past understanding, Much money is owing to them*, obsolescent *The house is building* (in all these the subject of *is* is the same as the zeroed object of *understand, owe, build*); *This is understandable*, which could be derived from *This is capable of one's understanding it* (as though from *This is able for one to understand it*). In all these cases the lower verb can also be said in the passive: *It is ready to be eaten, Much money is owed to them, The house is being built* (the last, only in recent English).

To say that the passive includes these three constructions means that it is the resultant of the application of all three to the source of the passive, i.e. to the active. Hence the passive (*The book was taken by John*) results from taking the simple active (*John took the book*) and placing it under a higher operator, *is -en*, which belongs to the set of (3) above whose subject is usually the same as the object of the verb under them. We then have *is -en* operating on the pair *book* and *John took the book* (which under *-en* is nominalized to *taking of the book by John*). Here *of the book* is zeroed as being a necessary or usual repetition of the subject of *is -en*; then the subject *book* of *is -en* looks like a permuting of the object *book* of *take*, whereas in the derivation we see that the former is just a repetition of the latter that then caused the latter (the lower occurrence of the object *book*) to be zeroed. The *-ing* of *taking* is not an independent accretion (an operator or argument) to the sentence, but merely a marker, an indicator (G41) that the operator to which it is attached has become nominalized as an argument of a further operator (in this case *is -en*). Therefore nothing is lost to the sentence when the addition of *-en* involves removal of the *-ing*, yielding *taken*. These steps, then, suffice to yield the passive. Using arrows for derivation we have: *The book was in the state of the taking of the book by John → The book was in the state of taking by John → The book was taken by John.*

As to the domains: since the passive is the resultant of three components, the domain of sentences to which it can be applied is the intersection of the domains of the components. Thus it can be applied only to sentences which are within the selection (3.2.1) of: (1) the *by*-nominalization of the lower sentence, (2) *-en* as higher operator on it, and (3) the subject of the higher operator being the same word (with same referent) as the object of the lower (nominalized) operator.

(1) Hence, there is no passive of *The book costs $5* and *A cat is a mammal* (no *$5 is cost by the book*, *A mammal is been by a cat*, just as there is no *I noted the costing of $5 by the book* (but only *I noted the book's costing $5*) and no *I noted the being of a mammal by the cat* (but only *I noted the cat's being a mammal*). To say that the passive is available only for transitive verbs is not precise. There are passives of intransitives, provided these carry the *by*-nominalization: *Everybody laughed at him* has both *I noted the laughing at him by everybody* and *He was laughed at by everybody*. But for *He walked near the wall* one is far less likely to say *I noted the walking near the wall by him* and even less likely to say *The wall was walked near by him*. And just as *I noted the taking of a ride by the boy* is more acceptable than *I noted the taking of fright by the boy*, so *A ride was taken by the boy* is more acceptable than *Fright was taken by the boy*.

(2) As is seen in transformation (2) above, *-en* means roughly 'state'. Therefore, when it operates on a pair of arguments consisting of a noun and a sentence, it selects those pairs in which the noun can be said to be in the state of that sentence. This is seen in the perfect tense, where the subject of the lower sentence is the same as the subject of the higher *has -en*: *Goethe has visited Lago di Garda*, equivalent to *Goethe has the state of his visiting Lago di Garda*, is no longer sayable because, as noted, the present tense *has* cannot now be used of Goethe's acts; one would have to say *Goethe visited . . .* or *Goethe had visited . . .* [3] And it is seen in the passive, where it is the object of the lower sentence that is the same as the subject of *is -en* (*Lago di Garda is written about by many*, equivalent to *Lago di Garda is in the state of the writing about it by many*). The fact that the *is -en* asserts a state of its subject explains, for example, why one can readily say *America was left by a whole generation of its young writers in the 1920s*, but hardly *America was left*

---

[3] But one can say *Goethe says this in many places* or *Goethe is one of the greatest German writers*, because these are continuing properties of Goethe's work. (He is still saying things in his books.)

by my neighbor last month; for the expatriation of a generation of writers could be described as constituting a state of America, but hardly so the neighbor's actions.

(3) The condition that the subject of the higher *is -en* have the same referent as the object of the lower sentence explains how it is that the main meaning of *Someone was blamed by everybody* (in the sense of 'There was someone such that everybody blamed him') is different from that of *Everybody blamed someone* (in the sense of 'Everybody had one or another person whom he blamed'). The explanation arises from the following: in the unreduced language base, as in logic, quantifiers (e.g. *every*) can be shown to act not on nouns but on sentences (G252). The referents are first determined in the source sentence *Somebody blamed someone there* before the quantifier is added, and it is on this that the passive *Someone was blamed by somebody* is formed; the quantification to *everybody* is added separately, on the source and on the passive (G263).

The illustrations above give some indication of how the different constructions in which a word is found and the different meanings the word has in them can be derived by a few given reductions from a standard unreduced (base) set of operator–argument constructions. Thus, a few pervasive constraints combine to produce the many forms of sentences, including ones that seem exceptional. These illustrations also show (as follows from 2.3) that data can be usefully analysed only in respect to a specified whole structure of a language, and not as selected examples in support of episodic grammatical rules or processes.[4]

1.2. *Mathematical properties*

There are reasons to expect that the structure of language has in some respects a mathematical character. The grammatical classification of words is defined in general by relations among the words, rather than by particular properties (phonetic, morphological, semantic, or logical categories) that each word has by itself. An indication of this is seen in the simple fact that

[4] It is for this reason that the examples in the present book are almost entirely from English, although a single language is far from sufficing for a survey of language structure. In the case of English, there are available above all the multi-volume *Oxford English Dictionary*, and Otto Jespersen's *Grammar of English on Historical Principles*; in addition, the present writer's *A Grammar of English on Mathematical Principles* (referred to here as G) presents English grammar in terms of the syntactic theory of chs. 3–4 below. The latter book not only provides examples for the discussion in the present volume, but also shows how the methods and theory stated here suffice to yield an orderly and detailed grammar of a language. The English grammar of G thus serves for systemic evidence, without which a theory in a field such as linguistics would not be adequately established. As noted at the end of 12.1, the interrelations and regularities of the data are essential to any theoretical formulation (on grounds given in 2.2), so that the episodic presentation of examples in support of one analysis or another is of little relevance.

every language can accommodate new words. For a language at any one time, we can think of its word combinations as being determined by specified lists of words, e.g. the list of words including *sheep*, *child*, *grass*, each of which can 'make a sentence' with a word of a list that includes *fall*, *sleep*, *eat*, but not with any word of its own list: *Sheep eat grass*, but not *Sheep child grass*. However, when a new word enters the language (i.e. comes to be used), the question is: on what basis can we say what list it belongs to? If the new word combined with other new words and not with the previously existing words, the language would develop separate encapsulated grammatical systems for different lists of words; this is not generally the case, although something approaching it is seen in the morphological structure of the English *perceive*, *conceive*, *persist*, *consist*, etc. If the new word combined with old words, but differently than the old words did among themselves, we might in time lose any systematicity over the whole language; this too is not the case. What we find in each language is that large subsets of words act in uniform ways in respect to other large subsets.

This means that the late-comers (whatever words they were) combined with other words in the same ways that existing words did. Such fitting into existing word sets means that the new words satisfied the condition of being usable in the existing types of word–set combination. Thus the overall criterion of classification is the condition that words enter into the same combinatory relations. Hence the general characterization of word classes is in terms of their common combining-relations to other classes. It will be seen that this opens the door to characterizing the entities of language by their relations rather than by their inherent properties—a necessary condition for relevant mathematical characterization of language structure.

Beyond the considerations above, language has to have a constructive and relevantly mathematical structure if it is an unbounded set of different activities (different utterances) carried out by finite beings (2.5.1), and if it was not built by anyone on some template or model (the existence of which prior to language would be inexplicable, 12.4.3–4). Also, if language is structured as an information carrier (11.5), we can expect it to have necessary properties related to those known from mathematical Information Theory (11.7).

This last is of central importance. If we begin analyzing language without any preconceived guides, we find that the lack of an external metalanguage in which to conduct the analysis leads us to describe the data in terms of departures from equiprobability of combination (2.1–2); these are the constraints of 1.1 and Chapter 3. It is then immediately apparent that these constraints create the kind of redundancies studied in Information Theory, so that the structure of syntax is a system of contributions to the total redundancy—both of language and of its information.

If we now consider the data of language, it indeed seems overtly to have possibilities of mathematical treatment, as in the transitional probabilities among the successive letters that make a word in alphabetic writing, or the successive words that make a sentence. However, the transitional probabilities do not yield useful characterizations of words and sentences (although particular stochastic processes can be so formulated as to yield word and sentence boundaries, 7.3, 7.7). Also, free semigroups are formed by the sequences of letters and by the sequences of words. But these semigroups do not represent natural language, and their structure is not relevant to the structure of language. For example, no simple relation stated on these sets can characterize which different sequences have which different frequencies of occurrence. Rather, these sequences constitute sets, of which the words and the sentences of natural language are undefined subsets. Indeed, to define the particular subsets which constitute language is the first and essential task of a theory of language.

The present theory will be seen to have various mathematical properties (ch. 6). The crucial mathematical property lies in the fact that the essential determination of operators and arguments is made by a single relation on their combinations. That is, in each language, the classes of operators and arguments are not each a collection of words which have been selected on syntactic or morphological or semantic grounds (such as having a tense, or a particular suffix, or a particular kind of meaning). Rather, words are assigned to classes entirely on the grounds of a single co-occurrence relation: We say that A is operator on ordered classes, say B, C (or: A depends immediately on B, C), to mean that A is in the set of all words which appear in an unreduced sentence only if some word of B and some word of C, in that order, also appear in that sentence; here B or C may be stated to be the same class as the other, or as A, or different classes, or the null set (3.1.1). In turn, membership in B or C is not given by a list of words or by some property other than this dependence. Rather, B is itself defined as the set of all words which depend, in the above sense, on classes D, E; and so for C; and so on (3.1.2).[5] The word classes are thus defined by their dependence on word classes which are in turn defined by the same dependence relation. An infinite regress is avoided only due to the existence of a class of words whose dependence is on the null set of words. The single relation among words is, then, dependence on dependence; and a sentence is any sequence of words—or other objects—in which this relation is satisfied. In accord with this, there are just a few dependence-on-dependence statuses that a word can have (i.e. types of argument that a word requires). Hence, languages have a set of words whose status is that they can be defined (for reasons given in 3.1) as depending on null (these are 'primitive arguments', 3.1.2, e.g. *John*, *sheep*), words depending on (one or more) words that depend on null (e.g. *sleep*, *eat*), and words depending on words that depend on something

---

[5] The reasons for this situation, going back to the lack of an external metalanguage, are reviewed in ch. 6 and 12.4. Sublanguages, which have an external metalanguage, do not have the mathematical-object property (10.51).

not null (e.g. *occur*, *is probable*); this last type includes words whose status is that one of their arguments depends on null while another argument depends on something not null (e.g. *know*, whose first argument may be *John* while the second may be *eat* ((a) in 1.1).

It may seem surprising that a practical grammar, able to produce the actual sentences of a language, can be built out of word classes defined so generally and indirectly, rather than out of linguistically determined word lists. This development was perhaps necessary, because if word combinations were in respect to specified word lists, it would have been difficult to reach public understanding on how new words are to combine. Greater stability is attained if once a new word is used in a particular dependence-on-dependence status, it is publicly clear what types of argument it can occur with (5.4, 12.4). In any case, the fact that we can characterize in this way the word classes that are sufficient for grammar means that words, whatever phonetic and semantic properties they have, have also as their defining syntactic property their position in a network of occurrence-dependences in respect to each other. Thus the co-occurrences (sequences) of words in unreduced sentences as created by the first constraint is a mathematical object: a system of entities defined only by their relations and closed in respect to them, with operations defined on these entities.

Showing that a given system in the real world is describable as a mathematical object means that any other real system which is also an interpretation of that mathematical object shares the relevant properties of the first. In particular it means that any other set of objects or events that satisfied the first (dependence) constraint would also have sentence-like properties. And indeed, mathematical notation will be seen below to satisfy the first constraint but not the second (10.6), with the similarities and differences between language and mathematics being indicated thereby.

In addition, the effect of various constraints has been to produce various mathematical structures in language—orderings, semigroups, mappings, and decompositions involving sets of words and of sentences. These are noted in 6.3, 7.3, 7.6–8, 8.1.5–6, 8.3. What is important is that these structures are not merely describable, but are relevant to the informational function of language (11.5). The dependence-on-dependence relation underlies virtually everything in language structure; and beyond that, the mathematical structures noted here suffice to define almost all the semantically (and grammatically) important classifications and relations; and even the kinds of information capacity that language has (11.3, 11.6).

1.3. *Science sublanguages*

Analyzing languages in terms of constraints on word co-occurrence makes it possible to recognize the existence of discourse-structures and sublanguages within a language. If the grammar of a language were stated in terms of objects unique to language, such as a particular list of words constituting the class 'adjectives', it would not be possible to apply the entities (classes) of a whole grammar when investigating any special classes such as might arise in special conditions of language, unless the special classes were built of the same entities. In contrast, if the grammar consists in stating a relation among words, we may obtain, as carriers of the same relation, one set of classes for the whole language and other, appropriately different, classes for distinguished subsets of the language.

One distinguished subset of the utterances of a language is the set of sentence-sequences, each sequence being a set of two or more sentences which have been said in succession, as a single discourse. All such sentence-sequences satisfy, overtly or else in their base reconstruction, a constraint which calls for one or another kind of repetition or subclassing of words among the sentences (ch. 9). The repetition constraint expresses—or creates— a connectedness among the sentences. For many kinds of discourse, this constraint produces, beyond the linearized partial order of sentence structure, a double array in which one dimension records each linear sentence while the other dimension gives the similarities and differences within corresponding positions among successive sentences (9.3). Even the structure of proof in mathematics is a special case of this constraint on sentence-sequences (10.5.4, 10.6).

Another, very important, type of sentence subset is that in which the connectedness among the sentences lies in their sharing a subject matter. Here the repetition constraint creates sublanguages, with grammars different from the grammar of the whole language (10.0).

Not all subsets of utterances lead to different grammars. Quite the contrary, analyzing various arbitrary collections of utterances in a language leads to much the same grammar, and the grammars of different collections increasingly approach each other the bigger the collections are. However, if the corpus of material which is being analyzed comes from a single, more or less encapsulated, subject matter, one arrives at a grammar somewhat different from that of the whole language. The same operator–argument dependence which creates sentences necessarily appears here too, but with different resulting word classes and sentence types. In the case of arbitrary collections of sentences, the word classes that are obtained are constrained only by the dependence properties of their arguments. But in the case of a restrictive subject matter, for example a subscience, we may find one particular small class of nouns (primitive arguments) which occur only under a particular list of operators, and another noun class which occurs only under another list of operators; and certain of the operators may occur only under a particular list of further operators.

For example, in immunology (10.4) we find certain operators including *appear in*, *are produced by*, *are secreted by*, having *antibody*, *gamma-*

globulin, *anti-ferritin*, etc., as their first argument (subject) and having *lymphocytes*, *plasma cells*, etc., as their second argument (object): *Antibody appears in lymphocytes*. And we find *was injected into* having *antigen*, *sheep erythrocytes*, *virus*, etc., as its first argument and having *rabbit*, *left ear*, etc., as its second argument: *Antigen was injected into rabbits*. The words within a class are clearly not synonyms, nor do they necessarily constitute an otherwise-known semantic class. Combinations cutting across these operator–argument structures are not just false, nor are they nonsense (as they might be for the language as a whole—e.g. *is produced by* as operator on the pair *left ear*, *rabbit*); rather they are simply outside the sublanguage. If someone inserted *The left ear was produced by the rabbit* into an immunology article, it would immediately be recognized by all readers of immunology as not part of the science, independently of whether it was considered a reasonable English sentence.

Furthermore, the grammatical analysis here need not go as far as it goes for the whole language, only far enough to separate out the segments which fall into operator–argument classifications in respect to each other in this material—i.e. only enough to obtain a least grammar for the given science sentences. In some cases grammatical analysis beyond this point would be detrimental to maximum regularity for sublanguage operator–arguments. For example, the operators on the pair *antibody*, *lymphocyte* include *appears in* and *is produced by* (in *Antibody appears in lymphocytes*, *Antibody is produced by lymphocytes*). If we analyzed *is produced by* into its grammatical components we would end up with *Lymphocytes produce antibody*, which would not readily look (e.g. to a computer recognition-program) like the same sentence type as in *Antibody appears in lymphocytes*. But in 10.3 it will be seen that *Antibody is produced by lymphocytes* belongs to the same sentence type as *Antibody appears in lymphocytes*, in the immunology sublanguage: both have the same word sets as subject and as object, though in inverse order.

When we consider the word classes and sentence types obtained from the operator–argument relations within a subject matter, we find that while they necessarily satisfy the gross grammar of the whole language to which they belong they do not do so maximally: compare the kind of classes mentioned above for immunology with the classes defined by dependence on dependence in 3.1.2. Rather, what they are found to satisfy in detail is the set of objects and relations of the science in question. The word classes and sentence types of a sub-science correspond well to the objects and activities of the laboratory, the events observed, and the entities and relations of the theoretical discussion. Thus the grammar of the sub-science is a structural representation of the knowledge and opinions in the field. In terms of this grammar, the sentences of articles and reports can be represented in a canonical way as formulas written in a structured notation, which exhibit the specific information of the sentence and also show its relation to the structure of knowledge in that science. Utilizing both the discourse and the sublanguage properties of scientific articles makes possible (in principle) computer processing of their information.

It will be seen in 10.3 that while the sentences of the science are a subset of the sentences of the whole language, the grammar of the science is not a subgrammar of the grammar of the whole language. Rather, it is a kindred system, different not only in detail but in some basic properties which go beyond whole languages. In their purest form, the word classes and sentence formulas of a science (independently of their original language), and in a different way the double array of a single discourse, constitute not just sublanguages but new language-like systems of notation. These systems, intermediate in specifiable respects between natural language and mathematics, show something of the nature of scientific knowledge on the one hand, and of discussion in general on the other (10.5).

1.4. *Information*

The relation of language structure to meaning was known to many, especially to logicians who developed categorial grammar (n.6). When structure is stated in terms of constraints, this relation can be seen explicitly, in several respects. First, when one sees what the syntactic relations consist of, one can see how they come to bear the· meanings they do. In the first constraint, many operators name properties or relations or acts while their arguments name objects or events. However, such typical meanings are far from constituting a semantic criterion for operatorhood or argumenthood, much less a precise or predictive one: in English, operators include *father*, while primitive arguments include *time*. But when the condition that saying an operator word depends on saying some argument word is conventionalized into a requirement (12.3.5), it has the semantic effect that the operator is being said 'about' the argument. The operator is like a predication, on one or more operands: *came* is said about *John*; *eat* is said about the ordered pair *sheep*, *grass*. The dependence is more than an artefact of speaking: it presumably arose from the usefulness of interrelating the word meanings in communicating information. Since major modes of thinking, especially such as are more reasoned and precise, are mediated by language, there is little doubt that this relational meaning of predication, which developed in language, has become a step in the process of thought or cognition. It would seem that the basic structural features of languages may have implemented cognitive development, while less universal features, such as tenses or conjugations, have little or no visible effect on cognitive structure, even in the languages that have them.

Second, the relation of language to information is more explicit than to thought. The fact that the constraint which creates sentences creates the meaning 'to say about' supports other evidence that language developed as a carrier of information (11.3, 12.3). It will be seen that other structural properties of language are explicable on this ground, and that, in contrast, little in grammar is structured to carry subjective expression, emotion and the like: indeed, such expressive features of speech as the intonations of anger, hesitancy, etc., are not included in grammar (hence, in effect, in language), because they do not combine with anything else in a way that forms further structures. That the content of language is seen here to be the communicating of information, rather than the expression of meaning in general, leaves room for the expression of other meanings, such as feelings, to be carried by other vehicles and behaviors such as music and the other arts (10.7).

There is another important result concerning information. Those transformations which do not bring further operators into a sentence ((b) in 1.1) have been found to be mostly either alternative linearizations of the original partial ordering of words, or else reductions in the phonemic (phonetic) composition of words in the sentence, under conditions in which no relevant information is lost or added in the sentence. The reduced form may lose some words which seem to have contributed information but did not really, as in *I expect him any minute* from *I expect him to be here any minute*. Or it may put things in a different grammatical relation than the source did, as in *The conductor, who was in a great rush, did not look back* from *The conductor did not look back; the conductor was in a great rush* (where only in the reduced form is the secondary sentence a modifier on *conductor*). There is something important which is preserved under transformation: it may be called the substantive or objective information, which is the same in the reduced sentence and in its source. The meaning difference between an unreduced sentence and its reduced form consists largely in the speaker's attitude toward the information (e.g. what is the topical part), or in the hearer's access to the information, and of course in the compactness of the information.

When we include among the constraints the choosing of particular words for a given sentence—in effect this is the second constraint (3.2)—the methods of the present theory isolate the first two constraints, which are universal and essential, as bearing all the substantive information and as creating a base sublanguage from which the whole language is derived. Within each sentence of the base, each application of these two constraints contributes accretionally to the information in the sentence. The other sentences are derived from the base by constraints which do not change the substantive information, though they may lose distinctions by degenerately deriving from two different base sentences the same word-sequence (which is thereupon ambiguous).

In the theory presented here every grammatical constraint (above all, the operator dependence) has a stated meaning which is the same in all its occurrences. And every word has stated meanings; its meaning is the same whenever it occurs with the same operators and arguments. That is, the association of words and grammar with phenomena of the world is made in a regular way. Hence it is possible to work toward orderly processing and retrieval of the specific information in sentences and discourses, on the basis of the syntax and words of sentences (cf. ch. 4 n. 1).

A major gain in describing language in terms of constraints—which are a construct of theory—rather than in terms of word combinations—which are the observables—is that each constraint can be associated with a meaning, in a way that each type of combination cannot always be. The relation of constraints to meaning is not surprising. On grounds of Information Theory we would expect the total departures from equiprobability, created by all the constraints together, to determine the total informational capacity of language. In the present work, it is seen that each constraint, acting on its predecessors, adds thereto its own departures from equiprobability and therewith its additional informational capacity (11.3.2). The constraints determine even what kind of information a system can carry: that mathematics can carry truth but not fact; that natural language cannot distinguish nonsense, but sublanguages can (10.5–6). A detailed stepwise correlation is thus created, in the base sublanguages, between form and content (11.4). The constraints—on sounds, on what constitutes words, on syntax—arise from efficiencies of communicational behavior (12.3.5) and from the speakers' experience of the world (11.6); but they appear physically as the structure of language events, as classifications of components of sound-waves. Hence, the relations within information can be processed on computers by carrying out the structural operations of language, although otherwise the attempts to mimic the brain and cognition by computer programs are not at present based on adequate or relevant methods. The relation of language structure to information thus makes possible the complex processing of information. In addition, it provides a stepping-stone toward understanding the structure of information, by showing that total informational capacity can be reached as the ordered product of specifiable contributions to information capacity, and by suggesting that all information involves a departure from equiprobability acting on departures from equiprobability (11.7).

We see here that there is a particular dependence of co-occurrence which creates the events which are called language, and that this dependence necessarily gives these events the power to carry information. More specifically, the underlying structure of language is the structure of information (11.5), and in 11.6.1 it is proposed that the co-occurrence constraints which create information are a reflection of co-occurrence constraints in the perceived world.

1.5. *The nature of language*

The theory of syntax presented here does not constitute a theory of language
in all its manifestations; it is not clear that there can be a single such theory.
However, what is shown here about syntax creates a framework which has to
be taken into account in many other questions concerning language, and
which casts new light upon these: for example, on the relation of language
change to language structure, on the differences among languages, and on
the difference between 'natural' language and language-like systems such as
discourse structures, science sublanguages, mathematics, and computer
languages. More centrally, certain considerations are established here with
respect to the structure of information, and with respect to the nature—even
the early development–of language. (Unavoidably, the sketch below is a
very inadequate summary of the discussion in Chapter 12.)

It must first be noted that the primitive elements are determined by
objective means, not by meaning or even by meaning-differences. The
phonemes, which carry no meaning, are obtained by a behavioral test of
what constitutes repetition in the speech community (7.1), and the words by
a stochastic process on phoneme-sequences in utterances (7.3). Given this,
it has been seen that the syntactic relations, which carry such fundamental
grammatical meanings as being predicate, or being subject, (a) do not have
to be invented or discovered as underivable *sui generis* properties of
grammar, but (b) can be derived from constraints on word co-occurrence.
In (a) such relations as being subject of a verb are necessarily unique to
language and cannot be compared with anything else. In contrast, for (b),
the dependence constraint is a kind of dependence that we can seek in other
systems and in other types of activity.

The syntactic and informational structure of language contributes to
understanding its development (12.3). The operator constraint gives the
sentential occurrence of each operator in terms of the occurrence of its
arguments, but not vice versa. Thus in the descriptive order of sentence-
making, the operator classes are 'later' than their argument classes. This
descriptive order has some bearing on the development of language, in
particular on the early development of sentences and syntax. It says nothing
about the existence of words independently of sentences. Many words which
later became arguments, and many which later became operators, must
have been used in pre-sentence times, said separately or in various combinations. But the coming into existence of sentences as explicit kinds of
word combination, with explicit sentential (predicational) meaning for the
combination, arose out of certain words coming to require, for their entry
into a combination, the presence of words from stated sets or else of null.
The order in which the definitions of syntactic classes refer to other syntactic
classes, which is the elementary order of sentence-making, may have a

relation to the historical order of the syntactic classes: words may not have become established at a particular operator level before words of their argument levels have become established as being such.

In this developmental picture, the dependence may well have arisen not as a structure, but as a preference: simply a matter of certain words A being more likely to be used in association with words of some particular set B than alone (or than with other words A). However, once this likelihood of occurrence became conventionalized, with some B present or implicit (via zeroing) whenever an A appears, the requirement became a (sentential) structure and gave rise to predication as the meaning of that word relation (12.3.5). Such conventionalization of a use-likelihood into a requirement, with an attendant grammatical meaning, is common in grammar, for example in the distinction between verbs and adjectives (G189).

All other syntactic relations can be obtained from the operator constraint. And since the meaning of that constraint is 'saying about'—the operator is said about its argument (1.4)—one can see that the constraint contributed a successful form of juxtaposing words so as to give information. This is of importance for understanding the development of language. For it is obvious that language has to have developed from a state in which there was no language. The problem is not resolved by attributing language to a specifically linguistic mode of neural processing, since it is even harder to understand how such a mode of processing arose before language existed. Furthermore there would remain the question of why the neural processing, and language itself, has the particular grammatical dependence-constraint that it universally has. In contrast, once that constraint exists, its learning requires no unique preconditioning: one need only hear a great number of sentences constructed by this dependence and thereupon construct further juxtapositions of words that satisfy the same dependence.

The same operation which creates the elementary sentences is then recursively applied to make extended sentences. Other sentences are further derived from existing ones by other conventionalizations of language use.

Sentences are obtained, and defined, as segments of speech within which the operator–argument relations, and most reductions, hold. Within a sentence there may be transformations. Phenomenologically, a sentence transformation appears as a relation (a constraint) among observed sentences. However, as a process it is merely a change of shape (primarily a reduction) within a sentence (8.1.7, 9.0), yielding not another sentence but another shape of the given sentence. Beyond this, there are also constraints among sentences. These are necessarily quite different from transformations: they are preferences or requirements for word repetition in certain operator–argument relations to each other within successive sentences (e.g. discourses) and within other special aggregates of sentences (sublanguages). Hence each sentence, discourse, and sublanguage is a structure, the locus of specifiable constraints (comparable in this sense to a biological organism). We may be able to say this also for the successive discourses over time in a particular science. In contrast, the set of sentences in a language, and the set of discourses in it, are only aggregates that are not defined by crucial relations over the members of the set (comparable, in a way, to community in biology).

Even such recent developments as sublanguages are obtained by existing linguistic relations operating in new circumstances. Throughout it is seen that language is not a single object or a one-time creation, but a system evolving by accretion, one which has grown out of self-organizing factors and out of the language-survival value of effectiveness in transmitting information (12.4.2–4).

We see here a few independent kinds of structure common to all languages: (1) the setting up of particular phoneme-sequences as words or morphemes, in a way unrelated to syntax; (2) syntax i.e. the partially ordered dependence in word co-occurrences, which can be broken up into dependence on word-class presence (which carries the meaning 'predication') and the individual word likelihoods within this dependence relation ('selection', which carries word meaning); (3) having phonemic variants (reductions) within words in the dependence relation; (4) the metalinguistic capacity created by the independence of (1) from (2) (cf. 5.1); (5) change; (6) analogic processes; and (7) room in each individual language (or language family) for special structures based largely on correlations between (1) and (2), or between (3) and the domains of words for which these variants hold.

It may be noted that the dependence relation and likelihoods of (2) above are the crucial steps that create sentences. They are not only universal but also semantically functional, and are learnable as a generalization or extension of simple properties which are exhibited in every sentential word combination (and overtly in unreduced sentences). The rest of syntax is not in detail universal among languages, not simple, and can only be acquired by learning from repeated hearing of sentences and sentence fragments. Such are the specific phonemic sound-sequences which constitute words with their approximate meanings, and the specific reductions with their specific conditions in each language. Finally, there are certain properties of language, generally neither universal nor directly functional or information-bearing, which may facilitate the speaker's learning to speak or the hearer's grasp of a sentence, but which do not have to be learned as such, as additional properties, by the users of a language. Such properties are: the distinction between morphology and syntax (a matter of certain particular kinds of similarities and of reductions), word-class distinctions other than between argument and operator (e.g. between verb and adjective, or between words of the same root), patternings of reduction which create conjugations and declensions, and the like.

In Chapter 3 it will be seen that given a set of base sentences of a language produced by the dependence and likelihood constraints (and normal lineariza-tion) of (2) above, the reductions and transformations (and alternative linearizations) of (3) above produce the rest of the sentences in a way that does not change the dependence and likelihood relations among the words, and thus does not change the substantive information of the source sentence. The processes that make sentences are thus separable into the fundamental informational constraints of dependence and likelihood, and the secondary paraphrastic constraints of morphophonemics, i.e. of the phonemic shapes and the relative locations of words without changing their dependence relations in the sentence. The dependence (and likelihood) relation is close to the view that language has an operator–argument structure, which is commonly held because of the pervasive verb–noun relation in the world's languages. But from the analysis of the other processes (excluding analogy) as morphophonemics it transpires that the complexity and patterning of languages are not due to the fundamental constraint.

First, only a part of the notorious complexity of language arises, as by-product, from the dependence constraint (i.e. from resultants of succes-sive applications of the operator–argument relation), or from likelihood (i.e. from similarities and contrasts among the selection-domains of each word). The major complexities in many languages arise from the institu-tionalization of differences (as in the verb–adjective distinction in some languages, 12.3.5), and in morphophonemic similarities as in the different declensions for different subsets of nouns. Second, the widespread pattern-ing that yields much of the special character of different languages turns out to be not a necessary component of information and communication, nor an unavoidable by-product of the operator–argument relation, but a systemat-ization of morphophonemic similarities and contrasts—something which has the properties of culturally set behavior and of standardized patterns of decoration (8.4).

Thus language structure has some universal properties relevant to speaking (e.g. phonemes, linearity), and some relevant to information (e.g. pre-dication), some not-universal properties which may be convenient for information but are not essential to it (e.g. grammatical structures for time, quantity, or aspect), and some properties, different for different languages, which may be irrelevant or even distracting for information (e.g. in some cases declensions and conjugations). These distinct types of structure arise in appropriately distinguished ways in the theory presented here.

In addition to all this there are major issues concerning language which are tangential to language structure as discussed here, but related to it. Such are: the regularities of language change (comparative, and more generally historical, linguistics); the relation of time-slice (synchronic) structure to change (diachronic); the differences between the speaker's grammar and the hearer's; the arising of geographic and social dialectal differences; the use of language for purposes of art.

There are also many areas of investigation which are related to language structure as presented here, but remain to be undertaken. For example:

Ongoing cases of institutionalization of language use;

Specifics on the motive forces and directions for the evolving of language structures;

The relation of the mechanisms of historical change to evolving structures;

Does the dependence-on-dependence relation indeed apply to all known natural languages, as it seems to?

How does the operator–argument structure compare with the old deaf sign-language which is not based on knowledge of spoken language?

Detailed descriptions, in terms of the dependence presented here, of major language families, and of little-known languages such as those of Australia and Borneo;

Detailed descriptions of reductions in various language families; many details remain to be worked out in English too;

Special problems in deriving morphemes from operator–arguments and reductions: affixes, pronouns, prepositions, particles;

Relatively recent development of such problem morphemes: e.g. articles;

The regularization of the double array in discourses (9.3);

Argumentation and consequence as constraints on sentence-sequences in science (10.5.4).

In sum, the most general theoretical finding reported here is that the word combinations which constitute sentences can be characterized not by a sequential relation but by a partial ordering relation on words, with all grammatical relations being defined in terms of this partial ordering. The most general grammatical finding is that if we take, as primitive events on the sentences of a language, a set of statable reductions in the shapes of words which have high expectancy in their sentences, then the residual form of each sentence, after undoing any reductions in it, is an overt fixed operator–argument relation among its words (the above partial ordering). The most general informational finding is that every grammatical relation has a fixed meaning, and that the constraints which create the partial ordering also create its information.[6]

---

[6] At various points, the conclusions which are reached here turn out to be similar to well-known views of language. For example, the operator–argument relation has similarities to the predicate structure in Aristotelian logic. The way it functions in the formation of sentences has similarities to functors (and more generally to the deep understanding of language theory) in the categorial grammar of S. Leśniewski, followed by K. Ajdukiewicz and others in the Polish School of logic; cf. particularly W. Buszkowsi, W. Mariszewski, J. van Benthem, eds., *Categorial Grammar*, Benjamins, Amsterdam, 1988; H. Hiż, 'The Intuitions of Grammatical Categories', *Methodos* (1960) 311–19 and 'Grammar Logicism', *The Monist* 51 (1967) 110–27; J. Lehrberger, *Functor Analysis of Natural Language*, Mouton, The Hague, 1974. Various conclusions about structure and meaning here appear similar to, or inverses of (this too being a similarity), arguments made by L. Wittgenstein. And the analysis of science sublanguages

addresses problems posed by R. Carnap concerning a language of science. Also, various statements about meaning (in ch. 11) accord with common knowledge or opinion. The intent of this work, however, was not so much to arrive at such conclusions, as to arrive at them from first principles, from a single method which is proposed here as being essential because of the lack of an external metalanguage. The issue was not so much what was so, as whether and how the essential properties of language made it so.

2

Method

2.0. *Introduction*

Linguistics has not in general been one of the sciences in which the relevance and correctness of statements are determined by controlled methods of observation and argumentation. It is therefore desirable to consider what methods are relevant in linguistics, in the hope of establishing criteria for investigation and analysis. Choice of method is not less important than responsibility in data, and the choice should be determined not by personal preference or current custom but by the nature and the problems of the data. Furthermore, it will be seen in 2.2 that it is not enough to go by the separate observable features of the data, such as their linearity or their carrying meaning, but rather that one has to consider the regularities of all the data, and their interdependence.

2.1. *The issue to be resolved: Departures from equiprobability*

Chapters 3 and 4 present a theory of language structure, sketched in 1.1. This theory is a conclusion which has devolved from analyzing language by a particular method. The choice of method arose out of the choice of problem, that is, out of choosing what issues were to be resolved. There are various issues, and types of data, which one might consider in seeking to investigate language. One is evidence as to the speaker's relation to what is said—his intent and how the sentence was put together. Another is the hearer's (reader's) perception of what is said—how it is decoded and what he understands in it. A third is the meaning of what is said. A fourth is the situational and behavioral context—what situations and events correlate with that which is said, and what are its effects, both immediate and in a broader cultural background. Lastly, there is the form of language—the flow of sound-waves which constitute each event of speaking; alternatively, any adequate representation or equivalent of it, either by instruments such as the sound spectrograph or by writing. This last type of data, as sequences of sounds or letters, is observable separately from the intent, perception, meaning, or context. It is sufficient for carrying the great bulk of information carried by language, as witness the adequacy of writing even though it lacks the direct support of speech properties or personal contact or extra-linguistic context. Within present knowledge, the form of utterances, i.e. the sound-sequences and written marks, is the one aspect of language that is observable precisely, fully, and independently of other features of speech and language.

There is reason therefore to address first of all the form of utterances, and to consider what methods are appropriate for studying it. Our objective is a precise theory which can analyze any sample of utterances (chs. 7–8), and can predict the structure of utterances (chs. 3–5) in a given language in a given period—and not predict more than is the case. It will be seen that in obtaining this we also obtain a record of the information in each analyzed utterance.

This is not to say that there is not additional knowledge which can be gleaned from other types of language data. If we knew all there was to know about the intent, perception, and context of each event of speaking or writing, we would doubtless learn more about the utterance, over and above what we can learn from analyzing the form. But we can obtain such knowledge only episodically and for the most part vaguely. And, when we analyze form, to use such knowledge only where we happen to have it interferes with reaching a systematic analysis of the form and the information it carries (e.g. if we explain the auxiliary verbs by the need to state modalities of actions, in contrast to combinatorial and likelihood factors as in G298).

It must nevertheless be noted that the data about the form of utterances are fuzzy in marginal cases. There are utterances for which it is not clear whether they are 'in the language' or not—speakers may disagree, or not be sure. And there are utterances which have such different meanings that the question arises whether they are not different utterances though with the same form. Here as in general, it will be seen that a theory constructed first of all to fit all such utterances as present no such problems can be then used to account for the marginal and problematic utterances.

2.2. *The relevant method: Least redundancy in description*

To investigate the form of utterances, we consider the entities involved and their change or combination. We note immediately that no matter what the entities, some combinations do not occur in utterances: *tpiks* and *musp* are not English words, nor is *slept the a the* an English sentence. Clearly there are departures from equiprobability in the combinations.

In language such departures from equiprobability are not only universal but necessary, because the whole set of human languages has no external metalanguage. This is a crucial consideration. For in order to describe the form of a language—or of any system—we have to identify its parts or properties, their changes, and the relations among them. In the case of logic and mathematics, the statements and formulas which are comprised in them have such explicit forms—of symbols and their combinations—that it is clear that statements able to describe logical and mathematical formulas do not themselves have the structure of those formulas. The statements that describe a system are in a different system, called the metalanguage—in this case the metalanguage of logic or mathematics—which is richer in certain respects than the system it is describing. In the case of natural language, we have no different system in which the elements and combinations of language can be identified and described. The identifications for a corpus of language utterances can indeed be made, but only in the sentences of a natural language (the same or another)—that is, in sentences which have already been built out of the same kind of elements and combinations which they are describing (5.1). We cannot describe the structure of sentences except in a system which itself has sentences, with predications and word-likelihood differences. Furthermore, no external metalanguage is available to the child learning its first language, nor to early man at the time language was coming to be.

In the absence of an external metalanguage, one could seek to identify the entities of language by extra-linguistic means, such as the occurrence of words in life circumstances that exhibit their meaning. Many words may indeed be identified on such grounds. But many other words, and the ordering which makes out of them sentences as against ungrammatical word collections, cannot be thus identified. In contrast, when only a small percentage of all possible sound-sequences actually occurs in utterances, one can identify the boundaries of words, and their relative likelihoods, from their sentential environment; this, even if one was not told (in words) that there exist such things as words (7.3). And when identified words combine with each other in relatively few regular ways which are used throughout the course of an utterance, one can recognize utterances, long and short, in distinction to the non-occurring sequences of words; and one can recognize within long utterances segments (sentences) identical to otherwise-observed short utterances. This holds even if one is not told that there exist such things as sentences (7.7). Given, then, the absence of an external metalanguage, it is an essential property of language that the combinations of words in utterances are not all equiprobable, and in point of fact that many combinations do not appear at all.

It follows that whatever else there is to be said about the form of language, a fundamental task is to state the departures from equiprobability in sound- and word-sequences. Certain problems arise. No one has succeeded in characterizing the departures from equiprobability by fixed transitional probabilities between successive words of an utterance; the reason turns out to be that the overall characterization of word-configurations is a partial ordering of words (3.1). Furthermore, since the set of all possible word-sequences in a language is in principle denumerably infinite while the grammar of a language must be finite (2.5.1), grammar as a catalog of the combinations which occur in utterances is not a practical program. Finally, many combinations (utterances) are marginal: it is not clear whether they are or are not in the language. All these difficulties can be circumvented by seeking out not a list of attested combination but a set of constraints on combination (1.1). A few constraints can together preclude unboundedly many word-sequences while allowing others; and constraints stated on different or partially fuzzy word domains can create marginal forms whose satisfaction of the constraints may then be in question.

If we obtain the occurring combinations, as departures from equiprobability, by stating the constraints that allow them, we are specifying (in the constraints) the various contributions to non-equiprobability that together make up the total non-equiprobability of the system. It is therefore clear that the constraints as formulated should not introduce any additional redundancy, over and above the least needed to account for what is actually present; for the non-equiprobability of language is precisely what we are trying to describe. As a simple example of appealing to fewest constraints, consider the difference between *walk*, *walked* and *go*, *went*. In *walk*, we have one of a large set of operators which occur with the present, future, and past tenses. In *go*, we have a single word having all properties of the operator set except that it does not in general occur with past-time adverbs such as *yesterday*; and in *went*, we have a word with the operator properties but which occurs only with past-time adverbs. This case requires us to partition the operator class into three subsets (*go*, *went*, and the others); this partition correlates with no different properties of that class. In contrast, we can say that all operators occur with all time-adverbs (except for certain quite other problems with 'auxiliary verbs' such as *can*, *used to*), but some words have different phonemic content in different environments, something which is known for quite a few words in different cases (5.4). In the latter case, the theory has to carry an intermediate ('morphophonemic') level of elements between phoneme and word: certain phoneme-sequences (e.g. *walk*, *go*, *wen-*) are bounded by word or morpheme boundaries, i.e. are word-size segments; and every word is a set of one such segment (e.g. *walk*) or of a disjunction of 'alternant' such segments (e.g. *go/wen-*). But this level, of morphophonemes and word-alternants (7.4), is needed for many words in various situations, and the constraint involved in specifying it for *go* affects a far smaller domain than the constraint involved in partitioning the operator class to suit *go*.

If, then, we have for a language different descriptions, adequate to characterize its utterances, but with different amounts of departures from equiprobability ascribed at various points in the course of the descriptions, we opt for the one with least such departures, since that one has clearly added least to the inherent departures in the language being described. The consideration of least non-randomness in description is involved in the discovery of phonemes, which underlies the whole development of a science of language, and it leads to many other methods in the analysis of language. The effect of the least-redundancy test, when applied at each point in the language description, is a grammar with fewest possible and most independent objects (elements and constructions), fewest and least-intersecting classes of objects, fewest and most independent rules (relations, transformations, etc.) on the objects, fewest differences in domain for the rules, and finally, fewest abstract constructs. Such a grammar is least redundant because each object (other than arbitrary ones defined only by their relations), and each rule or relation and classification, and each statement of domain for these, is a constraint on the equiprobability of parts in the utterances which are being described.

At each analysis of a linguistic form one would therefore seek to add as little as possible to the system describing the forms thus far analyzed. Certainly, one would not repeat information already given. For example, if a given phoneme- or letter-sequence (say, *attempt*) has been stated to have a particular meaning and particular other words as high-likelihood ('selectional') arguments, one would avoid giving the information twice: once when the word appears in verb position (*They attempt to stop smoking*) and again when in a noun position (*Their attempt to stop smoking failed*); the preferred alternative would be to take the second form as being the same sentence 'nominalized' under a further operator.

More generally, occurrences of a word-size segment in two different environments are not given two different grammatical classifications, if there is any regularity in these two environment-differences, that is, if we find other cases of a word-size segment in two comparably different environments (7.4). For in such cases, one could formulate a derivation that would carry each of these words from one environment to the other, without claiming that a given segment constitutes a different word in each different environment. This avoidance of multiple classification applies to many situations: one case is the noun–verb difference exemplified by *attempt*, above, where one can note that when a sentence (*They attempt to escape*) becomes the argument of a further operator (e.g. *succeeded*) its verb takes on noun form and the arguments of that verb receive adjectival forms (*Their attempts to escape finally succeeded*). That is to say, nominalization (i.e. occurrence in noun—'argument'—position) happens not to verbs (or operators in general) but to sentences: nominalized verbs are not a word class, but nominalized sentences are the form that sentences take when they are arguments of a further operator. Thus *The attempt succeeded* is a reduction from *Someone's attempt succeeded*, or the like. In another case, instead of saying that *eat*, *write*, etc., are each both transitive and intransitive (*John writes poetry*, (1) *John writes*), we say that under these transitive verbs the object is zeroable when it is an indefinite noun (*John writes things* is reduced to (1)); this zeroing is widely attested elsewhere.

There remain cases where certain occurrences of a word-size segment have no recognizable similarity to other occurrences in respect to their selection of high-likelihood neighbors; here we cannot avoid assigning the different occurrences to different homonymous (i.e. same-phoneme) words. This can occur when the segments are in the same gross syntactic positions, as in the two nouns *heart* and *hart*, or in different ones, as in the noun *sea* and the verb *see* (or *a prune* and *to prune*). Homonymy may be the most efficient analysis even in some cases where the different occurrences are etymologically identical but differ grossly in selection (and so in meaning), as in *a smart fellow* as against *a smart blow* and *He smarted from the blow*, or in *a humor magazine* as against *the four humors of the body*.

The same minimalizing considerations that go into identifying the elements go also into identifying the relations among them, these identifications being in any case two sides of the same coin. The analysis of a language form should make maximal use of analyses used for other forms, avoiding as far as possible recourse to new (*ad hoc*) classes, rules, domains. To explain a form, then, is to find its maximal similarity to the other forms, with the whole grammar thus becoming a best fit for the data. This means analyzing each form in respect to the whole relevant grammar—and the grammar of a recognizable linguistic community, not of an arbitrary speaker (since language is essentially a public structure).

Making a least grammar for an unbounded set of sentences entails that as many words and sentences as possible (and their meanings) be derived from others (by changes and additions), preserving whatever has gone into the prior making of those others. And it entails that the derivation be in small steps repeated in various combinations rather than large steps each of which accounts for a whole difference between starting-point and end-point. A large step, such as the totality of differences between the active and passive forms of the verb, can be used only at one point of the grammar—in this case, to derive the passive. But if we can obtain its effects from a succession of small steps (where the resultants of one step are in the domain of the next), we may find that the same small steps occur elsewhere in the grammar, in different combinations (for the passive: 1.1.2, G362). We thus seek to have fewest derivational steps, maximally repeated in different applications. An example is seen in the elementary transformations of 8.1.

Given these considerations, when we find a word—with more or less the same phonemes and with clearly related meanings—appearing in different grammatical statuses, we do not simply give it a multiple classification, but rather, as noted above, use one status as derivational source for all the others. Examples are the many words which appear in English as noun and verb (e.g. *to house* from *a house*, and *the day's take* from *to take*); or as verb with and without object, or with different object structures (e.g. *read* from *read things*, and *expect John* from *expect John to come*); and the many environments and uses of the prepositions (G80). Also, if words or morphemes

undergo fixed kinds of meaning changes ('lexical shift') in different grammatical conditions, we would try to attribute the meaning-shift to the
changed grammatical conditions. Examples are abstract nouns as being
nominalized verbs, more precisely as nominalized sentences whose subject
and object have been zeroed (e.g. *growth* from *the growing of things*);[1] and
metaphors (4.3) from paired sentences under the connective *as* (e.g. *I'll buy
your argument*, roughly from *I'll take your argument as one takes things when
one buys them*).

In all this, it should be noted that no assumption is made that there exists a
structure in language, and no appeal is made to any particular principles of
structuralism. Such structure as is found comes out in the process of making
a least description.

2.3. *Simplicity*

The need to minimize redundancy in grammatical description means that
what may informally be called the overall simplicity of the grammar is not
only a matter of elegance or science policy, but a methodological requirement. Furthermore, since language necessarily has a constructive characterization (6.0, 2.5), the simpler the grammar, the fewer different processes
are needed *in toto* to reach all sentences. The total of simplicity is in large
measure the outcome of specific contributions to simplicity, each due to
specific conditions in the data. For example, if a sentence undergoes several
expansions (additions) and reductions (or other transformations), it does
not have to be assumed that they all take place in one order or another.
Some may be defined on the same form and thus take place simultaneously,
independently of each other. Also, since language contains not only long
sentences but also short ones, in which the possibilities of combination
among the entities are fewer, one can seek a simpler system by trying to
analyze the longer sentences in the same way as the short ones, or as
combinations of the short ones.

Yet another example arises from the fact that since languages change
through time, there may always be some forms which do not fit perfectly into
the current grammar. The total system is then simpler to the extent that it
minimizes the disturbance due to such on-going changes. This suggests that
rules be so stated as to be able to house current derivations in repect to
combining-possibilities, phonemic shape, position, etc.—formulations in
which the relevant past or the immediate future of forms would be minor
changes in condition or domain. A major simplification can be achieved in

---

[1] In meaning, *growth* may seem to have also other sources, as when it means 'amount of
growing'. However, the other meanings can be derived from the same *growing* with zeroed or
implicit *amount* (*of growing*), *product* (*of growing*), etc.

many cases by considering not just a form but also its 'behavior', i.e. the environment and other factors which characterize its occurrences. It has been noted here that the general 'behavior' of words is such that it is easier to state the constraints which preclude certain types of word combination than to state all the types of word combinations that can appear in a language.

However, simplifications in any one part of the grammar have to be weighed against their effect in other parts of the grammar. To see this in practice requires a detailed example such as the following: The description of sounds is simplified when we collect sounds into phonemes, roughly on the condition that these sounds have complementary environing sounds (7.2.1). On this basis, English *h*, which never occurs at the end of a syllable, and the *ng* sound (as in *sing*), which occurs only at the end of a syllable, could be considered members of one phoneme, '*ng/h*'. (Syllables are phonetic, non-syntactic, restrictions on phoneme-sequences.) But this would disturb the phonemic similarity (7.3) of certain words when these occur in different combinations. For example, *long* ends in the *ng* sound, but *longer* can be pronounced either with *ng* alone or with the *ng* sound followed by *g*; and the related word *longitude* is pronounced only with *n* followed by *j*. If we take *ng* as a member of an *ng/h* phoneme, then *long* and *longing* and *longer* contain the *ng/h* phoneme, while *longer* (with pronounced *g*) contains different phonemes, *n* and *g*, with *longitude* containing *n* and *j* phonemes. If, however, we take the *ng* as a sound restricted to syllable end (with no relation to *h*), it can then be analyzed as a syllable-end composite of the *n* phoneme followed by the *g* phoneme. Then the two-phoneme sequence *ng* at word-end usually has the single sound *ng* (*long*, *longing*, *longer*), while the other pronunciation of *longer*, and *longitude*, can be related to the presence of the same *n* and *g* (or *j* replacing *g*) phonemes. The second analysis fails to maximize the collecting of complementary sounds into phonemes, but gives a simpler phonetic characterization of phonemes and a simpler phonemic composition ('morphophonemics') to words such as *long* in their various combinations. (Note that the reason against *ng/h* was not the lack of phonetic similarity between the sounds but the combinatorial simplicity at the next level, that of words.)

There are many other situations in which consideration of systemic simplicity affects the analysis of a given form. In many cases, it is possible to find an analysis in which both elements and the constraints on their combinations are simplified. Thus the classification of phonemic distinctions into phonemes, of word-size segments into words, of words into operator classes, are all done not only for reasons of taxonomy, with the general simplicity advantages that taxonomy provides, but much more, explicitly, because these classifications plus the constraints stated on the classes (phonemes, words, operators) are together far simpler and fewer than the constraints that would have to be stated on the individual phonemic distinctions, word-size segments, and individual operator words. The sets that are defined in the methods discussed here are thus characterized by relations among their members (cf. 2.5.5).

Nevertheless, there are also forms for which any simplification in one respect involves complicating some other part of the description. This means that the relevant simplicity has to be measured not in respect to the analysis of each form separately, but in respect to the whole system that describes or predicts the utterances of the language. It includes the simplicity of the primitive elements and their properties, and that of their operations or relations, and of the types of domains on which these latter are defined— and also the simplicity of the relation between the forms and their meanings, which is an essential concomitant of them.[2]

The simplicity of the system is greater if any property or analysis that is stated at one point in the description of a construction is preserved under later developments in it. Every preserving of a property saves having to state it separately for the different occurrences of the words. We try to find elements and relations such that our analyzing a construction in terms of them is not affected by further processes acting upon it.

In particular, no analysis of a construction should have to be corrected by the analysis of a later element operating on that construction. For example, (1) *A whale is not a fish* should not have to be derived from an assertion (under sentence intonation) (2) *A whale is a fish* to which is added (3) *not*, denying that assertion. Also, given the true (4) *Lead paint is dangerous*, we would not want this assertion about *paint* to be derived from the false (5) *Paint is dangerous* which is an assertion about *paint* in general, plus the true (6) *The paint has lead* (or the equivalent).[3] This does not mean that we forego relating (1) to (2) or (4) to (5). Rather, we obtain the sentence (1) *A whale is not a fish* from (3') *One denies* (or *I say the denial of*) operating on the unasserted intonation-less sentential construction (2') *A whale is a fish* (G321). And, for other reasons, we obtain *Lead paint is dangerous* from *Something, which is paint which has lead, is dangerous* from (5') *Something is dangerous* to which is adjoined (6') *This something is lead paint* from *This something is paint; this paint has lead* (G124). It is a contribution to simplicity to have each step in the making and analysis and interpretation of a language construction be accretional, as in the primed derivations above, and not require a reconsideration of what has been done up to that point.

A more complicated case of this parsimony of analysis is the avoidance, in analyzing a form, of going beyond what the language itself has distinguished. For example, in such pronouns as *he*, it is not possible, except in special

---

[2] Since each relation has a specifiable domain (in some cases fuzzy) and each acts only on the resultants of prior relations (down to the sounds), we can in principle measure the effect of each relation, or each proposed analysis, in the whole grammar (10.4.5)—and even which proposed grammar is 'least'.

[3] The truth of a sentence is a special case of its meaning, relating the information of a sentence (11.6) to axioms or statements believed to be indeed the case (10.5.4). Unlike meaning in general, it is not a known property for every sentence.

contexts, to determine uniquely from what is said who of several preceding individual human males is being referred to. A simplest grammar would not try to outdo language by creating a set of determinative readings, one for each possible antecedent, which may then be narrowed down by later contextual material. (Each reading would then have a definite antecedent in the way that the language as spoken does not have.) That such alternatives would take us beyond the structure of the language becomes evident when it is found (5.2) that in all cases where grammar has definitely located antecedents for pronouns, it has done so by special devices (e.g. bringing the pronoun to immediately after the antecedent). This means that fixed ante-cedents are provided for by creating special easily recognized relative locations; and where this has not been provided, it may be judged that the uncertainty as to antecedent is not due to there being a disjunction of determined antecedents, but rather to there not being determined ante-cedents at all in the given grammatical case. *He* then means 'someone mentioned' rather than 'this or that one or the other'.

By the same token, although language has in principle the addressing capacity of stating the relative linear location of every phoneme and word in a sentence, this capacity is not part of a grammar, because universally it is not used. Indeed, languages use words for referring to easily recognized locations, e.g. *latter*, without any system of overall addressing. To reject a description that overreaches language means that a theory and notation should at no point produce forms of distinctions beyond what language can contain, just as it must not produce less than what language contains. This suggests a theory fitted uniquely to language data (including its marginalities), neither more nor less, rather than a more powerful system which must then be whittled down by separately stated conditions.

A final consideration of simplicity: in addition to the internal simplicity of the grammar, one must require that there be least metalinguistic burden. This does not refer to the simplicity of the structure of the metalanguage. The metalanguage will be seen in 10.2 to be a sublanguage of the natural language, one whose grammar is partly different from that of natural language. We are here not speaking of its own structure, but of the discourses in it: how much has to be said in it about the language which it describes. We can avoid special metalinguistic comments about special events (forms and their 'behavior') in the language if we can obtain these events from the language grammar (which is in the metalanguage) with no addition to it, or from an addition which may be useful for other possibilities in the language. For example, *kerb* can be written (primarily in British English) for 'paving curb' but not for other occurrences of *curb*. This peculiarity can be included in the grammar if we say that by the side of the environmentally limited morphophonemic (sound) alternants of a word, such as *go* and *wen-* above (2.2), there may also be environmentally limited spelling variants, such as *kerb* in particular word-environments. This adds nothing to the elementary vocabulary of the metalanguage. For another example, Edward Sapir had once noted what would seem to be a life-situation restriction on word use: only a person who had studied mathematics at school can without dissimulation call it *math*. Here we can say that *math* is a morphophonemic alternant not for *mathematics* but for *mathematics as known from courses*; this provides the same restriction for the word use without leaving the confines of the ordinary grammar. In both examples the solution involves specification of the word's environment, something which is already known to be relevant.

Considerations of metalinguistic burden suggest cautions against both reductionist and teleological formulations in a science. In the case of reductionism, a given science is stated in terms of a prior science of it—one whose resultants yield the elements of the given science—without indicating that these elements have wider capacities than the prior science could have provided for. If indeed such wider capacities exist, that fact cannot be disregarded, and if not included in the formulations of the given science that fact would have to be left for informal expression in its metalanguage.

In the case of teleological terminology, the entities of a given science are named not solely by the processes that create them but rather by their end product, in particular by what they lead to in a successor science or field, one to which the given science is prior. While this end product may be of central import to the successor science, it may not play a role in the processes of the given prior science. Hence any naming or explanation made in end-product terms within a given science is a burden either on the grammar of that science or on its metalanguage.

2.4. *Can meaning be utilized?*

Language is clearly a carrier of meanings, both in its words and in the particular ways the words are combined. In considering methods for analyzing it, one might think that the forms—the words and constructions—of language, and the meanings which they carry, can be described each independently of the other. However, there is a feature of language which requires that the two be interconnected. This is again the absence of an external metalanguage. It has been seen in 2.2 that this lack makes it impossible to identify all of the elements in the flow of speech without noting the constraints on their combination. We will see here that this same lack makes it necessary to refer to constraints on combination also in specifying the meanings of some of the elements.

The reason is that there is no way of defining the meaning of all of the words, and of the constructions created by hierarchies of constraints on their combination, except in language. Other means, such as pointing to the referent of a word, or understanding the meaning of a word or sentence from the situation in which it is said, are undoubtably adequate, and essential, for grasping the meanings of many words and short sentences. But they cannot suffice for every word, nor for its different meanings in different combinations. The meaning of such words as *time*, *consider*, *the*, *of*, and the secondary-sentence meaning of the relative clause structure, cannot be adequately gathered from the real-world situations in which they are used. Also, any metalanguage in which their meaning could be given would have to have already for its own words the kinds of meanings which are being explained. Hence, these less obvious meanings can be learned—by speakers of the language and by analysts of language—only by much experience with the neighboring words and sentences with which they occur. Indeed, when one learns the meaning of a word from its dictionary definitions, one is learning from its sentential environment—the words of the definition and the examples of usage—and not from an extra-linguistic experience with the word. Furthermore, many words vary the combinations into which they enter, and with that their meaning. We cannot in general learn the change in co-occurrents from a prior change in meaning, but we can learn the change in meaning from seeing the change in co-occurrents: e.g. *meat* from earlier 'food', *fond* from earlier 'foolish', *provided* and *providing* in their new conjunctional meaning 'if' (11.1).

It follows that for an adequate knowledge of meanings in language, we need to know the constructions, in addition to an unspecified amount of semantic information garnered from the situations in which earlier-heard words and sentences had been said. Linguists have discussed when to bring meanings into the analysis of language—before, after, or simultaneously with the grammar. However, such a choice does not really arise. This is so because the meanings that are expressed via language (and even more so those which are inexpressible in language) are not in general discrete and definite items which can be mapped onto the words of linguistic constructions. There are a few sets of meanings which we can know precisely independently of language: for example, numbers, biological kinship relations, the spatial coordinates as we experience them. In these cases, we could start with the meanings and ask how a language expresses them. But this method cannot be extended to the vast bulk of meanings that correspond to words and constructions of a language. There, we cannot list a priori a set of meanings covering the relevant experiences of a given community. We have to see first what the vocabulary of the language is and then how it is used, in what combinations and in what constructions. We cannot say a priori what objects or actions or relations are included in one meaning as against another: we have to see what words combine with *chair* as against *bench*, or with *slip* as against *slide*, or with *from* as against *out of*.

In addition to being not sufficiently knowable independently of the word combinations in a language, the meanings are not precise enough to specify all combinations of a word, and are not sufficient to predict how words will combine. For example, from any purely semantic definition of *full*, *empty*, *equal*, one would not expect that the language would contain *fuller*, *fullest*, and *emptier*, *emptiest*, while *more equal* does not occur in ordinary English (though it is used in George Orwell's *Animal Farm*, without taking us outside of the English language). A person, e.g. a foreigner, who selects words for his sentences purely on the basis of even the fullest dictionary definitions will produce many combinations that everyone would consider wrong, and would miss many normal combinations.

Not only is it untenable to use meaning as the general framework for language analysis, we also cannot use it as an occasional criterion in deciding how to analyze linguistic sentence structure, for example, at points where finding a simple combinatorial constraint is difficult. For one thing, the decision where to use it would be arbitrary and would be different for various investigators. For another, as Leonard Bloomfield pointed out, appeal to meaning when formal analysis is difficult cuts off further attempts to find a formal 'explanation' (in the sense of 2.2). When one perseveres in purely combinatorial investigation, satisfactory explanations are reachable in various cases where meaning seemed an easier way out—for example, in the analysis of pronouns (5.3.2, 2.3) and of metaphor (4.2). More importantly, when the whole language is analyzed consistently in terms of constraints on combination, with no freedom for the investigator to beg off from that method, language is found to be built out of a few virtually exceptionless rules whose force does not appear when we look only at partial analyses of language, or when constraints on combination are referred to only sporadically. Also, the grammatical constructions that are obtained in a purely formal analysis turn out to have sharper correlation with meanings than do those recognized in ordinary partially semantic grammars (cf. the last paragraph of 2.4).

Meanings in language, then, are meanings of words and meanings of constructions—those words and constructions which are in the language, and which have been known before we ask their meanings. (And indeed, the boundaries and identities of words and sentences can be established by statistical means, with no knowledge of the meanings.) Thereafter, we find that the meanings are not additional properties unrelated to the syntactic forms, but are a close concomitant of the constraints on word-choice in the operator–argument relation and on the participation of words in various reductions and constructions (11.1, 3).

All this is not to deny the usefulness of considering meaning in formal investigation, let alone in studies of language use, language arts, and the like. Meaning, and especially the perception of difference in meaning— whether between two forms or between two occurrences of a form—may be used as a preliminary guide where to look for possible constraints. It is clear from all the above that in no way is meaning being excluded here from the study of language, but that it is not being used as a criterion in determining the analysis of forms. Indeed, general considerations of meaning will be obtained from considerations of form (11.4).

A final word about the relation of meaning to the investigation of forms: it has been seen that a description of the combinations of sounds or of words may contain two sources of redundancy: redundancy due to the way the classifications and rules are stated, and residual constraints due to the language itself. To the extent that meaning is connected with constraints on combination, it is connected to the residual ones, and not to those imposed by the form of description. Therefore, in any case, we have to reach the description with least redundancy (2.2) before we consider meaning and its interpretive relation to the constraints of form (11.4).

2.5. *Properties of the data*

2.5.0. *Introduction*

Language has several universal properties (4.4, 6.1), in addition to the constraints on word combinations on sentences. These too affect what methods are relevant for analysis.

2.5.1. *Unbounded language, finite speakers*

First is the fact that the set of utterances in a language (as describable in a grammar) is unbounded. There is no reasonable way of specifying what would be the longest sentence, and almost every sentence can have something added to it, whereas the speakers and understanders of the language are finite beings. Many things in language—what sounds are distinguished phonetically, what phoneme-sequences are pre-set as words or morphemes, what kinds of word combination are used—depend upon a finite body of public experiences on the part of speakers and hearers. This means that every language must be describable by a constructive (recursively enumerable) system, one in which there is a finite and reasonably learnable core upon which repetition or recursion without limit—and also change—can be stated (6.0), with possibly a finite learnable stock of 'idiomatic' material which is not subject to these unbounded operations.

2.5.2 *Discrete elements*

The success of writing systems, all of which use discrete marks, as carriers of language suggests that language can be represented with discrete elements. In addition, it will be seen in 7.1 that in the continuous events and changes in speech one can find ways of making cuts and distinctions which create the discrete elements of spoken language and of structural language change. It has also been found that the events which remain as continuous elements do not combine complexly with other elements, and so in effect remain outside the grammar. Once discrete elements are reached, the choice of relevant methods for further investigation is greatly narrowed.

2.5.3. *Linearity*

Speech has extension in time only, and its repesentation in writing is primarily a sequence of marks, along a line, although concurrent marks are used in some cases. These facts suggest that language can be described linearly, or at most multilinearly (with the observed event being a super-position of several concurrent linear events). The linear representation, however, does not mean that the elements of language are linearly ordered in respect to structure: they may be partially ordered with a linear projection, as indeed is proposed here. Not surprisingly, those linear analyses of language which succeed are not directly in respect to word-sequence, but are in terms of specially defined relations, commonly called grammatical relations, among the words of the spoken or written utterance. These grammatical relations represent special cases of an underlying partial ordering of the words.

2.5.4. *Contiguity*

Talk or writing is not carried out with respect to some measured space. The only distance between any two words of a sentence is the sequence of other words between them. There is nothing in language corresponding to rhyme, meter, or beat, which defines a space for poetry and music, or to the bars in music notation which make it possible, for example, to distinguish rests of different time-lengths. Hence, the only directly experienced elementary relation between two words or entities in a word-sequence is that of being next neighbors, or of being simultaneous (especially in the case of intonations). Any well-formedness for sentence structures must therefore require in the base a contiguous sequence of (discrete) objects, although later permutations and operators may intervene. The only property that makes this sequence a construction of the grammar is that the objects are not arbitrary words but words of particular classes. But the sequence has to be contiguous, it cannot be spread out with spaces in between, because there is no way of identifying or measuring the spaces.

By the same token, the effect of any operation that is defined in language structure, i.e. the change or addition which it brings to its operand, must be in its operand or contiguous to that operand. Of course, later operators acting on the resultant may intervene between the earlier operator and its operand, separating them; but this is a later operation and is in certain grammar situations avoided. In the description of the final sentence, such a separation (i.e. interruption due to the embedding of later operators) can be recognized. But in defining the action of the earlier operator on its operand, this separation cannot be identified; the separation can only have been due to a later event.

It follows that if language can have a constructive grammar, then there must be available some characterization of its sentences which is based on purely contiguous relations. The contiguity of the successive words is related to this situation, but does not in itself satisfy this requirement since the sentence has been found here to be characterized by a partial ordering of its words of which the linear order is merely a projection (3.1, 4). This contiguity can be utilized in designing a computer program to analyze the structure of sentences (ch. 4, n. 1).

2.5.5. *Variety within a class*

It will be seen in Chapter 7 that the search for a least redundant apparatus of description leads to many definings of classes, carried out in such a way that a property can be more simply stated about the class than about the individual members. This may happen because all the members have the same property, which we do not want to repeat for each member. Or some members may have only some of the class properties, so that they are only in some respects members of the class (e.g. the English auxiliary verbs, G300). And some members may have some properties while others have others, and it may be possible to say that two such members between them constitute a whole (composite, or disjunctional) member of the class (e.g. *go* and *wen-* above). In the last case, we can think of the composite member as being the true member, with the members that are included in the composite being regarded just as variant forms that the true (composite) member takes on in various situations. In all these cases it is enlightening to state a class property.

The forming of classes is thus not a 'similarity'-based grouping of the data but a complex way of finding the most regular way of attributing properties to entities, that is, of defining properties and their domains (2.3). Therefore, a consideration in forming classes is the variety, degeneracy, and other difficulties which are met with. For example, there is the question of how many and how unusual are the differences among the various sounds (allophones) grouped into one phoneme: this is equivalent to asking how similar the sounds of a phoneme are in its various environments. A comparable question is how similar are the phoneme-sequences that constitute a given word, in the various environments in which that word appears. We can say that *part* is a phoneme-sequence constituting a word; and *hart* is a phoneme-sequence constituting in some occurrences a word for deer and in others the word *heart*; and the two pronunciations of *economics* (with *e* and with *iy*) are two phoneme-sequences each constituting the same word; and *is*, *am*, *are* are three phoneme-sequences each of which constitutes the same word but over particular complementary arguments. Alternatively, using classification, we can say that *heart* and the deer name *hart* are two words whose phonemes are the same; in *economics*, the phoneme-sequence contains a morphophonemic variation between *e* and *iy* in the first position; and *be* has different alternant phoneme-sequences in specified complementary environments.

In classes of words there is the question to what extent they all take the same set of operations (relations) or environments. From the vantage-point of the operations or environments, the question is to what extent they have the same domain of words on which they act. Just as there are many classes with fuzzy membership, there are many operations and constructions with fuzzy domains—some being only slightly different from those of other operations, others being of indeterminate boundaries. In the latter case, one is not even certain that a given operation acts on a given word that is possibly in its domain, so that the resultant of the operation is a marginal sentence, such as *The baby took a crawl*.

An important question here is how much degeneracy arises when various operations are carried out on particular members in their domain, i.e. when one operation on one operand yields the same sequence of words—or of phonemes—as another operation on another operand. A major example is the case of sentences which are ambiguous because of different derivations: *Visiting relatives can be difficult* from *It can be difficult to visit relatives* or from *Relatives who are visiting can be difficult*; also *Frost reads smoothly* from *Frost reads in a smooth manner* or from *Reading Frost goes smoothly*.

There are yet other properties which have not so far been investigated in detail. Such is the distinction between what the speaker is constructing and what the hearer recognizes, and more generally between the synthesis and analysis of sentences.

2.5.6. *Synchronic single-language coexistence of forms*

The fact that there are various languages in the world and dialects within them, and that languages change through time, means that there are different possible objects for linguistic description and analysis. One could try to describe all languages together—either their intersection, namely what is universal and common to all, or their join or envelope, namely anything that is found in at least one language. One could also try to describe a single language or family through time, giving the succesive forms of particular elements and constructions. However, there is special importance to being limited to a single language community (disregarding mutually understandable, or hardly noticeable, differences of geographic and social-class dialects) during one 'synchronic' period (during which change is too slow to be seen as such, hence a time-slice). When language is described within these boundaries, all its attested forms, meanings, and distinctions coexist for its users. They can occur together in the same constructions and sentences and discourses, made and understood by the same community of users. Not surprisingly, it is under these conditions that the description of language yields a fully intertwined grammatical system, for it is here that we survey the forms and the constraints that are used together.

2.5.7. *Diachronic change*

Natural languages are open: new sentences and discourses can be said in them. Furthermore, natural languages change in time. As far as we can see, they do so without at any point in time failing to have a largely coherent grammatical structure.

At any moment in the history of a language, it is possible to make as complete a grammar of the language as we wish. No item of the language need be left out as undescribable; any item which is not a case of existing rules of the grammar can be listed separately or fit in (as a special case under special conditions) to some existing rule in respect to which it can be described. That is, we can describe a language at any time t_1, giving as complete a description as we wish and necessarily including a large number of individual facts (operations whose domains are individual words). At any sufficiently distant time t_2 we can do this again, and in all likelihood there will be some difference in the two descriptions, in respect to some items X, due to changes in the language in the intervening period.

Since t_1 and t_2 are arbitrary, and since there are few if any discontinuous points in a language history (although there may be such in the description of language history), the description used for X at t_1 must be valid up to some period t_i, $t_1 < t_i < t_2$, and the description used for X at t_2 must be valid from t_i and on, without there being a discontinuity at t_i. We conclude that during the period t_i, the grammatical items X must have been describable in two different ways. Before t_i, the item might be described in one way, to fit it into certain features of the grammar. After t_i, it will have changed sufficiently so as to require a different description, fitting it into some other features of the grammar. At t_i, both descriptions must have been possible, i.e. at t_i the amount of change in X must have been sufficient to make X fit the t_2 features, but not so great as to make X no longer fit the t_1 features.[4] To this

---

[4] It does not matter whether an individual changes his speech during t_i, or whether the items X are used differently by people who began to speak after t_i as against those who began to speak before t_i.

extent at least, some forms in language may have non-unique analyses. The situation at t_i is indeed often observable in detailed grammars, e.g. in the case of transformations which are in progress. Thus in *He identified it by the method of paper chromatography*, we have two sentences connected by semicolon (secondary stress): *He identified it by a method* and *The method was of paper chromatography*. In *He identified it by the means of paper chromatography*, we can attempt a similar analysis into *He identified it by some means* and *The means were of paper chromatography*. This is, however, not very acceptable, and a more acceptable analysis would be to take *by the means of* as a new preposition similar to *by* itself. The latter analysis is already inescapable in *He identified it by means of paper chromatography*, since *Means were of paper chromatography* does not exist.

The possibility of a structural static grammar at time-slices of a changing system is due to the fact that there are similarities and other relations among the various domains stated in a grammar and among the various operations stated in the grammar. In terms of these higher-level systemic relations, the exceptional domains of an operation can be shown to be extensions or modifications of one of its regular domains.

This means that a description of language has to provide for the existence of items which don't quite fit into the rules for the rest of the language, but can nevertheless be related to those rules as extensions of their domain or small modifications of their operation.

2.6. *From method to theory*

The methods presented above were determined by the issue of constraints posed in 2.2, and by the nature of language data as seen in respect to that problem. Much of the nature of the data, such as its linearity and contiguity, is observable in speech before any analysis is carried out. But the possibility of defining discrete elements is observable without further ado only in writing, especially alphabetic; in speech it is not seen until after certain cuts and partitions are made in the set of speech sounds.

Carrying out these methods does more than provide an analysis of the data. It narrows the range of possibilities toward forming a theory of the constraints which were sought in 2.2. Such selective effect of the data methods upon the ultimate theory happens when the methods are scrutinized for their relevance (here, to the constraint issue), when they are carried out over the whole of an adequate sample (here, a sample of the whole language), and as fully as possible—so that the entities reached by the analysis are maximally independent of each other (i.e. are least constrained combinatorially), and when the analyses are made by these methods alone with no appeal to external considerations (such as phonetic similarity, meaning, intention, analogy, or history). This is not to say that such considerations may not have great weight, and may not even be the creators of particular forms and relations; but for purposes of approaching a theory we must first see to what extent a single relevant method can explain the forms, and to what (if any) extent forms due to other influences can be domesticated—reinterpreted—in terms of the central method.

It is especially possible to consider the analysis as laying the groundwork for a theory, when every set of elements and operations or relations that is finally reached is structurally related to a single language-wide property, rather than some of the sets being merely residues created by setting up the other sets. An example of the latter case would be if the set of inter-sentence derivations were not characterizable as reductions but were simply the set of whatever changes one needed to obtain one sentence from another, rather than a motivated, structured set of reductions and transformations.

The reason that the methods help to suggest a theory is that they are not simply empirical or descriptive. They necessarily organize the constraints into a maximally simple system, so that the description of language comes to consist of maximally unconstrained elements and rules (2.3).[5] Such a system may not be identical with the way the speaker produces his sentences or the way the hearer figures them out. But it has a unique least status in respect to the data of the language (2.2). And it has a crucial status in respect to the information carried by language, in that the information is certainly related to the departures from equiprobability in combination; and a most parsimonious grammar reveals most closely just the individual departures from equiprobability in a language (11.4).

It should be noted that the method leading to a least system is not reductive in a simplistic sense. It organizes the constraints, but it does not lose them. As an example: when constraints on certain entities (say, sounds) are used to define a 'higher' and more powerful class of entities (in this case, phonemes) which are freed of these constraints, the old constraints still apply to the members themselves within the new classificatory entities, but these old constraints do not interfere in the simpler relations that can be stated among the new entities.

---

[5] This is not to say that we have here a discovery or decision procedure for grammar. However, given a particular metatheory of language—the need to map the departures from randomness (2.1), which arises from the essential lack of an external metalanguage—we can propose procedures that lead to a least grammar, which constitutes a best fit of the data in the given direction, though not uniquely the best fit. It is nevertheless the case that the departures from randomness in the various languages are so massive and so similar that all grammars exhibit a strong family resemblance, no matter in what style they are stated.

II

Theory of Syntax

A Theory of Sentences

3.0. *Constraints on equiprobability*

It has been noted (1.0, 2.1–2) that a central problem for a theory of language is determining the departures from equiprobability in the successive parts of utterances. It has been found that in natural languages it is not merely that not all combinations of parts are equiprobable, but that not all combinations occur. That is, given even very large arbitrary samples of the utterances of a language, some word combinations have zero probability of being found. For the given language, then, we make a distinction between the combinations that have zero probability and those that have any positive probability. It will be seen below that the zero probability characterizes what is outside of syntax (3.1), while the differences in positive probability characterize the differences in word meaning within syntax (3.2).

Hitherto, we have been considering only the directly observable data of word combinations in a language. However, simply listing the combinations is not only impossible because of their number, but also fruitless because it leads to no relevant principled classification. Furthermore, any such listing would be imprecise since, for a great many word combinations, speakers of the language are uncertain or do not agree, as to whether they are in the language: given that *take a walk* and *take a run* are in, are *take a climb* or *take a jog* or *take a crawl* also in? Finally, since the details of word combination change faster than anything else in language, no exact list would be correct over any long period.

In contrast with list-making, it is possible to characterize the word combinations by stating a set of constraints each of which precludes particular classes of combination from occurring in utterances of a given language (1.1). These constraints will be found to hold over long time periods. They can be so formulated as to act each on the product of another constraint (beginning with a null constraint, i.e. with random word-sequences), so that a few types of constraint will suffice to preclude very many types of word combinations and produce the others. It will be seen that the constraints which produce the combinations found in a language can allow for many exceptional forms and changes, by minor fuzziness in the domains over which they are defined: e.g. why *America was left by me on July 6* is odd, but *America was left by a whole generation of young writers in the twenties* is natural (1.1.2, G365). The regularity of the constraints, and the simple relations among them, fit in with a mathematical formulation of syntax; and their direct specification of departures from equiprobability gives them a direct informational value.

Characterizing sentences in terms of constraints on combinations rather than of word combinations directly is a matter of method. But when the sentences are described as resultants of constraints, one can consider the constraints to be constructs in a theory of sentence construction.

Since sentences have not been defined at this point, these constraints are not, in the first place, stated in respect to the set of sentences. They can be said to hold over the set of utterances provided that, first, a small number of frozen expressions (such as *goodbye* and *ouch*), each with unique constraints, are excluded, and second, the period punctuation between sentences (or, in speech, between successive sentence intonations) is considered a conjunction roughly like *and*. In 7.7 a procedure for finding sentence boundaries will be presented, permitting us to define sentences. The constraints of Chapter 3 will then suffice for sentences. Further constraints, among sentences of an utterance, will then be presented in Chapters 9 and 10.

3.1. *Zero vs. non-zero likelihood*

3.1.0. *Introduction*

The crucial property of language is that the presence of words in a sentence depends on how other words in the sentence depend on yet other words in it. This dependence is the essential constraint on equiprobability of words in sentences. Though more or less empirically come by, this dependence can be considered a construct of the syntactic theory. Here, we first define the dependence (3.1.1), and then state what it means that the dependence is carried out in respect to further dependence (3.1.2). To a first approximation, we are speaking here, for example, of how the presence of a verb in a sentence requires the presence there of some noun as its subject.

3.1.1. *Dependence*

To begin, we note that in the sentences of a given language we can find a distinguished subset—called base—in which this relation of dependence is overtly exhibited (4.1). We will then see that all other sentences in the language can be derived in a regular way from those of the base, without altering the word-dependence, though with some loss of its visibility (4.2). In the base sentences, most words are morphologically simple (unimorphemic, i.e. roughly affixless).

We define the dependence as follows: If A is a simple word, and $b, \ldots, e$ is an ordered set of classes of simple words, then A is said to depend on (or, require) $b, \ldots, e$ if and only if for every sentence in the base, if A is in the sentence then there occurs in the sentence a sequence of simple words $B \ldots E$ which are respectively members of $b, \ldots, e$. Within the given sentence, A may then be said to depend on the word sequence $B \ldots E$. If in the given sentence there is no other word G such that A depends on G and G depends on the given occurrence of $B \ldots E$, then A depends immediately on that occurrence of $B \ldots E$. A is then called the operator on that $B \ldots E$, which in turn is called the argument of A in the sentence; B may be called the first argument, and so on. When the sentence consists only of A and its $B \ldots E$, the operator is necessarily contiguous to its arguments (2.5.4)—before, after, or between them. When another word F depends on A as A depends on $B \ldots E$ (3.1.2), the constraint would be simplest if the F is similarly contiguous to (the resultant of) A together with its $B \ldots E$; this is indeed seen to be the case. The contiguity of operator to argument in the base makes it easy to check if no word G as above intervenes in the operator–argument relation of A to $B \ldots E$. In *I know sheep eat grass* (1.1), *know* is the operator on the pair *I*, *eat* as its argument, with *eat* in turn as the operator on the pair *sheep*, *grass* as argument.

Deriving the remaining sentences from the base sentences is carried out by means of the reductions and other transformations, which are changes in the phonemic (alphabetic) shape of words in an utterance. In terms of them we can make a more general statement about dependence, without referring to the base set or to sentences at all: A depends on (or requires) $b, \ldots, e$ if and only if for every utterance in the language, if A is in the utterance then there occurs in the utterance a sequence of simple words $B \ldots E$ (members of $b, \ldots, e$), any of which may be in a reduced (including zeroed or permuted) form as provided by the reductions of 3.3. (The base contiguity described above is then modified by these reductions and permutations.) As an example: in *John expects Mary to come*, which is approximately in base form, *expect* depends on a word of the class of *John, Mary, coffee*, etc., and on a word of the class of *come, go*, etc.; *come* in turn depends on a word of the class of *John, Mary, coffee*. In *John expects Mary, Mary expects coffee* (which are not base sentences), it can be shown (3.3) that *come* is still present in both sentences though in zero shape (*John expects Mary to come, Mary expects coffee to come*), so that the dependences remain as before.

Dependence is a relation between a word and an ordered set of word classes (or, in a sentence, between a word occurrence and an occurrence of a sequence of words belonging to those classes). Over a whole language, most or all dependences hold not for a single word such as A above but for any word of a set, a; so that dependence can be considered a relation between a word class (e.g. a) and an ordered set of word classes ($b, \ldots, e$).

As examples, we take a few words and their dependences in short sentences.

First, *sleep*. We find *child* (as in *The child slept*—for *the* and tense see 4.2), *tree* (*sleeps all winter*), *earth* (*sleeps under the snow*), *night*, *universe* (*will then sleep until the next Big Bang*), rarely *algae*, *vacuum*, etc. In contrast, we do not find *go*, *eat*, *up*, *because*, etc. (not *Go sleeps*, *Going sleeps*, *Eat sleeps*, *To eat sleeps*). At this stage, the distinction between words that may be found, even very rarely, with a particular word, as with *sleep* here, and those that are never found is imprecise, and any characterization of such two sets of words would be *ad hoc*. However, when we compare, in any large corpus of data, the lists for *sleep* with the lists for certain other words, such as *fall* or *sit*, we find similarities, though with different frequencies, in what words do not co-occur. With *fall* we find *child*, *tree*, *book*, *night*, *earth* (*can fall into the sun*), rarely *universe*, perhaps never *vacuum*; and clearly not *go*, *going*, *eat*, *to eat*, *up*, *because*. With *sit* we find *child*, *book*, *night*, *earth*, *universe*, *algae*, almost never *vacuum*, and not *go*, *going*, *eat*, *to eat*, *up*, *because* (no *Because sits*). Even so, at an empirical level some problems remain: we may perhaps find *Going doesn't sleep* (though not *Go doesn't sleep*) and perhaps even *To eat doesn't sleep*. However, these differ in the circumstances of their use from *This child doesn't sleep*, *Algae don't sleep*, *Vacuum doesn't sleep*. For example, for the almost-never words in the set that includes *child*, *algae*, *vacuum*, we find different frequencies among them in respect to *sleeps* or *falls* (where they may be almost never in varying degrees) and in respect to *doesn't sleep* or *doesn't fall* (where all are appreciably less infrequent). In contrast, in the set that includes *of*, *eat*, *because*, all have the same frequency (zero) with respect to *sleeps* or *falls*, and all have the approximately same zero or vanishingly small frequency with respect to *doesn't sleep* or *doesn't fall*.

All this is not quite enough to establish a distinction into a few large classes: *child*, *vacuum*, etc. and *sleep*, *fall*, *because*, *up*, etc., both with respect to *sleep*, *fall*, etc. But it is enough to suggest a postulate that there is such a distinction. This postulate, then, holds that there is (a) a class of words (*up*, *because*, *sleep*, *fall*, etc., called O below) which have zero likelihood (or probability, 3.2.3) of occurring with words of a subset of them (*sleep*, *fall*, etc. called O_n below), and (b) a class of words (*child*, *tree*, *universe*, *vacuum*, etc., called N below) which have some positive likelihood no matter how small of occurring with that subset (O_n). In (b) any member of N may conceivably be found with any member of O_n, e.g. *The comet slept*, *Space slept*, *Time fell*; and indeed one cannot say which of these sentences, if any, is absolutely excluded (especially in fairy-tales and the like). It will be seen that the distinction into these two classes O and N, and such subsets of O as O_n, will show up throughout the grammar, and will conform to various informational differences.

Next, *wear*. We find here for the most part not single words occurring with *wear*, but pairs: *child*, *coat* (*The child wore a coat*), and *night*, *hue* (*The night wore a dark hue*), even *tree*, *bark* (*This tree wears a heavy bark*), or very rarely *child*, *bark* (*The child wore bark*); but not *child*, *eat* (no *The child wore eat*, or . . . *wore eating*), or *go*, *bark* (no *Go wears bark*, or *To go wears bark*), or *eat*, *because* (no *Eat wears because*). Much the same exclusion holds for *eat* (*child*, *egg* or *acid*, *copper*, etc. occur, but not the pairs *go*, *child* or *go*, *because*: no *Go eats because*), although more pairs can be readily found with *eat* than with *wear*. One might think that pairs like *child*, *egg* or *acid*, *copper* that are not found with *wear* should be excluded there. But although we do not find *The child wears eggs*, we find *She wears a set smile*, *She wears perfume*, *She wore roses in her hair and fruit on her hat* and we may find *The child wore egg on his face*. In any case, the set of words definitely excluded under *sleep*, *fall* (namely *because*, *go*, *sleep*, *fall*, etc.) is also excluded under *wear*, *eat*. And the words that are not excluded under *sleep*, *fall*, either because they are found there or because they could conceivably be found there, are similarly not excluded in both positions under *wear*, *eat*.

We thus find support for a class of words postulated to have zero likelihood of occurring with the *sleep* and *wear* set in the shortest sentences of the language—shortest either observably, or after correcting for the reductions presented in 3.3. This class will henceforth be marked O (operator), while the residue class, whose members have some positive likelihood no matter how small under the *sleep* and *wear* classes, will be marked N because most of them are simple (unimorphemic) nouns. The *sleep* set, a subset of O, will be marked O_n, and the *wear* subset O_{nn}, to indicate that they occur with one or two members of N respectively.

We next note that in any shortest sentence in which *is probable* or *is unlikely* occur, they are found with a member of the O class (including the O_n and O_{nn}) and not with a member of the N class alone (except, rarely, due to reductions). We have *wear* under *is probable* in (1) *That John will wear a tie is probable*, but we don't have *John is probable*. More precisely, the only members of N that are found in those sentences (e.g. *John* with *probable* in (1)) are the ones already accounted for as occurring with the O_n, O_{nn} members in question, as in the case of *John*, *tie* occurring with *wear* when *wear* occurs with *is probable* in (1). The case of *is unlikely*, *is probable* appearing only with N (e.g. *A boy is probable* in guessing about a birth) are accounted for by the reductions of 3.3 (in this case from e.g. *For the baby to be a boy is probable*), as are the cases of an O without its N (e.g. *To win is unlikely* reduced from *For us to win is unlikely* or other such sentences). The class of *is probable* is then marked O_o, since its co-occurrent (argument) is an O (which then brings its own co-occurrents into the sentence). Note that the O argument under O_o may be not only O_n (like *wear* above) but also O_o, as in *That John's coming is unlikely is quite probable*.

Words like *assert*, *declare*, *deplore* are found to have two co-occurrents with them: first an N and second an O, as in *John deplores your arrival*, where *John*, *arrive* are an ordered pair occurring as arguments of *deplore*, while *you* is accounted for as argument of *arrive* (an O_n). Hence the class of *deplore* is O_{no}.

Similarly, *entail* and *because* are O_{oo}: we may find *Your arrival entails John's departure* or *John departed because you arrived*. We do not find *You entailed John* or *Your arrival entailed John*, nor *John because you* or *John because you arrived*. As with O_o, the O_{oo} and O_{no} can occur on O_o or O_{no} or O_{oo} as well as on O_n, O_{nn}: in *John asserts that Mary asserts that you departed because I arrived*, the first *assert* (an O_{no}) has *John* (*N*) together with the second *assert* as its ordered co-occurrents (arguments); the second *assert* has *Mary* (*N*) and *because* (O_{oo}) as its co-occurrents; *because* has *departed* (O_n) and *arrived* (O_n); *departed* has *you* (*N*), and *arrived* has *I* (*N*).

The presence of an O_n thus depends on an *N* word as argument, since an *O* co-occurrent is excluded. And each O_{no} word depends on the ordered pair of *N*, *O*, since *O* words and *N* words respectively are excluded in the two argument positions. And so for the other words.

3.1.2. *Dependence on dependence*

We now consider the word classes on which a word depends, i.e. what classes appear in argument position. The crucial finding here is that each of these argument classes is characterizable not by a list of members (which would be difficult, since words can be added to a class or lost from it), but by what that argument class in turn depends on. For English, each argument class is characterized by whether it depends on something at all (i.e. on the presence of yet other words in the sentence) or on nothing: that is, by whether it in turn has an argument or not. In *John asserts Mary came*, we say that *asserts* (O_{no}) depends on the pair *John* (*N*), *came* (O_n); and *came* depends on *Mary* (*N*). In turn, *John*, *Mary* can be said to depend on nothing; they can also be said to depend on nothing except the operator that depends on them (reasons for the first formulation: see below and 3.2.1 end). Thus the presence of *asserts* in a sentence depends on the ordered presence of a word (e.g. *John*) that depends on nothing further, and of a word (*came*) which does have something further on whose presence it depends (in this case, *Mary*). No property (e.g. meaning or subclass) of the pair *John, came* other than their dependence status is essential for the presence of *asserts* in the sentence. The presence of a word, then, depends not on particular other words but on the dependence properties of the other words.

The status of *N* is clearly different from that of all the *O* classes. The question is what should be the systemic status of this difference. Here it may be relevant that the ability of an *N* word to occur alone as an utterance differs from that of all the *O* words. When a word occurs alone as an answer, that is only because its environing words were zeroed as repetition of the question:

Who was there? John from *John was there*; *What did John do all day? Study* from *John studied all day*. When an *O* word appears alone, in imperative intonation, it can be shown to have been reduced from a particular sentence structure (G351): *Stop!* from *I order you that you stop*. But when an *N* word appears alone, with or without exclamation intonation, it is far less clear what are the specific environing words if any that have been zeroed: e.g. the various uses of *John!*, or *A hawk!*, or the numbers (in counting). True, these grounds do not determine any single way of characterizing *N*. But they support the preference for the first formulation above, that an *N* word be defined as depending on the null class rather than on the operator over it.

Then by the side of the primitive arguments *N*, which require nothing, all *O* are operators, with non-null requirement. All $O_{n\ldots n}$, e.g. O_n and O_{nn}, require only words that themselves require null. All $O_{\ldots o \ldots}$, e.g. O_o, O_{no}, O_{oo}, require in at least one argument position a word that requires something, which in turn may require something or null. Such are, in English O_o (e.g. *probable*), O_{no} (*assert*), O_{noo} (*attribute*, in *John attributes his success to my helping him*), O_{oo} (*entail, because*). There is also a class O_{on} (*astonish*, in *John's winning astonished me*); and within O_{on}, the prepositions (e.g. *without* in *John left without me*) as operators almost all of whose members can also appear, due to various reductions, as if they belong to other *O* classes.

3.1.3. *Partial order*

Here, as at various points in the syntactic theory, we find a partial ordering, that is a relation ($\geqslant$) on a set, of being 'higher' or 'greater' ($>$) or else equal ($=$), in respect to which, for all *a*, *b*, *c* in the set we find $a \geqslant a$ (reflexive), and if both $a \geqslant b$ and $b \geqslant a$ then $a = b$ (antisymmetric), and if $a \geqslant b$ and $b \geqslant c$ then $a \geqslant c$ (transitive); but there may be members *a*, *b* in the set which are not comparable, where neither $a \geqslant b$ nor $b \geqslant a$.

The dependence relation imposes a partial ordering of word classes in respect to their definitions. The definition of *N* ('zero-level') includes no reference to any particular class. The definition of the $O_{n\ldots n}$ ('first-level') classes has reference only to *N*. And the definition of the $O_{\ldots o \ldots}$ ('second-level') classes (i.e. operators having an operator as their argument) includes reference to *O* (of whatever argument) and in some cases to *N*. This yields a partition of the words of the language into the three levels named above, and within the first and second levels into the argument-requirement subclasses O_n, O_{nn}, O_o, O_{no}, etc. (6.3(d)).

When this relation is applied to the words of a sentence we obtain a partial ordering of them. In each sentence, in the set of word-occurrences $A, B, \ldots$, Z we have a transitive relation: $A > B$, when *B* is in the dependence chain of *A* (that is, *B* is the immediate argument, or argument of the argument to any depth, of *A*). When *A*>*B* and there is no *C* such that *A*>*C*>*B*, we say that *A* is the operator on *B* (*A* covers *B*), and *B* is the (immediate) argument of *A*. When *A* covers both *B* and *D*, we call *B* and *D* co-arguments of *A*. The co-arguments of an operator occurrence are linearly ordered: *A* on *B*, *D* does not form the same sentence as *A* on *D*, *B*. In *John expects Mary to come* we have *John*, *come* as co-arguments of *expect*, and *Mary* as argument of *come*. The ordering is a semilattice; and no word-occurrence can be an immediate argument of two operator occurrences (6.3).

If one wishes to think of the making of a sentence as a process, one can view the semilattice, that is the dependence order of the words in a sentence, as the partial order of entry of the words into the making of the sentence. No word can enter into a sentence, which is being constructed out of words and previously made (component) sentences, unless its prerequisite is available—this means, unless its required argument is present and is (under the lattice condition) not serving as immediate argument of any other operator in the sentence. The prerequisite is any one word of each of these classes on which the given word depends. Then each entry of an operator makes a sentence, out of a non-sentence, namely *N*, or out of a sentence (that is, something that has been made into a sentence by a prior-entering operator). Thus *come* made a sentence by operating on *Mary*, while *expect* made a sentence by operating on the co-argument pair *John* and *come*. The linear ordering of the co-arguments means that *expect* may not be defined (and indeed is not) on such a pair as *come*, *John* (no *Mary's coming expects John*). And though both *John saw Mary* and *Mary saw John* are sentences, they are not the same sentence, nor is one derivable from the other or paraphrastic to it. The similarity of operators to functors in categorial grammar within logic is evident.

3.1.4. *Predication*

The major semantic yield of the dependence relation is that it carries inherently the meaning of 'predication'. If there is a word *A* which is not said except in company of one or another word *B* of a set *b*, then when *A* is said with some *B* it has perforce the effect of saying something about that *B*. Certainly if a word's only meaning were a property or event relating to certain things, it would only be used together with the words for those things (or with pointing or otherwise referring to those things, which can in turn be covered by metalinguistic predications, 5.2). Sentences are, first of all, word combinations in which something is said (predicated) about another thing. Predication inheres not in co-occurrence *per se* but in the co-occurrences which are required. The idea of such predicating, or the way of doing it, cannot have existed in the minds of early speakers without the existence of sentences. There may well be concepts or mental events that exist in the mind independently of their expression in language, but this could not be the case for concepts which are essentially about language. Therefore sentence-hood can only have arisen not as a vehicle for an otherwise conceptualizable relation of predicating, but as a concomitant of some other, more behavioral, relation which then yielded predication as an interpretation. We may suggest that this underlying relation was the practice of not saying certain words of a set a except in the company of certain others of a set b, because the a-words were not specific enough except when used only with, hence about, a b-word.

3.2. Inequalities of likelihood

3.2.0. *Introduction*

Within the dependence relation of operator word to argument class there is room for a 'fine-dependence' relation of operator word to argument word (or word-sequence), which makes a further contribution to the departures from equiprobability of word-occurrence. This possibility exists because the dependence of a word on a class states only that every member of the class has a positive likelihood, no matter how small, of occurring with that word. Nothing says that the likelihood has to be a priori determinate, as being random, or equal, etc., for all members of the class. Hence, for a given O_n, different N can have different likelihoods of appearing as its argument; equivalently, the O_n word has different likelihoods of occurring on different N words. (Just what the likelihoods consist of will be considered in 3.2.6.) For example, *sleep* has reasonable likelihood of appearing on *John, the child, the dog*, but lower likelihood of appearing on *the city* (as in *The city sleeps*), or on *trees* (in *Trees have to sleep each winter*, which is not really metaphoric) or *earth* (*The earth sleeps under a blanket of snow*), and lower yet on *window, chaos*. If an operator has two or more positions in its argument (e.g. subject and object), its likelihoods of occurring on the individual words in one position may be independent of its likelihoods on the words of the other position; if they are not independent, we have to speak of the operator's likelihood of occurring on the word m-tuples formed from the m ordered classes in the argument. For example, the O_{nn} operator *see* has greater likelihood of occurring on *John, dogs, fish* than on *space, water, the blind* (taken as noun, e.g. in French *un aveugle*), in its first argument position, and more or less independently it has greater likelihood on *John, dogs, fish, water, the blind* than on *space, vacuum* in the second position of the *NN* argument. In contrast, the likelihood of occurrence of the O_{nn} operator *eat* can best be described in respect to pairs of individual words in its *NN* argument, e.g. with greater likelihoods on *children/cereal, goat/ paper, virus/host, acid/copper, revolution/its children, car/gasoline, Venus fly trap/insects*, than on *children/gasoline*, *virus/paper*, *Venus fly trap/cereal*. These likelihoods can change through time, and can differ in different subsets of speakers, but they are roughly stable over short or long periods, so that the likelihood inequalities of arguments in respect to particular operators and vice versa can be used in analyzing language.

3.2.1. *Likelihood constraints*

If we consider the various words which have the same argument class, e.g. *see*, *eat*, etc. in O_{nn}, we find that they have different inequalities of likelihood in respect to the individual argument words in the class: *see* has greater likelihood, but *eat* less likelihood, on *John/water* than on *car/ gasoline*. In fine, we can state the inequalities of likelihoods of each word in an argument class in respect to the words in the operator class over that argument: e.g. the likelihood of *water* to be in the second N position under *see*, under *eat*, and under other O_{nn}.

The distinction between argument-requirement and likelihood is not immediately observable in the data. It is a choice of analysis to isolate the combinations whose frequency not only is zero in the given data, but can be expected to remain zero in any further data, and whose distinction from the likelihood-based combinations is expected to be useful for constructing a theory. The dependence of an operator on an argument is thus a construct. It is based on a particular criterion, of what arguments they in turn depend on, which has the merit that the arguments in turn can be characterized by the same criterion. In contrast to this construct, the likelihoods of an operator in respect to its argument words are the direct data of word-combination frequency within the dependence structures, in (a sample of) the sentences of a language (3.2.3, 3.2.6). These likelihoods form a vast body of data, which cannot be surveyed except in limited sublanguages (ch. 10).

The inequalities of likelihood of a given operator word in respect to an argument class amount to a rough grading of likelihood, though not in general to a specifiable partial ordering thereof. If we check the likelihoods for *sleep*, for example, we find many N words which have reasonable likelihood (much higher than the average of their class) of appearing as its argument. These have been called, e.g. for *sleep*, the normal co-occurrents, or selection, of that operator in respect to its argument. Some of these words may occur frequently, e.g. *you, I, the baby*, in *How did you sleep?*, *The baby slept well*, etc. Others may be very infrequent, as in *A thrips was sleeping on the leaf*; but even these are immediately acceptable if they are included under a classificatory word which in turn is a reasonably frequent argument of the given operator (see below). For other argument words the operator *sleep* may have, in various degrees and respects, less than selectional likelihood: *trees, flowers, bacteria*. Yet other N may be even more infrequent as arguments of *sleep*, or may appear only when the *sleep* is under particular further or neighboring operators. For example, *The saucer slept* may occur (not even metaphorically) in a fairy-tale, in which neighboring sentences will be of appropriate restricted types; *The universe slept*, *The winter slept* involve metaphor, which is analyzed at the end of 4.2 as reduced from a sentence-sequence under the operator *as*. And almost any *N* can be the argument of *sleep* when *sleep* is under *not* (because it is on *say*, 5.5.3), as in *Rocks don't sleep*. All the latter *N* would not be included in the selection of *sleep*.

The statuses of the various argument words under an operator do not in general suffice for setting up any precise subsets within the argument class. It is even impossible to define a set of 'human' or 'human-like' *N* as the first-position selection under *say*, *know*, *believe*, *expect*, and other such O_{no}, since we find not only *I know that he left* and *The dog knows he left*, but also for example *The virus knows when it is in an environment containing material for replication* and *The dog said yes with its tail* and *Here are ten dollars that say you are wrong* (in laying a bet) and *These rocks can expect to last another 5 billion years*.

Because of its imprecision and instability, as well as its vastness, the data concerning selection, or likelihood in general, has not been used directly in defining the structures of language. But there are stable properties and structures, which can be defined in terms of the likelihood inequalities or gradings. For one thing, since reductions will be seen not to affect the likelihood gradings among the words of a sentence, these gradings are preserved under reductions and transformations (the latter being mostly products of reductions, 8.1.2). Indeed, transformations can be phenomenologically defined as subsets of sentences throughout which the likelihood gradings are preserved (8.1.3). More important, the reductions will be seen to be based on extremely high likelihood (or, equivalently, low information) in the operator–argument relation (3.3.2). Similarities and contrasts in likelihood grading also yield many of the grammatical categories, semantically important though somewhat fuzzy as to membership, that give texture to each language grammar: e.g. durative and momentaneous verbs, indefinite pronouns, etc. The more or less stable likelihood differences—the fine dependence—thus constitute an additional constraint on word-co-occurrence, beyond the dependence of 3.1.

While the likelihood constraints differ from the dependence constraint of 3.1, they do not conflict with it, because the likelihood of an operator is postulated not to reach permanent zero in respect to any word in the argument class. Also, the likelihood constraints do not repeat any of the dependence constraints, which are stated on argument classes; the likelihoods are stated in respect to the argument word, as a further restriction on what the dependence admits. While virtually everything in language structure will be found to be encased in the relation between an operator and its argument, the likelihood of an operator can be affected not only by its immediate argument but also by some distinguished word further down in the dependence chain—e.g. an argument of its argument to some depth. As an example of one such situation, *melt* is not a likely operator on *house* (as in *The house melted*) but it has some likelihood on *house made of snow*. In this kind of situation, a word with modifier (the latter being reduced from a conjoined sentence) can carry a different selection, which may approximate the selection of a different word, in this case *igloo*, as in *The igloo melted*, *The house made of snow melted*. The consistency of selection breaks down in negation and question: *virus/paper* is not in the selection of *eat*, but one can readily say *Can viruses eat paper?* and *Viruses do not eat paper*. The explanation is that negation and question are not operators (on *eat* in this case), but are part of the metalinguistic performatives in which each sentence is encased (5.5).

Perhaps every word has a unique selection of normal co-occurrences in the argument positions under it or in the operator position over it, and almost every word has unique inequalities or grading, from most likely to least likely co-occurrents in each of these positions. Nevertheless, words that are close (or opposite) in meaning have largely overlapping selections, as to a lesser extent do words that belong to the same universe of discourse (e.g. *walk, run, climb, jump*). That is, they have many normal co-occurrents in common. What is more important, even the divergent co-occurrents are not unrelated, for their selections, in turn, have large overlaps. Thus the operator *climb* but not *jump* has *crocodile, snake* in its first-argument *N* selection (*The crocodile climbed the river bank*) and *jump* but not *climb* may have *flea, robin* in its *N* selection; all of these nouns *N*, however, are in the selection of the operators *eat, die, sleep*.

Without attempting the impossible task of describing the network of selections among the words of a language, we can recognize in a rough way that each word has (or shares in) what we may call a clustered ('coherent') selection. Operators share in a clustered selection if their argument selection has a large overlap and if many words in the selection of one which are not in the selections of the others have, in turn, large overlaps in their operator selections; similarly, many arguments have a coherent selection of their operators. We see in this that the occurrence likelihoods of arguments under a particular operator, or of operators over a particular argument, are neither arbitrary nor unrelated to those of other words. A cladistic similarity relation here can give coherence to word co-occurrences and to the domains of reduction; we may also see here some properties of the sociometric charting of how social contacts cluster.

The importance of coherent selection is that it helps characterize semantic word classes and also the internal structure of such classes (e.g. of a generic name and its species names). It also enables us to tell when a given word consists of two homonyms (namely, when its selection contains two or more distinguishable coherent selections), or when a particular co-occurrent of the word is not really in its coherent selection, but is rather a result of selectional extension or metaphor that happened with the given word and not with other words.

It is clear that similarity of selection is closely related to similarity of meaning, and this for two reasons. In part, the meaning of an operator affects what argument word the speakers will use it on. But in part, the combination in which a word already occurs affects the meaning it has for the speakers of a language. These combinations, i.e. the selection of each operator or argument, arise not only from directly semantic considerations— what is useful to say about a given argument word—but also from the complex morphological, analogic, and historical factors that determine word frequency and word replacement in combinations, as well as meaning-change.[1] Extending the selection of an operator, i.e. increasing its likelihood on particular argument words, is not based on semantic considerations alone. For example, when airplanes began to be talked about, why did people come to say in English *The plane flies* rather than *the plane floats* or *the plane glides*? Of balloons one is likely to say both *fly* and *float* (but usually in different circumstances). Consider also that one usually says *to boil eggs* rather than *to cook eggs*, although this changes the meaning of *boil* from that in *to boil water*. Such selectional peculiarities can be found in much of the vocabulary of a language.

Words often have somewhat different meanings when they are in combination with different words—a further indication that co-occurrence can create meaning, not only the converse (11.1). This may be more noticeable for operators in respect to their arguments than for arguments in respect to their operators: *boil* has different meanings in *boil eggs* and *boil water*, but *eggs* has the same meaning in *boil eggs* as in *buy eggs*. The greater variety in $O_{n\ldots n}$ meanings supports our formulating the likelihood and dependence relations as being in the first place a restriction on the operator in respect to the argument words rather than the converse (3.1); this formulation is convenient since the restriction is a further constraint within the dependence relation which had been formulated for the operator in respect to argument classes.

3.2.2. *Likelihood types*

In each language there are words whose selection can be summarized in ways that are of interest for the structure and meaning capabilities of the language. The major types of selection will be noted here.

One case is that of unrestricted selection.

[1] Henry M. Hoenigswald, *Language Change and Linguistic Reconstruction*, Univ. of Chicago Press, Chicago, 1960.

There is a rather small set of words, e.g. *say*, whose arguments carry no likelihood restriction. Thus, *say* can be said of any sentence, hence on every operator, as in *I say he's late* or *He said she saw me*. This holds also for *deny* (source of *not*), *ask* (source of the interrogative), and *request* (source of the imperative), although some stative operators such as *exist* may be rare under *request* (G321, 331, 351). This set of words will be found to have the status of metasentential operators (5.5). The further likelihoods, of operators to their arguments, under *deny*, *ask*, *request*, may differ from those under *say* itself: *A stone sleeps*, *Vacuum sleeps* may be unusual under *say*, but less so in *A stone does not sleep* (from *deny*), *Does vacuum sleep* (under *ask*). But the partial-order constraint of 3.1 holds throughout: *Going does not sleep* and *Does going sleep* are about as unacceptable as *Going sleeps*.

There is also a small closed set of 'indefinites' whose union has all operators in its selection. For example, *something* is in the argument selection of very many operators, and any operator under which *something* does not normally occur will be found to have *somebody* (and *someone*) in place of *something*. Since they have to fit under every operator, such words as the *something–somebody* set unavoidably carry the meaning of indefinites: they do not fit one operator rather than another; hence they mean no more than being the argument of whatever operator is on them. (Although *he*, *she*, *it* seem also to constitute such a set, this is because they arise as referential reductions of all arguments (5.3), whose union necessarily has all operators in its selection.)

Another case is broad selection, when an operator has in its selection not all the words in its argument but very many of them, far more than most operators have. This is the case for some O_{oo} (operators on a pair of sentences) such as *because*, *entail*; also operators with mostly space-relation meanings originally, such as *near*, *to*, *before*. It is also the case for certain operators of exceptionally broad applicability, such as *in a . . . manner* (an earlier English form for which was reduced to *-ly* as in *He walked slowly*, *He spoke agitatedly*), or such as *the condition of* (an earlier English form for which was reduced to *-hood* as in *childhood* 'the condition of being a child'). Also, operators of time and aspect (i.e. the onset and durational properties of an event) occur readily on almost all operators, or on all those which are durative, and so on.

A special kind of broad selection is seen in quantifiers, and somewhat differently in classifiers. Words for numbers and amounts appear readily as arguments under almost every operator (*Two came late*, *He ate two*, *Much gets forgotten*), where they seem first to adjoin and then paraphrastically to replace the words whose count or measure they are giving (*Two men came late*, *He ate two apples*, *Much that is known gets forgotten*). But in addition these words have a selection of their own, such as *plus*, *more than*: *Two plus two equals four*, *Much is more than little*. Words for sets, containers, and fragments are similar: *The team played well* from *The players (in the team) played well*, but in addition *The team was formed last year*; also *The whole room applauded* from *The audience (in the room) applauded*, but in addition *The room is dark*; also *This piece is too sweet* from *This (piece of) cake is too sweet*, but in addition *This piece is only one-sixteenth of the cake* (G260).

Somewhat differently, classifiers are words each of which has (with certain modifications) the selection of the union of a unique subset of other words (its classificands). Thus the selection of operators over *snake* includes the selections of operators over *viper, copperhead*, etc. In addition, however, there is a small selection of other environments (not applicable to their classificands), as in *Snakes constitute a suborder*. When all these classificatory words are analyzed it is seen that they are co-arguments, under quantifying and classificatory operators, of the words whose selections they can take over (e.g. *A viper is a snake*).

In addition to unrestricted and broad selection, there is 'strong' selection, which is seen when a word occurs exceptionally frequently with a particular operator over it or argument under it: e.g. verbs under particular prepositions (as operators on them), in *wash up* as against the less frequent *wash down* and the rare *wash under*; also the frequent *stop in* and *stop over* as against the rare *stop out*; and, differently, the frequent occurrence of *hermetic* with *seal* as argument (*hermetically sealed* from *The sealing was hermetic in manner*).

An extreme case of strong selection is the narrow or quasi-idiomatic, as seen for example in the operator *loom*, where we have by the side of *loom* itself (as in *It loomed out of the dark*) also *It loomed large* (roughly derivable from an artificial *It loomed into its being large*) but not *It loomed small* nor *It loomed dangerous*, etc. We can consider as strong selection also those words that do not combine grammatically with anything, and so occur only under metasentential and metalinguistic operators (ch. 5): e.g. *ouch, hello* in *He said Ouch, Hello is an English word*.

The major case of narrow selection is seen in idioms and frozen expressions. These may have unusual word combinations as in *hit upon* (*He hit upon a new idea*), or unusual reductions as in *make do* or *The more the merrier*, or unusual meaning of the whole expression as in *old hat* or *kick the bucket* in the general sense of 'die' (these are often due to metaphor (4.2), which is a product of reductions).

There is also a strong 'negative' selection. Comparable to the demand that certain words or repetitions appear under a given operator is the demand that certain words not appear, that is, the rejection of certain argument choices. Thus while one can readily say *He is truly coming, He is certainly coming, He is undoubtedly coming*, one cannot say *He is falsely coming, He is uncertainly coming*, and perhaps not *He is doubtfully coming*. The reason is that *He is truly coming* is a reduction of the sentence-pair *He is coming; that he is coming is true*; hence (1) *He is falsely coming* would be a reduction of (2) *He is coming; that he is coming is false*, which would generally be rejected. The difficulty lies not in *That he is coming is false*, which can readily be said, but in conjoining the two sentences, which is not likely when one sentence denies what the other sentence asserts. The contradictory combination can physically be said, as in (2) above; but because of its unlikelihood (negative selection) it does not participate in the reduction to adverb, *falsely* in (1) above (G306).

3.2.3. *Reconstructed likelihoods*

The types of likelihoods seen above, from unrestricted, broad, and strong to narrow and negative, are all observable directly in the data. There are also other situations, in which the regularization and simplification of the description of the data (2.3, 2.6) make us reconstruct certain word-occurrences which in turn show that certain word combinations had high likelihood in the reconstructed source.

For example, as noted above, *expect* in most cases has the argument pair N,O, as in *I expect Mary to win*, where the O_n word *win* has *Mary* as its argument in turn. When we find a shorter second argument, apparently just N, as in *I expect Mary*, the problem is to derive it from the more common O second argument, since in almost all cases when a word has two forms of argument one can be derived from the other (2.3). In this case it is found that any further likely operators on *I expect Mary* are the same as are likely on *I expect Mary to come* (or *to be here*). Also the likelihoods of the various N to appear alone as object of *expect* are graded in about the same way as the likelihoods of those N when they are subjects of *to come* (or *to be here*) (alone, or when these last are objects of *expect*): *I expect time* is less likely than *I expect Mary*, as *I expect time to be here* (or just *Time is here*) is less likely than *I expect Mary to be here* (or *Mary is here*). We conclude that *I expect Mary* is a reduction of *I expect Mary to be here*, etc. In the source forms, once every N second argument under *expect* is reconstructed to N *to come* (or N *to be here*), the locally synonymous set *come, be here*, etc., now has greater likelihood as second argument of *expect* than any other word or locally synonymous set in that position (G161).

Another example of high likelihood in reconstructed sentences is seen in the operators that are called reciprocal verbs. We consider *John and Mary met*, by the side of (1) *John met Mary and Mary met John*, whereas there is no *John and Mary saw* by the side of (2) *John saw Mary and Mary saw John*. We note first that when a sentence like (2) exists in the language then a sentence like *John and Mary saw each other* also exists, with *each other* being taken as a reduction of *Mary and John respectively*. Then from (1) we have *John and Mary met each other*. What is special about *met* is that the *each other* is zeroable. This happens under all operators of a subset O_{rec} (reciprocal verbs) of O_{nn}, which are definable by the condition that given (3) $N_1 \, O_{rec} \, N_2$ (where N_1 and N_2 are in the argument selection of the O_{rec}) in a text one can always add (4) *and* $N_2 O_{rec} N_1$ without affecting the likelihoods of the further operators, i.e. without affecting what the rest of the text says about (3). This means that we can reconstruct every occurrence of (3) $N_1 O_{rec} N_2$ into an occurrence of (3, 4) $N_1 O_{rec} N_2$ *and* $N_2 O_{rec} N_1$ without skewing the structural description, or distorting the information, in the utterances of the language. With O_{rec} operators, the contribution of (4) is implicit; introducing it does not add information and is indeed not commonly done (hence the rarity of (1) *John met Mary and Mary met John*). Every occurrence of *John met Mary* (and so for every O_{rec} sentence) can then be considered as a case of *John met Mary and Mary met John*, which is transformable to *John and Mary met Mary and John respectively*, which is always reducible to *John and Mary met each other*. The *each other* is thus an alternative which is available on every occurrence of *John and Mary met*, and is zeroable (3.3.2). Note that when (4) is not likely, as in *John met his doom*, where *doom* is not in the first-argument selection of *met* (i.e. not a normal subject of *met*), so that *His doom met John* is most unlikely (though one cannot claim that it has no meaning), we rarely if ever have the reduction to *John and his doom met each other*, and hardly ever the further reduction to *John and his doom met* (G314).

A major type of high likelihood in reconstructed sentences is to be found in the reconstruction of word-repetition. A simple example is seen in repetition under *and*, *or*, *but*. To see it, we first consider the conditions for *-self*. If we compare *I introduced myself*, *You introduced yourself*, which occur in English, with *I introduced yourself*, *You introduced myself*, *I introduced me*, *You introduced you*, all of which do not, it is clear that *-self* is adjoined to a pronoun (which is a reduced repetition of a neighboring 'antecedent' argument) as object (second argument) of an operator if and only if that pronoun repeats the subject (first argument) of that operator: *-self* is adjoined to *I introduced me* and not to *I introduced you*. If now we wish to derive (1) *I came up and introduced myself*, contrasted with the non-occurring *I came up and introduced yourself*, simply by extending the conditions for *-self*, we would have to say that the antecedent could also be the subject of the verb preceding the verb whose pronominal object receives the *-self*. This is not a simple extension, because many other words can intervene between the antecedent subject and the pronoun with *-self*. Instead, we can reconstruct a subject for the second operator (*introduced*) by assuming this subject to have been zeroed on grounds of its having repeated the first argument (*I*) of the first operator (*came up*). The reconstructed sentence would then be (2) *I came up and I introduced myself*, where no extension of conditions is needed to explain the *-self*.

This reconstruction is justified independently. In all sentences consisting of a subject followed by two operators connected by *and*, *or*, *but*, both operators have the class of that subject as their required subject class (e.g. no NO_n and O_o: *John phoned and is probable*); and both operators have the given subject word in their subject selection. If the second operator does not have the first subject in its own selection the unlikelihood of the sentence is immediately noticeable, as in *John is unmarried but pregnant*. Put differently: the likelihood grading of operators after *John is unmarried but . . .* is approximately the same as after *John is unmarried but John . . .* (e.g. *is pregnant* would be equally unlikely in both cases). Description of data such as the occurrence of *-self* and the likelihood of second operators above would require additional grammatical statements unless we assume that the missing subject of the second operator is a zeroed repetition of the subject of the first operator. By the criterion of least grammar (2.2) we reconstruct this repetition, obtaining (2). When we do this, (1) *I came up and introduced myself* is a case of (2) *I came up and I introduced myself*. Partly similar analyses can be made for missing second operators (e.g. *plays* in *John plays piano and Mary violin*), or objects of the second operators (e.g. *the new keys* in *John misplaced, and Mary found, the new keys*), under *and, or, but*. When these missing words are reconstructed as repetitions of the correspondingly placed word before the *and, or, but*, we obtain filled-out second (component) sentences after the conjunctions, without any missing words. In the set of sentences, including component sentences, which have no missing words, it is then seen that under these conjunctions it is more likely that at least one of the positions after the conjunction is filled by the same words as in the corresponding position before the conjunction, than that every position after the conjunction differs from the corresponding position before. This property, of expecting some word-repetition in corresponding positions, is part of the selectional characteristics of *and, or, but*. As such, it contributes to the meaning of these conjunctions, which are readily used to collect the various acts or events done by a given subject or to a given object, or the various subjects and objects which participate in a given act (G139).

Another type of repetition preference, not restricted to corresponding positions, holds under many other conjunctions. When we consider bi-sentential operators (O_{oo}), we find what seem to be semantic differences, between arbitrary-sounding or nonsensical sentences such as *I sharpened my pencils because the snow was soft on Tuesday* and ones that sound ordinary, like *People are corrupted by (their) having access to power*. It is notable that in many cases the latter, more 'acceptable' kind of sentence has a word repeated in its two component sentences (e.g. *people* and *they*) while the former kind does not. The relevance of this observation is supported when it turns out that the former lose their eyebrow-raising effect when a third component sentence is adjoined which creates a word-repetition bridge between the original component sentences: e.g. *I sharpened my pencils, having promised him a pencil sketch of a snowy tree, because the snow was soft on Tuesday*. True, there are many bi-sentential sentences which lack repetition but are nevertheless immediately acceptable, such as *I will stay here when you go*. But in such cases we can always adjoin a third component sentence which creates a repetitional bridge, and which is zeroable by 3.3.2 as involving merely dictionary definitions or other knowledge common to speaker and hearer. For example, *I will stay here and not go, when you go*, or *I will stay here, which means that I won't go, when you go*. We see here that the selectional characteristic of these bi-sentential operators is that their component sentences be bridged by repetition of some word, with common-knowledge bridges being then zeroable. Indeed, it is this selection that gives them their connective meaning (9.1).

Yet a different repetition-preference is seen in many O_{no} and O_{nno} operators, whose first one or two arguments are N and last is O (this lower O carrying its own arguments in turn). When one of the arguments of the lower O is the same as one of the earlier argument words under the 'higher' operator (the O_{no} or O_{nno}), the argument of the lower operator is zeroable. Which lower argument is zeroed in respect to which higher argument depends on the particular higher operator. For certain higher operators there is apparently a greater likelihood that a particular lower argument should be the same as a particular position of the higher argument, and it is this lower argument that is zeroed. When one of the arguments (the subject or the object) of the O argument (the 'lower' O) is missing, the question is, then, whether or not the missing lower argument is a preferred repetition of an N argument of the higher O_{no} or O_{nno}. Under some of these higher operators, e.g. *observe, hear*, there is no omission for repetition: we have *I observed John shaving himself* and *In the mirror I observed myself shaving myself* but no *I observed shaving myself*. Similarly, when one says *I heard shouting* and *I heard the shouting at me* the missing subject of *shout* is not a repetition, but a zeroing of indefinite N (3.3.1) from *I heard someone shouting at me*. Note that if we start with *I heard myself shouting at myself*, there is no repetition-omission to yield (1) *I heard shouting at myself*; the latter is non-extant presumably because of the non-existence of *I heard someone shouting at myself*, from which alone (1) could have been derived.

As to the cases where lower arguments are zeroed:

First, omissions of the subject of the lower operators: under *prefer, want, report, admit*, etc., we have (2) *I prefer for John to shave himself* and (3) *I prefer for me to shave myself* and also *I prefer to shave myself* (with omitted repetition of the subject of *prefer* in (3)), but no *I prefer to shave himself* (there being no grounds for omitting *John* in (2)). Similarly, there is *I reported having introduced myself falsely*, which can be reconstructed only to *I reported my having introduced myself falsely*; also *I admit to having introduced myself falsely*. Under the O_{nno} *offer, request, ask*, we have such sentences as *I offered (to) John to introduce myself* with the subject of *introduce* omitted as repetition of the subject I of *offer*, but also in some cases *I offered John to introduce himself* with the subject of *introduce* omitted as repetition of the second argument of *offer*. However, under *order, tell, advise, prohibit, prevent*, we have only *I prevented John from introducing himself, I ordered John to introduce himself* (from *I ordered John for him to introduce himself*), and no *I ordered John to introduce myself*; and if we can say *I ordered John for me to be included in the list* we apparently cannot omit the *for me* as a repetition of the subject *I* (to obtain *I ordered John to be included in the list* as meaning 'for me to be included').

Second, omission of the object of the lower operator: under the O_{no} *deserve, merit, accept, avoid, escape*, we have *You deserve (for you) to see yourself in this costume*, and *John merits (his) receiving the prize*, where the omittable (parenthesized) subjects of *see, receive* repeat the subjects of the higher operators *deserve, merit*, etc. But we also have *John deserves punishment* and *John escaped sentencing for his crime* and *John merits our help (to him)* or . . . *our helping him*, where it is the omitted objects of *punish*, etc., that repeat the subjects of the higher operators. In contrast to *deserve*, under the O_{no} *suffer, undergo*, and *is -en*, (1.1.2), we have only *Napoleon suffered defeat* (or . . . *Wellington's defeat of his forces*), *John underwent an operation (on him) by a team of three surgeons, I withstood their attack (on me)*, where it is only the object of the lower operator (*defeat, operate, attack*) that is omittable as being a repetition of the subject of the higher operator. In contrast, under the O_{nno} *defend*, the object of the lower operator is omittable as repetition of the object of the higher operator: *I defended John from attack (on him)*. Note that the apparent ambiguity in French *défendre*, in *défense de la patrie* compared to *défense de fumer*, lies in *défendre* admitting both the omissions of the English *defend* and the omission of the English *prohibit, prevent, order*. Should the question be raised as to the evidence that the verbs under *suffer, undergo*, and *defend* indeed have objects (which are then zeroed), note that only transitive verbs (which have objects) appear as lower verbs under these: there is no *I defended John from sleep*. In contrast, under *prefer*, etc., the lower verb can be transitive or intransitive: *I prefer for John to wear it, I prefer for John to sleep*.

In the O_{oo}, O_{no}, and O_{nno} cases here we have seen that various omissions in the lower sentence have to be reconstructed (by such evidence as *-self*) as being repetitions of an earlier argument of the higher operator. When we add these frequent omissions to the occurrence of observable repetitions, it is seen that the (original) repeated material has a high likelihood, higher in the positions noted than any one other word in those positions. The positions of this high-likelihood repetition differ under different higher operators, in a way that conforms to (or affects) the meaning of the higher operators (and that does not in general vary with the meanings of the lower operators). Thus under the higher operator *withstand*, the object of *attack* (and of every other lower operator under *withstand*) has high likelihood of being a repetition of the subject of the higher operator, but under *defend* (in English, but not in French) it often repeats the object of the higher operator; under *prefer* the subject of the lower operator often repeats the subject of the lower operator, while under *prevent* it often repeats the object of the higher operator. One could say that this is just a reflection of the meanings of these higher operators. But almost all selection can be considered a reflection of meaning. Nevertheless, if we start with the meanings—especially given our lack of an effective universal framework by which to formulate such subtle meanings—we could not reach the particular classification of words which we have found here, in respect to particular grammatical processes such as the position of omission: how zeroings under *offer* differ from under *order*, and how French *défendre* has the repetition likelihoods and zeroings (and meanings) of both *defend* and *prohibit* in English.

3.2.4. *Likelihood restricted to zero*

One type of likelihood remains to be discussed. All the likelihood types noted up to now were assumed to describe combination probabilities larger than zero. This enables us to say that the likelihood differences act on the resultant of the operator–argument relation, since the latter excluded all combination probabilities equal to zero. Hence no matter how strong the selection of particular arguments under a given operator, the other words in the argument class of that operator were assumed to have more than zero probability there; and indeed the occurrence of very low-likelihood argument words does not make an impossible sentence, only a nonsensical one (e.g. *The vacuum conversed with the ether*), or a non-reducible one (e.g. *He is coming; that he is coming is false* not reducible to *He is falsely coming*, 3.2.2 end).

Nevertheless, borderline cases arise. For example, under *order* the omitted subject of the lower operator can be reconstructed as a repetition of the second argument of *order*, as in *I ordered John to shave himself* reconstructed to *I ordered John for him to shave himself*. But it is questionable whether any other subject can appear here for the lower operator: can anyone say *I ordered John for you to shave yourself*? True, one can perhaps say *I ordered John for you to be admitted*, but this appears limited to those lower sentences which may be transforms of a source in which the subject is the same as the object of *order*, e.g. *I ordered John (for him) to admit you* or . . . *(for him) to allow your being admitted*. It may therefore be that all other subjects of the lower operator have zero probability under *order*. Similarly, it is difficult to find *deserve* and especially *merit* (above) without an argument of the lower operator repeating the subject of the *deserve* or *merit*.

This situation exists explicitly in English, in the auxiliary verbs *can, may, shall*, etc.: *John can play violin*. We could analyze this as *John plays violin* with an auxiliary on the verb, or else as a two-verb expression; with *John* as subject. These analyses would involve the serious cost of introducing new and *ad hoc* entities into English grammar, and would violate the fundamental partial ordering of 3.1. Alternatively, we may say that *can* and the others are O_{no} operators of the *prefer*, *want*, type above, in which the subject of the lower operator is omitted when it repeats the subject of the higher operator. Then the only serious difference is that under *can*, etc., for the subject of the lower operator to be different from the subject of the higher operator has not just a lower probability but zero probability: one does not say *I can for John to play violin* as one can say *I prefer for John to play violin* and even *I want for John to play violin*. In earlier periods of English, some of the auxiliaries indeed could have different subjects for their lower operator, so that we have here not a principled restriction on these higher operators but a lowering of probability to zero (G298). The result in modern English is the same whatever the history, but the history provides support for saying that in addition to the massive zero-probability constraint that creates the operator–argument classes (3.1), a few words within an established operator class may lower to zero the probability for part of their argument class (this being then viewed as a limiting case of the likelihood constraint).

The fluidity and secondariness of this zero grade in the likelihood constraint, in contrast to the large-domain zero-probability constraint that underlies the argument relation, is seen in a few words such as *order*, *merit*, which approach the zero grade seen in *can*, but fall short of the definite grammatical exclusion that we find under *can*. The fluidity of the *can* situation is also seen in the attempts to circumvent the exclusion, as in the case of *try*. Many speakers of English would consider subject-repetition to be required under *try*: *I will try to shave myself*, but only questionably (1) *I will try for him to shave himself;* however, the questionable form is circumvented by a virtually meaningless use of *have*: *I will try to have him shave himself* can mean the same as (1) in the sense of *I will try that he should shave himself*, aside from its other meaning of 'I will try to arrange for him . . .' (the latter from *I will try* on *I will have him shave himself*).

3.2.5. *Word subsets*

The various special selections noted above yield certain subsets of words, mostly such that the words in a subset have selections that are similar, complementary, etc., but also of those words which together comprise the selection of particular interesting operators (3.2.2). For example, there are operators whose argument selections are virtually identical. Such are the various operators expressing duration, onset, and the like, such as *continue*, *last*, *throughout*, which appear on approximately the same selection of arguments (which are themselves operators, such as *sleep*, *exist*, *grow*): *He slept throughout the night* but not *He arrived throughout the night* (G77).

This situation has the effect of creating a set, very fuzzy in the case of English, of durative verbs.

There are other words whose selection is similar to (mostly, has a large intersection with) the selection of some other word, or includes the other selection, e.g. classifier words as compared with their classificands. Also, there are words whose selection contrasts with, or is virtually complementary to, the selection of certain other words: *sleep*, *grow*, are said normally under *was throughout*, *lasted*, etc., and not so readily under *was sudden*, *occurred at*, etc.; but *arrive*, *depart*, *say* etc., are readily arguments of *was sudden*, *occurred at*, and rarely of *was throughout*, *lasted*. Such contrastive selections may partition the words of a class into two (or more) selectional subsets (e.g. in some languages imperfective or durative as against perfective or momentaneous verbs, G275). There are also words such as *speak*, *think*, which occur as normally with *lasted* as with *was sudden*. Hence 'aspect' in English is a graded property, which does not easily create subsets of operators, and does not partition them as having sharply either durative or else momentaneous selection.

In almost all cases the selectional subsets are fuzzy—imprecise or in various degrees unstable as to membership; some words have some but not all the properties of the subset, or new words enter the subset, and so on. Thus the pair *something*, *somebody* (or *someone*) can be considered as variants of a single indefinite noun; the word *people* has some of the properties of *somebody* but not all. The auxiliaries *can*, *may*, etc., seem at first like a well-defined English subset; but, for example, *need*, *ought* have only some of the properties common to the set (G300). The property of *falsely* above (3.2.2 end) is shared by *uncertainly*, *improbably*, *doubtfully*, and a few others, but not with equal definiteness. Setting up such subsets is not necessarily precluded by such complexities of conditions. For example, an operator which is not in the selection of *throughout* may enter that selection if it has a plural subject (first argument): there is no *The guest arrived throughout the afternoon*, but there is *The guests arrived throughout the afternoon*. This does not mean that the 'momentaneous' operator *arrived* has become 'durative', but that a momentaneous operator with a plural subject (*The guests*) can constitute a durative sentence. Durativity is selectional; it is thus a property of a word together with its arguments; the definition of selection allows it to be affected by lower parts of the argument chain. But the collecting of operator words into (fuzzy) subsets is still possible if we choose controlled conditions with maximal distinction: English durativity would be defined on operators with subjects in the singular (with certain durativity restrictions on their objects too, such as between the continuous or 'imperfective' *line* and the completive or 'perfective' *circle* as objects of *draw*).

All the word-class names used above (other than the N, O_n, O_o, etc. defined in 3.1) are definable in respect to selectional (3.2) or morphological

(3.3) properties on the N, O_n, etc. This holds, for example, for metasentential operators, indefinites, pronouns, adjectives, prepositions, auxiliaries, reciprocals, and durative-momentaneous verbs.

The difficulty in establishing word subsets shows that in general languages have few if any definite subsets in respect to likelihoods (3.2). In science sublanguages (10.3–5) it will be seen that selectional subsets are possible and of great importance, when subject matter is restricted.

3.2.6. *Frequency; low information*

We have seen (3.2.2) several types of special likelihoods in operator–argument occurrence: unrestricted and exceptionally broad selection, and preference and rejection for certain combinations, down to the zero grade of likelihoods (3.2.4). It remains to see just what features of likelihood are needed in the present theory. The different likelihoods within the operator–argument relation, which are roughly stable over short periods, would be needed only for describing the data of actually occurring sentences within the set created by the partial order of 3.1, and for indicating the (mostly fuzzy) word subclasses formed by similarities of selection (3.2.5). However, the upper and lower extremes of likelihood determine the domains of reduction (3.3), which is a universal process in language. In order to characterize—one might almost say motivate—reduction it is necessary to specify likelihood difference in a way that is measurable at least in principle.

If there are n words in an argument class, say N, then for a given operator on it, say *sleep*, some members of N will occur, in many non-specialized collections of sentences, in much more than $1/n$ of the occurrences of *sleep*, while other members of N occur in much less than $1/n$. The occurrences for each N will not tend toward a common $1/n$ 'average' as the size of the collection increases. We thus have not a random distribution of N as argument words of *sleep*, but inequalities of likelihood, and to some extent a grading, of N under *sleep*. These inequalities of likelihood are much the same for given N in any very large corpus of data. The set of higher-than-average-frequency members of N constitutes what was called (3.2.1) the selection (in N) under, for example, *sleep*. A comparable formulation can be made for the likelihood, and the selection, in the various O_n (including *sleep* and all other members of O_n) as operators on a particular member of N. In some cases the likelihood of each member (or of certain members) of an operator class to occur on a particular argument word depends on the identity of the latter's arguments words in turn.

Another way of characterizing the selection under a given operator is by seeing the variety of further contexts (i.e. further operators and co-arguments) when the argument is on a particular argument word. The occurrences of, say, *sleep* on a particular argument word may have many different and unrelated further operators on *sleep*, or different co-arguments of *sleep* under these further operators, up to considerable distances in the text from the position of *sleep*. Thus *child* as argument of *sleep* may occur in a personal letter, in a hospital report, in a newspaper account of daily life, in a novel, etc.—all with different further contexts in the same sentence and with different neighboring sentences. In contrast, *saucer* as argument of *sleep* may occur in fairy-tales, stories told to young children, possibly some science fiction (many of these sources having similar kinds of further operators and co-arguments on *sleep*). Also, this and many other combinations occur primarily under *not*, and question (*I ask whether* ——, *I wonder if* ——), and similar further operators (e.g. *Saucers don't sleep*). We can say that argument words which occur under a given operator in a wide variety of further contexts are in the selection (or normal likelihood) for that operator, but ones such as *saucer* under *sleep*, which occur under that operator only in a narrow range of further contexts, are not.

To a first approximation, likelihood is relative frequency. It is not absolute frequency in utterances of the language, since the total rarity of a word (e.g. *thrips*, an insect) has no general relation to its meaning—or to its reducibility (as in *The thrips crawled out and flew away* from *The thrips crawled out and the thrips flew away*). Nor is the likelihood measured relative to the whole sentence; if it were, we would have the impossible task of considering each sentence as a different environment in which to count the word likelihoods at each position. Rather, likelihood as used here is the frequency of each word in a given argument or operator class (N, O_n, etc.) relative to a given operator word over it or argument word under it (or co-argument under *be*, as in the classifier *is an insect*, below). In the great bulk of cases, it is relative to the immediate operator over it or argument under it, but in some cases an operator or argument further up or down the chain may be relevant to a word's likelihood. One example was noted above in the further operators which characterize children's stories. Another is the expectable frequency of *melt*, above, as operator on *house*, which, as noted, is very low (e.g. in *A house can melt*), but higher if *is of snow* operates secondarily on house (in *A house of snow can melt*). (In the source, *melt* does not operate on *house* alone but on *house* plus modifier: the source here is *A thing which is a house which is of snow can melt*, from *A thing can melt; the thing is a house; the house is of snow*, G124). The appeal to a secondary operator (i.e. a modifier, e.g. *of snow*) is not needed if we have a single word whose selection is that of *house of snow*, e.g. *igloo*.

The considerations listed above are corrections on our first approximation to what constitutes likelihood. In addition, the frequency has to be counted in respect to the reconstructed sentences (3.2.3): when the *-self* in (1) *I prefer to shave myself* shows that (1) has to be reconstructed to (2) *I prefer for me to shave myself*, we count *I* as occurring twice in (2) but also twice implicitly in (1).

Finally, certain words of low frequency relative to a given operator or argument may have a classificand or synonymy or definitional relation to words that are more frequent relative to that operator or argument. For example, we may have a word which is so infrequent that no grading of its observed argument or co-argument words is of interest: e.g. the number in *The rock weighed 3.073 kilograms*. But while *rock*—and everything else— may be very rare with this particular number, it has selectional likelihood under normally frequent words of the same subclass, numbers, as in *The rock weighed 2 kilograms*. There are many *N* which we may suspect were never said before with a particular operator or co-argument, but once said are recognizable as comparable to normal selection: e.g. (1) *The very young thrips maneuvered itself into a narrow crack in the gray bark*. The immediate acceptance of such infrequent material as a normal sentence is due to its component sentences having special relations of classification, synonymy, and the like to sentences of ordinary frequency. Thus, *is an insect* is in the selection of the infrequent word *thrips*, as in the classifying sentence *A thrips is an insect*; and the common *move* is in the classificatory selection of the somewhat less frequent *maneuver oneself*. These selectional relations classify the word combinations of (1) under the word combinations of a sentence which is far less infrequent or unusual than (1), namely *The insect moved into a crack*.

The term likelihood is used here, then, for the frequency of occurrence of a word as specified above: relative to a particular operator or argument word of it, in respect to the reconstructed base form of any sentence which is in reduced form, and with allowance for classifier relations and the like. This frequency is not in principle unmeasurable, since it is primarily a word-pair relation, between argument and immediate operator, rather than a relation between a word and its whole sentential environment. In addition, what makes this relative frequency manageable is that its use in defining word subclasses and reduction domains requires only that a few grades be distinguished—roughly, frequency estimated to be higher than for any other word in the given position (i.e. highest likelihood); selectional (higher than average); selection that is exceptionally broad or narrow (in various ways); and rejection ('negative').

These various likelihood grades carry different amounts of information, in the sense that is known from the statistical structure discussed in Information Theory. In this sense, a word that has broad selection and high total likelihood carries less information than other words, and a specified word (or locally synonymous set words) that has highest likelihood in a given environment can be considered to contribute no information in that environment. Since various kinds of likelihood are characteristic of different subsets of words, and since these have informational properties, one might wonder if some other factors, specifically semantic ones, might not underlie the various frequencies observed. This is a reasonable supposition, but the semantic factors that we can distinguish are more varied yet, and are not organizable in any objective way, and not measurable. Among such factors in determining frequency are: the informational usefulness of particular word combinations and the non-language occasions for them to be said, the particular aspects of experience that it is useful or customary to note (time, place, evidentiality, etc.), euphemisms and avoidances of particular expressions, expressive and communicational exaggerations (e.g. a preference for stronger terms for time, such as may have led to development of the periphrastic tenses, e.g. from *took* to *has taken*), preference for argot and in-group or colorful speech, and so on. Aside from these more or less semantic factors determining likelihood and the change and extension of likelihood, there is the massive effect of analogic extension and change in selection which is determined by complex grammatical factors (of similarity, etc.), only part of them semantic.

3.3. *Reduction*

3.3.0. *Introduction*

In reaching the syntactic theory presented here, the core of the work was to find a small set of relations between sentences in terms of which we can say that one observed sentence has been derived in some explicit and non *ad hoc* way from another. Here, to 'derive' is to find a constant difference between two formally characterizable sets of sentences. The intent in establishing such derivations is that the meaning be preserved in corresponding sentences (or sentential segments) of the two sets: the meaning of *John phoned and wrote* is the same as in *John phoned and John wrote*, and the meaning of *John phoned* is the same in the sentence *John phoned* as in *I deny that John phoned*. It was found that such derivation can be effected by reductions and transformations which have an important property: if we undo these, the residue, consisting of the pre-reduction ('base') sentences (4.1), has the transparent structure of dependence and likelihood (3.1–2). Furthermore, it was found that the reductions which suffice to produce this residue are not simply an arbitrary conglomeration of inter-sentence relations which had been so chosen as to leave this residue of transparently constructed sentences. Rather, most of the reductions fall into a coherent and not very large set of reductions, merely in the phonemic composition of words, which take place when those words contribute little or no information to their sentence. The fact that language is the resultant of two coherent systems, each with its own structural definition (and not one being just whatever is left as the residue of the other) gives credibility to each of the systems. It is therefore important not merely to note a few examples of reduction, but to present a survey of the major reductions of a language (English), sufficient to show the character of the reduction system.

The constraints of 3.1–2 create both the form and also the information-bearing grammatical relations of the base sentences. The reductions neither add nor remove information, and are thus information-preserving. In addition, while reductions create the vast bulk of sentences outside the base, a language may have non-reductional transformations of the base sentences (8.1); and other non-base sentences may be formed by analogic processes and by borrowing.

3.3.1. *The reduced shapes*

The most widespread reduction is to zero. In *John plays violin and Mary piano* or in *John left, and Tom too* it is clear that *plays* has been zeroed after *Mary*, and *left* after *Tom*; similarly *I will go if you will* is produced by zeroing *go* after *you will*. If we compare *I plan to see for myself* with *I plan for Tom to see for himself* it is clear that *for me* had occurred after the first *I plan* and had been zeroed there (3.2.3). It can be shown that *Never send to ask for whom the bell tolls* is in principle from *Never send someone to ask . . .* ; and *I don't have time to read* from *I don't have time for me to read things* (since elsewhere *read* has a second argument, its object, as in *He reads comics*). The zeroing of *to come*, *to be here* under *expect*, as in *I expect John* from *I expect John to be here*, has been noted in 3.2.3. Complicated forms, such as metaphor, the passive (1.1.2), and the zero causative (3.3.2, as in *He sat us down* from *He made us sit down*), can be derived from attested kinds of zeroing. That these are indeed zeroings, not merely 'implicit meanings', can be seen in languages which have gender agreement. Thus in a café one asks for *un blanc*, from *un verre de vin blanc*, but in a wine shop for *une blanc*, from *une bouteille de vin blanc*.[2]

Another widespread reduction is to affixes. These are short morphemes attached to words. In English and in many languages most prefixes have the meaning of an operator whose second argument is the word to which they are attached, their 'host': *He is anti-war* comparable to *He opposes war*. In some cases the prefix is visibly a reduced form, as in *mal-* (in *maltreat*) from a word meaning 'badly', or *be-* (in *because, beside*) from *by*. As to the suffixes of English, almost all have the status of operators, or rarely of co-arguments on their host, as in *earthly* from *earth-like*, from *like earth* (leaving aside here the specialization of meaning in *earthly*). For most English suffixes, there is no evidence of their having been historically reduced from free-word operators; but some suffixes are visibly such reduced forms, as in *sinful*, or in *kingdom* from Old English *dom* 'law, jurisdiction', or in *childhood* (older

---

[2] Noted by Maurice Gross.

child-had) from Old English *had* 'state, condition'. Note also the adverbial *-ly*, as in *slowly*, from the early *lice* (the *like* above) in dative case, meaning originally 'with form, with body'. Reduction to affix involves reduction in the stress relation to the host and in phonemic distance from it (called grammatical juncture), as in *-dom*, *-hood*, or in *childlike* from *like a child*. (For the position of affixes see below.)

Less common in English than the above are shortenings of exceptionally widespread words. This is seen in the prepositions, e.g. *of* from *off*, and *beside* from *by (the) side of*, *about* from earlier *on* plus *by* plus *out*. Most English prepositions have been exceptionally short words throughout their known history, but some have been reduced, even in recent times. Other recent colloquial shortenings are, for example, *will* in *He'll go soon*, *is* in *he's here*, *has* before verbs in *He's taken it* (and before nouns, only marginally in America, in *He's two jobs*), *going to* before verbs in *They're gonna make it this time*.

Another kind of reduction is seen in the pronouns. These have been considered to be words whose meaning refers the hearer to the meaning of some other word in the discourse. But such a characterization would constitute an addition to the grammatical apparatus, these meanings being different in kind from those of non-pronominal words. For a simpler system, we seek to derive pronominal meaning from the kinds of meaning found in other words. To this end, we note that in any case we have to recognize the existence of 'metasentential' words (ch. 5), and among these such as cite something in the same discourse in which the metasentential words occur; this holds both for words like *latter* which refer to a location in the discourse, and for words like *say* followed by quotes, which refer to the material in the quotes. The pronouns can be derived from words of this kind. As noted in 5.3, we would consider, say, *I met John, who sends regards* to be a reduction from, roughly, (1) *I met John; John—the preceding word has the same referent as the word before—sends regards*. Here *John—the preceding word has the same referent as the word before—* is reduced to *who* (or rather to *-o*, the *wh-* being a connective, 3.44). The word referent here simply carries its dictionary meaning, and is not itself a referential. Of course, no one ever said (1). This complex sentence is not the historical source of the pronoun. However, (1) shows that the referential relation of *who* to the preceding word (the first *John*) can be represented by a repetition of that preceding word, plus a secondary metasentential 'sameness statement' (between the dashes above). All the apparatus needed for (1), namely for the secondary sentences, and the metasentential words and sentences, is attested else- where in the language (5.2–3).

All the above are reductions *in situ*: prefixes are in many cases reductions of operators which had preceded the host word (the host being their second argument). Most suffixes can be obtained as reductions or equivalents of the last word in a compound word (G233); in many cases the suffix is originally the operator or predicate classifier on the word to which it is attached (now its host), as in *waterproof* from *water-proof* from *proof against water*, *childhood* from *the state (had) of a child*. Many of the changes of word-order in sentences are not due to any permutation process on the words within a sentence. Rather, some are alternative linearizations of the partial ordering which created the sentence (3.4). Others are complex additions and reductions which produce the appearance of a permutation process: *The letter was written by him* from *He wrote the letter* (1.1.2).

Nevertheless, English has a few constructions which seem to be due to transposing a segment of a sentence from one position to another (3.4.3). An example of this is seen in the modifiers which appear (in English) before their modificand. A modifier is a relative clause (G120), i.e. a sentence-like construction having at or near its beginning a relative pronoun (*which*, *who*, etc.) which refers to the word or sentential construction that preceded the relative clause, and is the modificand of the given modifier: e.g. *I met John, who sends regards* (above), or *The money which is needed is unavailable*, or *I phoned Mary, which was after I met you* (here the modificand is the sentence *I phoned Mary*). Here the *which is*, *who is*, can be zeroed, whereupon in English certain types of modifiers can or must be moved to before the modificand (G115, 229): *The money needed is unavailable*, *The needed money is unavailable*; *I phoned Mary after I met you*, *After I met you I phoned Mary*; but from *The money which is new is not counted* only *The new money is not counted* and not *The money new is not counted*. The full relative-clause modifier with its *which is* never appears before the modificand. Presumably it was not formed there; and the shortened *after I met you*, *needed*, *new*, etc. before the modificand were presumably transposed to that position after the *which is* was zeroed.

The same transposition is found in certain word-sequences which are constructed like modifiers. If we consider *He sells books which are (specifically) for school*, we have a zeroing to *He sells books specifically for school* (in which both *books* and *school* have full stress). Here a transposition can replace the preposition *(specifically) for*, yielding *He sells school-books*. This is the important compound-word structure, in which the first word carries full stress and the second has secondary stress. The last word in the compound is the modificand (host): *school-books* are books, not schools. A similar processs appears in *his writing of letters* (from *He writes letters*), *the fall(ing) of rain* (from *Rain is falling*), *This lock is proof against burglars*, and *his state* (earlier: *had) of (being) a child*, yielding respectively *his letter-writing*, *the rainfall*, *This lock is burglar-proof*, *his childhood*. In certain cases the last word in the compound is reduced *in situ* from having secondary stress to zero stress, thereby becoming an affix (e.g. *childhood*, *kingdom*, above). In rare cases the last word is reduced to zero, as in *red-cap*, which has to be derived from something like *a red-cap-person* since *a red-cap* is not a cap but a person.

Another transposition is seen in residues of zeroing under *and: John and Mary (too) left early* is from *John left early and Mary (too)* from *John left early and Mary left early (too)*.

We have seen that certain word-occurrences are reducible to zero, others to short words, yet others to affixes (when their juncture, i.e. their distance in time or space, or their stress status, in respect to their argument or co-argument—their host—is reduced). Also, that certain modifier-like occurrences of words are transposed ('permuted') from being after their arguments or co-arguments to being before them. These few changes create fixed differences in shape, accompanied by no difference in substantive meaning, between certain sets of sentences: so in the examples above, or as between *I'm expecting John, I'm expecting dinner*, and *I'm expecting John to come, I'm expecting dinner to arrive*. Many languages also have complex transformations between certain sets of sentences. These are larger fixed changes which are accompanied by no difference or by fixed differences in meaning (8.1): e.g. the passive, between *John will send the letter, Mary will find a stamp* and *The letter will be sent by John, A stamp will be found by Mary* (1.1.2); the relation ('zero causative', 3.3.2) between *He caused them all to sit down* and *He sat them all down*, or between *He treated the salad with pepper* and *He peppered the salad* (G317); and metaphor, between *He treated his speech with jokes as one peppers things* and *He peppered his speech with jokes* (G405). In many cases, these are composed of successive reductions and transpositions, in some cases with further operators acting on the original sentence. Thus the large transformations do not in general constitute a new process in sentence formation, over and above the reductions surveyed here.

The net effect of all the reductional and transformational possibilities is that in many languages the grammatical relations among words in a given sentence can be recast into many other grammatical relations among them, preserving only the fact that those words are, at a given depth, grammatically related, i.e. related in the partial order. This, without change in substantive information (or in the word-semilattice of the sentence): compare *He interdicted their transmitting the story* with *The interdiction was his in the transmission of the story by them*.

3.3.2. *Likelihood conditions*

3.3.2.0. Introduction

Reductions are made in particular circumstances. The necessary but not sufficient condition is that the word or sequence being reduced have exceptionally high frequency of occurrence, or contribute little or no information at the point it occupies in the making of the sentence. Only rarely is reduction occasioned by absolute high frequency independently of the sentential environment: e.g. in *goodbye*. More common is the reduction of words whose high total frequency is due to having an exceptionally broad selection, i.e. having normal (selectional) frequency to an exceptionally large number of other words. Lastly, there is the reduction of words which are the most frequent in respect to the environment (usually the specific operator or operand) with which they are appearing.

In these terms, we can understand that in the word-sequences of sentences, a word-choice which has very high likelihood (in the environment in which it occurs) needs less phonemic information to make it identifiable by the hearer, and so may be shortened (including reduced in stress, or reduced to affix, both of which mean reduced in phonemic distance from its environment). A word-choice which is determinable from its environment can even be reduced phonemically to zero. This last applies not only to words that have recognizably the highest likelihood in a given environment, but also to a word (or to any of a set of words whose difference is irrelevant in the given environment) which is conventionally accepted as the zeroable word there: e.g. *come* (or also *arrive*, *be here*, etc.) under *expect*. The issue here is not even whether the *come*, etc., can be shown to be the most frequent there but merely whether any hearer will supply one of these words as what was zeroed there.

These various high-frequency conditions for reduction are related to low information in terms of Information Theory. In the sequence of symbols that constitutes any language-like message, each sequential choice of a symbol (out of the total set of symbols) is a contribution to the information in the message. In a given environment of symbols, the more expected (i.e. likely) a particular choice is, the less that choice contributes to the information in the message. In the extreme case, a choice that is determinate from the environment contributes no information in that environment. In such a case, if such a symbol is omitted (with evidence that it has been omitted), it can be reconstructed from the environment: we can say that the choice had been made (by virtue of the environing symbols) but zeroed—i.e. its physical shape is zero, so that it cannot be seen except by inference from the environment.

The low-information status of these reducible words is seen in their meanings. In the case of words whose selection is broad, the increase in their selection is accompanied by a decrease in the specificity of their meaning. Compare the meaning of *about*, close to the meaning and selection of its etymological source, in *He walked about the room*, with its meaning and its far-extended selection, as in *He spoke about this problem* (G80); and the meanings of *give*, *take* in *He took a book*, *He gave me a book*, with those in their extended selection, as in *He took a good look at it*, *He gave it a good look* (G302). When the extension of selection, whether by metaphor or analogy or otherwise, is recent, the decreased specificity of meaning in the extended selection is obvious, as in the conjunctional use of *provided*, *providing* (as in *I will go, provided you agree*, G195).

In the case of words which are zeroed for having high frequency in a given environment (which in most cases is the immediate operator or argument under it), the fact that the word can be reconstructed by the hearer, on the basis of the environment, shows that the word contributed no information at that point.

3.3.2.1. Broad selection

A major example of broad selection is the set of prepositional operators (G80), extended from their original selection of time and place words (indicated by their etymology) to general relations, as in *about* above. Many of these words have been shortened, as in *beside*, *off* (3.3.1). A further reduction in English and generally in Germanic languages is replacing the preposition by the 'compound word' construction (G233). In this construction we begin with a preposition, which is an operator, between two words, its arguments: e.g. *books for school*, from *books which are for school*. The operator is zeroable if its second argument (*school*) has special likelihood in respect to the first (*books*), primarily as stating a characteristic property of the first argument. When the preposition is zeroed the second argument has the status of a shortened modifier of the first argument, and is transposed to before the modificand, as adjectives are in those languages (3.4.3). The trace of the zeroed preposition is having the main stress on the first word of the resultant: *school-books*. The process can be repeated, yielding *public-school-books* from *books (specifically) for public schools* (where *public schools* is not itself a compound), and repeated as in *earth-moving-machinery* from *machinery for earth-moving* from *machinery for the moving of earth*. Another example is *stone-grey* from *grey like stone*.

Other words having very broad selection get forms that relate them more closely to their arguments. This is seen in the interrogative and imperative intonations over a given sentence. It can be shown that these are derived from saying that one is asking or demanding the given matter, i.e. from *I ask* and *I request* over the given sentence: the *I ask* or *I request*, which can occur on almost any sentence, can be replaced by the respective intonation (G331, 351). A different situation is seen in the tense suffixes (specifically, the past *-ed*) which can appear on every verb (disregarding its morphophonemic variants such as in *stood*, *sang*), and the plural suffix, which can appear on most nouns. These can be analyzed as though derived (not historically, but in the grammatical system) from operators on the whole sentence: roughly, *I state that his going is before my stating it* to *I state that his going is before* to *I state that he went*, to *He went* (G265); and *I bought a book and I bought a book, up to five* to *I bought five books* (G252). Such derivations are offered here only to show that the operator–argument structure of 3.1–2 suffices to state all the information, within its grammatical universe, in the sentences of the language, including such categories as time and number which are not expressed overtly in operator form. In any case, these categories, which have very broad selection, are expressed by short morphemes attached as affixes to the words on which they act.

As to the other affixes, it is relevant that almost all have one or another of just a few very general meanings, such as the verbalizing set including *-ize*, *-ify*, *-ate*, and *en-* in *enrich* (all of which mean 'cause to be') or the nominalizing set including *-ness*, *-ment*, *-ion*, *-ure*, *-hood*, *-ship* (all of which mean 'condition' or the like). In the case of the past and plural, it is generally considered now that the various forms of each are morphophonemic variants of a single morpheme: a single past in *leaned*, *went*, *stood*, *sang*, *cut* (here, zero) and one plural in *books*, *children*, *sheep* (here, zero).

In *-ness*, *-ment*, *-ion*, etc., it is commonly thought that there are various nominalizing morphemes. However, it is very hard to state a difference in meaning, or in host domain (which words get which affix), or in sentential environment (i.e. further operators) that would distinguish all occurrences of one of these from all of another. Even in the relatively few cases of a word taking two different affixes of a set, yielding two different meanings, as in *admission* and *admittance*, the general meanings of the two affixes (in all their occurrences on *admit*) are close. The main difference is the grammatically uncharacterized choice of word per affix: *advancement*, *improvement*, but *progression*, *admission*; or *purify*, *solidify* but *realize*, *materialize*. The selection in respect to higher operators is roughly the same for all affixes within a set, except for special problems as in *-ion*, *- ance* on *admit*. Hence we can consider *-ify* and *-ize*, for example, to be complementary variants of a single variant-rich morpheme *A* whose combined selection (i.e. all the stems on which all variants of *A* occur) is much broader than the selection of any one verbalizing affix. Seen in this way, virtually all English affixes either themselves have exceptionally broad selection of host (their argument), or else are variants of a suffix-alternation (e.g. the plural or the verbalizing above) which in total has such selection. It is this broad selectional property that gives the variant-rich morpheme a low information status in the sense of 3.3.2.0, which justifies its reduction to affix (in varying forms). Many problems nevertheless remain in analyzing the selection and source of the individual affixes.

The broad-selection word *set* and its synonyms can also be zeroed in certain situations, when they occur with words for their members. For example, (1) *Gilbert and Sullivan wrote operettas*, cannot be formed from (2) *Gilbert wrote operettas and Sullivan wrote operettas*, since anyone acquainted with the matter would not say (2); this, differently from *Mozart and Beethoven wrote operas*. However, from (1) we can reconstruct (3) *The team* (or *set*) *containing Gilbert and Sullivan wrote operettas*, from *A team wrote operettas; the team contained Gilbert and the team contained Sullivan*. In (3), *the team containing* is zeroable, as being something that could be said in almost every sentence in which *Gilbert and Sullivan* is said (especially in that order, with the stress pattern of a single name). Then, in the reconstructed sentences, *the team of* (or *the team containing*) is very frequent before occurrences of *Gilbert and Sullivan*; it also adds no information, being common knowledge to people who speak about Gilbert and Sullivan, especially in the stress pattern stated.

Zeroing of broad-selection words in fixed positions is seen in the indefinite nouns. *Something* and *someone* are, between them, in the selection of every operator: *I ate something*; *I read something*; *Someone spoke*; *Someone*, or *something, fell*. In the second-argument position, the indefinite is zeroable under many operators (*I ate* from *I ate something*; *I read all night* from *I read things all night*). But it is not zeroed under certain operators, such as *wear*, where perhaps the indefinite is less frequent as second argument (no *I wear*, which would be from a perhaps less common *I wear things*).

Further, these indefinites, and classifiers (i.e. words naming a class) with more limited but still very broad selection (and hence broad meaning), can be zeroed when they are the last word in a compound. This case was seen in such compounds as *red-cap*, which differs from other compound nouns in the status of its last word (the modificand): as noted, a red-cap is not a cap. But if we assume the source to be *one* (or: *person*) *who specifically (by definition) is with a red cap* reduced to *red-cap-one* (or *red-cap-person*), then the properties of compound nouns would be satisfied, and a zeroing of the low-information, broad-selection indefinite or classifier host (last word of the compound) would then produce *red-cap*.

A similar zeroing would explain the 'zero causative' seen for example in *The nurse walked the patient*, *He leaned the broom against the wall*, which must be from, roughly, *The nurse enabled* (or: *helped*) *the patient to walk*, *He caused the broom to lean against the wall*. The transformation would have to be via length permutation (3.4.3), yielding e.g. *He caused-to-lean the broom*, and then compounding (3.3.1), yielding a non-extant *lean-cause*, and then zeroing the broad-selection *cause* (3.3.2.1), leaving *lean* and *-ed*. The trace of this transformation is seeing an intransitive verb (O_n, *walk*) appear as transitive, i.e. with a second argument, and one which could be the subject of that verb.

3.3.2.2. High frequency

Reduction also occurs in some high-frequency set phrases, such as *Goodbye* from earlier *God be with you*, and *'s, 'm, 'll, 'd* from *is, am, will, would*; compare also the Spanish *usted*.

The most varied and complicated reduction is to zero, always in word-occurrences that are reconstructible from the environing material (occasionally with some ambiguity due to degeneracy, i.e. to different possible sources). In almost all cases the high-likelihood words that can be zeroed, and their positions, are fixed. The difference between reduction to zero and reduction short of zeroing is seen in the treatment of expected word-repetition. In situations where the occurrence of repetition is frequent, and expected, but not specifically in any one position, a repeated word is reduced to a fixed form, called pronoun. Within a connected discourse there is a considerable likelihood that some of the referents will be repeated (9.1), though mostly in unpredictable positions: if some person or things are referred to, it is highly likely that one or another of them will be referred to again later on. When such repetition occurs, i.e. when a word is repeated with a statement that it has the same referent as another occurrence of that word (cf. *who* in 3.3.1), then the whole sequence—the repeated word together with the sameness statement—is reduced to a pronoun which states only the fact of repetition. Thus *John came in, but I was busy. So he left* would be considered a reduction of (roughly) *John came in, but I was busy. So John (which has the same referent as a preceding occurrence of John) left.* It is this kind of derivability that shows *he* to be a reduction and referential, not a new word.

In contrast, when a repeated word is in a fixed position where repetition has high frequency and where the word is reconstructible *in situ* by the hearer, then it is reducible to zero. This was seen in the O_{no}, O_{nno} operators where the zeroing depends on the higher operator—specifically on whether the first or second arguments of the lower operator is likely to be the same as one of the noun arguments of the higher operator. As was seen in 3.2.3: under *hear*, *observe*, where there is no special likelihood of repetition, there is no zeroing of an argument if it is repeated (e.g. no *You heard shouting at yourself* as a reduction of *You heard yourself shouting at yourself*). Under *prefer*, the subject (first argument) of the lower operator (which may be intransitive or transitive, i.e. with either one or two arguments) is zeroable if it is the same as the subject of the higher operator (*I prefer to stay* from *I prefer for me to stay*). Under *order*, the lower subject is zeroable if it is the same as the higher object (second argument) (*I ordered you to shred it* from *I ordered you for you to shred it*). Under *suffer*, the object of the lower verb (which is here necessarily transitive, i.e. has an object) is zeroable as repetition of the higher subject (*I suffered defeat* from *I suffered someone's defeat of me*). Under *defend*, the lower object is zeroable as repetition of the higher object (*I will defend you from attack* from *I will defend you from anyone's attack on you*). The high frequency of fixed-positioned repetition permits these repeated words to be reduced to pronouns in most cases, but they can also be reduced to zero, which was not possible when the positions were not fixed, as they are here.

Another zeroing of word repetition in fixed position was seen in 3.2 when a word in the second sentence under *and*, *or*, *but*, *more than* is zeroable when it is the same as the word in the corresponding position of the first sentence: *John plays piano and Mary violin.*

A less obvious case of zeroing the repetitions in fixed position is seen in the bi-sentential operators (O_{oo}) whose selectional preference is that in their arguments there be some word in the first sentence occurring also in the second. It is seen in 3.2.3 and 9.1 that such conjunctional sentences which seemed nonsensical became acceptable in an ordinary way if we interposed a bridging sentence that related a word from the first component sentence to one from the second: if in *I sharpened my pencil because the snow was soft* we interpose, say, *since I had promised a pencil sketch of a snowy tree.* These observations suggest that when we see conjunctional sentences that are acceptable despite their lacking word-repetition we should postulate an interposed bridging sentence that was common knowledge to speaker and hearer and hence reconstructible and zeroable: e.g. for *I'll stay if you go* we reconstruct an intervening component sentence like *which means, I won't go.*

There are various fixed positions in English in which we find zeroing of a particular word which may be the most frequent and is certainly the most expected, in that it, or in certain cases also synonyms to it in the given position, can be readily reconstructed by hearers of the sentence in agreement with the speaker. Among the examples, presented below, are: *which is*; also *for* and its synonyms before time and distance words; *come* and its synonyms under *expect*; *each other* under 'reciprocal' verbs.

When a sentence under a secondary intonation (which can be marked by semicolon) follows a preceding sentence, or enters as interruption after a word in the preceding sentence (3.4.4), the first word of the interruption is frequently *which*, *who(m)*, etc., as below. This *wh-* word is a pronoun (i.e. a repetition) of the preceding noun. The great frequency of the *wh-* pronouns means that more interruptions had begun (before reduction) with a repetition of the word or sentence before them than with any other word. This high likelihood explains why the repetition is reduced to a pronoun, and also why the *which*, *whom*, is zeroable when not separated from its antecedent by a comma: *A boy whom I knew ran up* to *A boy I knew ran up.* One verb, *is*, occurs much more commonly in interruptions than does any other verb, and the sequence *which* or *who* plus *is* (even if introduced by comma) is zeroable: *The list which is here is incomplete* reduced to *The list here is incomplete.*

Given that the zeroability of *which is* arises as noted above, the decision as to whether a particular occurrence of *which is* is zeroed depends on an interesting further consideration of likelihood, which creates the ordering of adjectives before a noun. The consideration here is that given $A_n \ldots A_1 N$, i.e. a noun with ordered preceding modifiers (mostly adjectives or adjectival nouns) that have accumulated before the noun, a following *which is* is zeroed only if the adjective following *which is* has good likelihood of occurring as an additional modifier, i.e. as interruption after the whole $A_n \ldots A_1 N$. This is an understandable condition, since upon zeroing this *which is* the $n+1^{th}$ adjective which it carries moves automatically (in English) to before the $A_n \ldots A_1 N$ and thus enters a closer modifier relation to the whole sentence. *A new red car* is said of a red car that is new (a new one of the red cars), while a *red new car* is a reduced form of a *new car which is red* (a red one of the new cars). This is clear semantically, and also from checking the relative permanence as against accidentality of modifiers which occur closer or farther before the noun (the generally most accidental or changeable are the farthest). If $A_a N_i$ *is* A_b is more likely than $A_b N_i$ *is* A_a then A_a is more permanent than A_b, at least on N_i. Since no property or modifier is more permanent than its modificand, the fact that each adjective A_{n+1} is a modifier of the whole following sequence $A_n \ldots A_1 N$ means that the adjectives before a noun recede from the noun in monotonically descending order of permanence and of likelihood. This is the adjective order wherein the adjectives closer to the noun are more durative or relevant classifications at least in the given occurrence, as we see for example in *a dented cheap old red car*. This permanence order decreasing with distance results simply from restricting the zeroing of *which is* to the case when the adjective following it has high likelihood to be an interrupting predicate, i.e. a further modifier (3.4.4), on the whole $A_n \ldots A_1 N$ before it.

On this analysis, the pre-noun (attributive) adjective order gives indirect support for the attributive adjectives being reductions of relative clauses (by zeroing the *which is*), and for the informational grounds of reduction (of the *which is*). For here the unusual ordering seen in the pre-noun adjectives is obtained simply by repeated application of the *which is* zeroing on each $n+1^{th}$ relative clause if it has high likelihood in respect to the host noun together with its previously reduced relative clauses. If a person wants to refer to the red one among the new cars, and has already said *red car*, he will not reduce a following *which is new* (from *; the car is new*) to obtain *new red car* (except perhaps by using contrastive stress on *red*).

The analysis given here builds up the adjective order by reducing each successive relative clause to an adjective as long as it is a likely further modifier on the noun as modified so far. This does not mean that all sequences of secondary sentences must be reduced to relative clauses and further, in this order. Quite the contrary, it is possible to build up two sentences, each with one or more relative clauses (from secondary sentences on it), and make the whole second sentence (with its relative clauses attached) into a large relative clause on the first sentence. An example was seen in the derivation of the 'restrictive' relative clause in 3.2.6, where *A house of snow can melt* was derived from *A thing can melt* carrying as a secondary sentence *The thing is a snow house* from *The thing is a house which is of snow* (where *the* is a reduction of the sameness statement on the second *thing*).

There are various other cases in which words with highest expectability in respect to their argument or operator are zeroed. For example, operators expressing duration have exceptionally high likelihood on arguments which refer to periods of time and are then zeroable in appropriate situations: *He talked during* (or *for*) *a full hour* is reducible to *He talked a full hour*, and *I saw him during one morning* (or *this morning* or *mornings*) to *I saw him one morning* (or *this morning* or *mornings*); but *I saw him during the* (or *a*) *morning* is not reduced in English. Similarly for space-measuring operators on measured-space arguments, which is a high-likelihood or preferential combination: *He ran for a mile* reducible to *He ran a mile*, and *He paced through the halls* to *He paced the halls*; but *He circulated through the halls* is not reduced, since *circulate* is less likely to be here a measured activity.

More specifically, certain words are conventionally the favored or 'appropriate' arguments or operators of certain others, and are zeroable when they occur with those others. An example is *come, be here*, or the like, under *expect* discussed previously. Another example given in 3.2.3 is the zeroing of *each other* under 'reciprocal' verbs, as in *John and Mary met each other* (where if John met Mary it is almost certain that Mary also met John), but not elsewhere, e.g. not in *John and Mary saw each other* (which is not reduced to *John and Mary saw*). Also scale-operators (G245) are zeroable in the presence of scale-measurements: e.g. *These chairs begin at $50* from *The costs* (or *prices*) *of these chairs begin at $50*, or *This chair is $50* from *This chair costs $50* (where the operator *cost* is zeroed, leaving the tense *-s* which is then carried by *be*). Such favored-word zeroing can also arise marginally in nonce-forms, colloquialisms, or special subject-matters: e.g. *at the chemist's* from *at the chemist's place* or the like, *She's expecting* from *She's expecting a baby*, *Coming?* from *Are you coming?*, and *Coming!* from *I'm coming* (the latter two despite the fact that the subject of a sentence is not otherwise zeroed in English). These words are reconstructible by the hearer from the environment, whether or not they are the most frequent there. Their meaning is assumed even in their absence, hence their presence contributes no information.

A different kind of low-information zeroing is seen in sentence-segments which engender in the sentence an event that expresses their information, making them superfluous in the sentence. This zeroing is seen in the performative segments *I say* (but not *I said* or *He says*) on a sentence, where the saying of the sentence carries the same information as *I say* on it (5.5). There are many kinds of evidence that a sentence as said is derived from its form plus a preceding *I say* or the like. In the case of other performatives, such as *I ask*, where again saying *I ask* on a sentence constitutes asking it, one can consider *ask* to be derived from *say* plus *as a question* or the like, with *I say* zeroable and *as a question* imposing the interrogative intonation. Note that the performative is zeroable only in its performative occurrences:

He will be angry if I ask: Are we losing? has no associated *He will be angry if are we losing?*

In contrast to all the high-frequency reductions above, exceptional low frequency can block a reduction that would otherwise take place. Thus it was noted at the end of 3.2.2 that the secondary-sentence intonation (the semicolon conjunction) has negative selection toward having its second argument directly deny its first argument (e.g. *He is coming; that he is coming is false*), with the result that the usual reduction of such secondaries to adverbs does not take place here (hence no *He is falsely coming*), even though the unreduced form is sayable within English.

The survey above of the main types of reductions shows them to be based, roughly, on exceptionally high frequency or expectancy, or exceptionally broad ranges of higher-than-average frequency (selection), within the operator–argument relation. Irregularities and mistakes in the application of this criterion can arise. For example, in *He's away, because his car is not in the garage* there is indeed a zeroing of *I say*, but that *I say* was merely a part of the first argument under *because*, and it should not have been zeroed; the source is *I say he's away, because his car is not in the garage*, where *I say* is not zeroable. (His being away was not caused by the absence of the car, unless he went to report it stolen.)

Aside from these likelihood conditions there may be, in one language or another, positional conditions for reduction, especially for zeroing. Thus in English the first argument of the highest operator is not normally zeroed, although this is done in the colloquial *Coming!* above. As to repetitions, what is reduced is usually the occurrence that is second (linearly), or lower in the partial ordering of words in the sentence; but reductions in the first occurrence can be found, as in *He had liked it from boyhood, so John majored in geology*. However, by the side of *John was emboldened by his winning the argument* we do not find (for the same meaning) *He was emboldened by John's winning the argument*. The positions in which pronouning and repetitional zeroing take place differ considerably, as also the positions of repetitional zeroing in contrast to other high-expectancy zeroing. For example, the final (host) noun of a compound can be zeroed, though only rarely, if it is an indefinite, as in *red-cap* above. But it is not zeroable on grounds of being a repetition, as can be seen if we compare *He is manager of the office and she is manager of sales* (where the second *is manager* is zeroable) with *He is office-manager and she is sales-manager* (where neither *manager* nor the second *is . . . manager* is zeroable).

The likelihood condition for reduction is practicable because it involves not the absolute likelihood of occurrence, nor even the general inequalities of likelihood, but a few grades: extreme expectability, extreme unexpectedness (rejection), and higher-than-average likelihood.

3.3.3. *Required reductions*

To speak of required reductions might seem to contradict the demand for a least grammar since it means that in addition to positing a particular form we have to say that the form does not exist as is but must be reduced. And indeed the great bulk of reductions are optional. That means that the unreduced forms exist, and can be constructed in the grammar by 3.1–2, and that the reduced forms also exist and can be constructed by making changes in the unreduced sources (3.3). Obtaining the reduced form out of the unreduced, by a reduction satisfying specified conditions, makes for a far less redundant grammar than constructing the reduced sentences *de novo*, since the reduced and unreduced forms are otherwise similar.

In various languages, however, there exist forms for which the least-grammar demand makes us first construct a non-extant form (which is similar to all other forms of its type) and on it impose a required reduction (of some known type) which yields the form that is extant. The most obvious case in English is that of the auxiliary verbs *can*, *may*, etc., where we first form the non-extant *I can for me to shave myself* with required repeated subject (3.2.4) and required zeroing of the repeated subject (with *for . . . to*), comparable to the preferred but not required zeroing of *for me* in *I want for me to shave myself* and the optional one in *I prefer for me to shave myself*. Had we constructed *I can go* directly (as verb acting on verb), not by reduction (from verb acting on sentence), we would have added to the grammar a set—fuzzy, at that—of operators whose second argument (e.g. *shave*) does not keep its own arguments (here, *I*) with it; it would then be impossible to define a sentence as a sequence all of whose words have their argument dependence satisfied.

Another required reduction in English is zeroing a repetition of the word that carries the comparative: *John is older than Mary is* (from . . . *than Mary is old*), *John bought more books than Mary bought* (from . . . *than Mary bought books*); but *John likes books more than Mary does* (from . . . *than Mary likes books*) where the *more* is on the whole component sentence *John likes books*, hence on its operator *like*, so that the second *like* is zeroed leaving its tense *-s* to be carried by *do*. Yet another required reduction in English is moving a post-*which is* adjective to before the preceding noun (more exactly, the preceding *A . . . AN*) when the *which is* is zeroed: *a new red car* from *a red car which is new* (we do not find *a red car new*). Also, under *and*, *or*, the moving of a verb whose object has been repetitionally zeroed, to before the antecedent object: *John composes pop-music and John plays pop-music* reduced to *John composes and plays pop-music*, without the intermediate *John composes pop-music and plays*; also *John composes light classical music and Mary plays light classical music* to *John composes, and Mary plays, light classical music*.

Other than this, almost all reductions in English are optional.

3.3.4. *How reduction takes place*

The fact that reduction is based on operator–argument likelihood rather than on features of the whole sentence means that its applicability depends on a subset of word-pairs rather than on the unmanageable set of all sentences.

A reduction or other transformation cannot take place before the conditions specified for it are satisfied, i.e. are in the sentence. For example, *plays* in *Mary plays piano* cannot be zeroed before that sentence is connected by *and* to *John plays violin*. Furthermore, the zeroing of a correspondingly positioned repetition under *and* (3.3.2) does not take place when the *and* interrupts its first component sentence too early. Since this zeroing compares the partial-order status of a word in the second sentence with that in the first, it can only be carried out after the partial order of the first sentence is available for comparison. In *John—and John is very insistent—thinks we should not go* the second *John* is not zeroable (even though the first has already occurred), because the partial-order (i.e. grammatical) status of *John* in the first sentence is not yet known at the point when the second *John* is said. Note that in *John thinks—and is very insistent—that we should not go* the second *John* is zeroable (from *and John is very insistent*) because the preceding *John thinks* suffices for the status of the first *John*. Had the first sentence been *John I don't like*, the interruption *and John is very insistent* would not admit zeroing (no matter where it entered) because the two occurrences of *John* do not have the same status: from *John—and John is very insistent—I don't like* we cannot obtain *John—and is very insistent—I don't like*.

It follows from this that words are reduced only in their reducible, i.e. high-expectability, occurrences. Consider the colloquial reduction of *going to* to *gonna*. The *going to* has selectional (normal) likelihood on virtually every verb, and hence very high total likelihood on verbs: *They are going to sleep*, *They are going to reject it*, or *to go*, or *to roam*. But it does not have selectional likelihood on every noun: *They are going to* is normally likely on *Rome*, *the river*, *the attic*, a bit less on *the chair*, *the fox*, still less on *the teaspoon*, vanishingly little on *the electron*, *time*. Hence the reduced form *gonna* occurs before verbs (*They're gonna come soon*, *They're gonna stampede*), but not before nouns (no *They're gonna the attic*, *They're gonna Stampede, Washington*). More generally, this shows that it is not a particular word but a particular word-in-combination that gets reduced.

Given these considerations as to when a reduction cannot take place, there is also evidence that if it does take place, it does so before any further operators act. For example, given *John knew that Mary came quickly*, we also have paraphrastically *John knew that Mary quickly came* (or . . . *quickly Mary came*), but not *John knew quickly that Mary came* or *John quickly knew that Mary came*: in the latter cases, *quickly* cannot refer to *came*. The only simple and non *ad hoc* explanation is that the permuting of *quickly* (from *Mary came quickly* to *Mary quickly came* and *Quickly Mary came*) took place before *knew* operated on the pair *John, came*; hence this *quickly* could not move into the then non-present *John knew* segment.

Furthermore, *John slowly knew that Mary would recover* is rare, and if it is said its *slowly* refers to John's knowing, and not to Mary's recovery; it is not a permutation of *John knew that Mary would slowly recover* or *John knew that slowly Mary would recover*. That is, if *slowly* is said on (1) *John knew that S_i* (where S_i is *Mary would recover*), it can be permuted among the arguments and operator in (1) but cannot enter inside an argument of (1). Reductions can thus be located between fixed points in the semilattice of word-entry—i.e. of operator–argument relations—in the sentence (6.3, 8.3).

Since reductions are carried out before any further operator acts on the sentence, any further operator acts on the reduced sentence. This does not affect the content on which the further operator is acting, since the reductions are substantively paraphrastic, and indeed any zeroed material is reconstructible and can be considered to be still present though in zero form. But it may alter the formal conditions for the further operator, and indeed some operators cannot act on particular reduced forms. If there are cases in some language of an operator not being able to act unless its operand sentence has been reduced, that would compromise the ability of the base to carry all the information in the language (4.1, 11.3).

As to how reductions are ordered with respect to each other:

Two or more reductions may be applicable as a result of a single operator entry. For example, when *and* operates on the pair *John phoned Mary*, *John wrote Mary* (with sameness statements, as in 3.3.1, on both *John* and *Mary*), it simultaneously makes both *John* and *Mary* zeroable in the second sentence, yielding *John phoned and wrote Mary*. These two zeroings are independent of each other; neither, either, or both may take place, with no indication of their being ordered in respect to each other. Somewhat similarly, in (1) *John came home, which was because John was tired* (with sameness statement on *John*) we have available a pronouning of the second *John* (to *John came home, which was because he was tired*), and independently of this a zeroing of the *which is* (to *John came home because John was tired* or ... *because he was tired*), followed by a possible permutation to *Because John was tired John came home* or *Because he was tired John came home*. In the former case, pronouning of the second *John* is still available, now yielding *he* in a different position: (2) *Because John was tired he came home*. (Pronouning the first occurrence, as in *They were as yet unsure of themselves, when the doctors first suspected a viral cause*, is so limited as to require more stringent conditions than the second-occurrence pronouning discussed here.) It is not necessary to state a time order for these events, which would require that pronouning be able to take place both before and after the permutation. A least-restriction grammar can obtain the existing forms by saying merely that both the pronouning and the *which is*-zeroing plus permutation are available in (1) above until some further operator acts on the whole of (1). Such a further operator is seen, for example, in *John's coming home, which was because he was tired, interrupted our conversation*. In (1), the permutation, if it takes place, creates a new pronouning-possibility in *John comes home*, if pronouning had not already taken place in *because John was tired*, yielding (2). All we need say, then, is that the conjunction which brought together *John came home* and *John was tired*, expressed in (1) by commas, admits two independent reductions which are available, unordered among themselves, until a further operator enters the sentence.

In contrast with these, there are cases of reductions which are ordered, simply because the conditions for one of them to take place require the result of the other. An example is moving an adjective to before the host, after its *which is* has been zeroed, as in *costs which are allowable* to *costs allowable* to *allowable costs*; the permutation requires zeroing of *which is* (there is no *which-are-allowable costs*). Another example is zeroing a high-expectability operator, or a classifier noun, after it has become the last (weak-stressed) member of a compound: *a quick drinking of coffee* to *a quick coffee-drinking* to *a quick coffee* (in *Let's stop for a quick coffee*); *a fly like a blue bottle* to *a blue-bottle-fly* to *a blue-bottle*.

A reduction is described as altering the shape of a word, even if to zero, but nothing is said about its altering the presence of the word in the given occurrence, and indeed the word continues to be present in the sentence. This is evident in the meaning, where a blue-bottle is a fly. It is also evident in various grammatical features, such as the presence of modifiers of zeroed words, as in the *momentarily* of *I expect John momentarily* from *I expect John to come momentarily*; *momentarily* is clearly on *come*.

In view of all this, it is not merely a figure of speech to say that reduction does not affect the basic syntactic status of the words which have entered into the partial order that creates a sentence. When we consider reductions as variant forms of a sentence, then, the reductions do not add constraints on equiprobability except, slightly, in the case of permutations. But if we want to account directly for the observable combinations of words that constitute sentences, then pronouning (which replaces many words by the same he, or it, etc.), and especially zeroing (no matter what traces are left), have complex effects on the departures from word-equiprobability. In any case, a transformed sentence has to be described by describing its source sentence, plus the change; if we try to describe the transformed sentence *de novo*, we have to repeat for it almost everything that was said for the source sentence.

3.4. *Linearizations*

3.4.0. *Introduction*

In all languages, speaking is an event in time only, and thus overtly linear, even though the linear form can contain simultaneous components, and can even contain long components which are simultaneous with a sequence of short ones.[3] The partial order established in 3.1 is descriptively prior to the linear order; that is, it is independent of the linear order in a description of language structure, whereas the linear order is not useful toward a structural description without reference to how it embodies the partial order (or the grammatical relations, which can be defined in terms of that partial order). This is not to say that the partial ordering exists in time or in human activity prior to the linear. There is no evidence against saying that sentences are constructed as word-sequences, with the partial ordering bringing in an additional degree of freedom (and so an additional dimension) in the choice of words, recognized now by educated speakers as 'grammatical relations' among the sequenced words. Within the sequence, the partial order is realized by two linear orderings: that of an operator to each part of its argument, and that between co-arguments (*John ate a fish* being not the same sentence as *A fish ate John*).

The relation of linear to partial order differs for speaker and hearer. The speaker chooses certain words in a partial order which satisfies their dependences, and says them linearly in time. Since much of the partially ordered set, particularly the arguments of operators said so far, is available to him when he is saying the linearly earlier words, there is room for his making the kind of slip called anticipatory contamination, such as Spoonerisms (as notoriously in *our queer old dean* for *our dear old queen*). The partial order, which underlies the linear, suggests that slips which are unintentional should appear chiefly in operator words drawing upon their linearly later arguments, or in argument words drawing upon their linearly later co-arguments; this is indeed found to be the case. We would not expect slips in a word to draw as frequently upon linearly later words that are also higher in the partial order, for these higher words are not necessarily in the speaker's mind when he says the earlier word. Intentional (jocular) contaminations do not show this distinction, since the speaker is then drawing upon a complete sentence or segment.

The hearer receives the words in linear order of time or writing and has to reconstruct their partial order as grammatical relations. In this he meets the well-known problems of degeneracies, especially those due to homonymy and zeroing. Upon occasion he also has to reinterpret an earlier word on the

[3] Z. Harris, *Methods in Structural Linguistics* (pb. title: *Structural Linguistics*), Univ. of Chicago Press, Chicago, 1951, pp. 45, 50, 92, 117, 125, 299.

basis of its linearly later operators or arguments, as in *She made him a good husband because she made him a good wife*. Here the expectation that the second *made* has the same partial-order relation to its following *him* as does the first one is counteracted by the gender of *wife*: the first *made him* is from *made him into*, the second is from *made for him*. Somewhat similar is K. S. Lashley's example, 'Rapid righting with his uninjured hand saved from loss the contents of the capsized canoe', where the hearer normally assumes *rapid writing* until he hears the last words.[4] In reconstructing the partial order, as grammatical relations, out of the linearly ordered words, the hearer uses the residual traces of each reduction which has taken place in the partially ordered material. In addition, certain changes are made in respect to the linear order: phonetic assimilation and dissimilation in neighboring words.[5]

3.4.1. *Normal linearization*

Thus the major departures from word-equiprobability in the making of sentences are made in a partial-order relation. This means that speech linearization has yet to be stated, over and above the partial order. One might think that the linearization could constitute a complex change in the constraints on equiprobability, possibly with different linearizations for different parts or kinds of the partial order. This is not the case. It seems that each language has simple normal linear projections of the partial order. In English, the operator is said normally after its first argument: the O_n *sleep* appears in *John sleeps*, the O_{nn} *eat* appears after *John* in *John eats fish*. In short sentences, which contain only one operator, the arguments have to be next to their operator, and this is the order in which they appear. In longer sentences, if the same projection is to be retained with least modification, the further operators O_{no}, O_{oo}, etc., acting on O arguments, would have the same linear order in respect to their arguments; and this is indeed the case. The O_{oo} *cause* acting on two O, e.g. *phone* and *leave* (each necessarily with its arguments next to it), produces *Mary phoning John caused his leaving*; and the O_{noo} *attribute . . . to* on the triple *I*, *leave*, *phone* would yield *I attribute John's leaving to Mary phoning him*. (If we expect the partial order in long sentences to be the same as in short sentences, it follows that if there are several N in a sentence under O_{noo}, the N which is its first argument is likely to be the one before it, just as is the first-argument N under O_{nn} in the short sentence *John eats fish*.)

---

[4] K. S. Lashley, 'The Problem of Serial Order in Behavior', in L. A. Jeffress, ed., *Cerebral Mechanisms in Behavior*, The Hixon Symposium, Wiley, 1951.
[5] The relation of speaker's grammar to hearer's will be touched upon, though very slightly, in 4.3 and 12.1.

In English and in many languages an operator (with its arguments) which is appearing as an argument of a further operator receives a marker (affix or word, e.g. *-ing*, *'s -ing of*, *that*, *for* . . . *to*) to indicate its new status as argument: *That John arrived surprised me, John's buying of the book surprised me, I know that John arrived* (where *that* is omittable), *I prefer for John to stay, I prefer John's staying.* This argument indicator obviates certain ambiguities. Consider three sentences connected by two conjunctions: SCSCS;. In *He came over because she was leaving because they had quarrelled*, there are two ways of grouping and understanding the sentence: *SC(SCS)* is not the same as *(SCS)CS*. But in *His coming over was because of her leaving being because of their having quarrelled*, which is *SC(SCS)*, as against *His coming over being because of her leaving was because of their having quarrelled*, the two groupings above are differentiated. Spoken language has no parentheses, but the argument indicator does their work. If the linearization places the operator before or after its full argument, instead of between the parts of the argument, this ambiguity does not arise, and the argument indicators are then not needed (though they would still be a convenience), as parentheses are not needed in Polish notation of logical formulas.

3.4.2. *Alternative linearizations*

The fact that there is a normal linearization of the partial order does not preclude there being alternative linearizations of it at different points or in certain situations. Indeed, one of the advantages in describing sentences in terms not of a direct linear ordering, but of a partial order which is then represented linearly, is that a limited freedom in the linear representation can account for important differences in sentence structure without constituting a difference in the grammatical relations which characterize the sentence: In many languages certain parts of a sentence which are normally not at its beginning have an alternative linearization at the front of the sentence, without changing the grammatical relations: *This I will say* (or *I this will say*) instead of *I will say this.* There is no evidence that this alternative linearization ('fronting') is a permutation taking place after the normal one has been formed; rather, it may be simply a different linear representation of the partial order (here *say* as operator on *I*, *this*), used for stylistic reasons or to make the front word stand out as the 'topic'.

That which is subject to alternative linearization (i.e. that which is linearized) is not simply particular words of the sentence, but rather particular items of the partial order, as will be seen. Furthermore, certain of these items have their adjuncts (modifiers, 3.4.4) and even their second arguments with them in any alternative position. In English, a noun always has its adjuncts (*Long books I cannot read*; never *Books I cannot read long*); an operator optionally (*Read I will immediately*, *Read immediately I will*).

Prepositional operators always have their second arguments (*Near the lab he has a house*, never *Near he has a house the lab*); verbs optionally (*Sign I will any and all petitions*, *Sign any and all petitions I will*). The distant items do not conflict with contiguity: arguments and their operators and their co-arguments all remain contiguous (except for later inserts). Since adverbs are secondary-sentence operators on their host verbs, they remain next to either the verb or the arguments of the verb (which were present with the verb when the adverb acted on it).

The main alternative linearization in English is the front position of non-first arguments and of tenseless operators: specifically, the second argument (*This I say*), the third argument (as in *On the table I put the book*), or the operator with its non-first arguments (*Explain this I will not*, *Tired he was*). There is a rarer placing of the second argument before the operator (*I this doubt*), and a limited interchange of the two arguments without losing their identity in the partial order (i.e. in the grammatical relations, as in *This say we all*). In longer sentences there are certain partial-order positions which do not have the alternative available to them: they are too far grammatically (too complexly down in the partial order) to appear in the front. This applies to certain positions in sentences which are second arguments of certain conjunctions: for example, the last argument (*Hiroshima*) does not have the choice of occurring up front in *I didn't mention Nagasaki because of your detailed description of Hiroshima*; but it has that choice when it is deep in a chain of arguments, as in *I remember his describing how they made the decision to bomb Hiroshima*, where one can say, if awkwardly, *Hiroshima I remember his describing how they made the decision to bomb*.

Another set of alternative linearization has certain adverbs, especially those (such as *scarcely*) which operate on the metasentential *I say* (5.5.3), appear at the front of the sentence, followed by the tense: *Scarcely had I taken it*. Slightly different is the order in *Tired though he was*, and in *Here arises an important question*, and also in the moving of tense when certain front words are dropped: *Were he there I would go* as compared with *If he were there I would go*, and *I ask: Will he return?* from *I ask whether he will return*.[6]

3.4.3. *Permutation*

There is in English one other important type of alternative linearization. This is the permuting of neighboring long and short constructions at the end of a sentence. For the major case, in English: an adjoined (secondary)

---

[6] Both fronting and interruption (3.4.4), together with their semantic effects, can perhaps be obtained without assuming a special added freedom by deriving them from adverbs on *say* (5.5.3): e.g. *This I will do* from *I say of this that I will do this*, with zeroing of the second *this* (3.3.2) and of *I say* (5.5.1).

sentence which has been shortened can move to before a linearly preceding item from the partial ordering (with any adjuncts, and possibly with its arguments) usually on condition that the preceding item is longer, and above all is not shortened to a pronoun. Thus *John clamped the lid down*, where *down* is reduced from a secondary sentence *John's clamping the lid was down(ward)* (secondary to *John clamped the lid*), has an alternative order *John clamped down the lid*, provided *the lid* is not pronouned to *it* (no *John clamped down it*). In contrast *John slid down the pole*, where *down* is simply an O_{on} operator on the pair *slid, pole*, has no permutation for *down*, and permits pronouning of *pole* (to *John slid down it*).

Somewhat differently, when *the car which is new* (where *which* is a pronoun for the preceding *car*) has its *which is* zeroed, the residual *new* moves necessarily to before *car*: *the new car*. Permuting the residue after zeroing *which is* holds for all uni-morphemic adjectives and for stated other words, including the compound-word structure (after zeroing the charac- teristic *for* or the like, as in *books for school* to *school-books*, and *grey like stone* to *stone-grey*). When *which is* is zeroed in *I have decided to leave now, which is because my time is up*, we obtain *I have decided to leave now, because my time is up*; and *I have decided, because my time is up, to leave now*; also *Because my time is up I have decided to leave now*. Another moving of a shortened conjoined sentence can follow zeroing under *and*: in *John phoned and Mary phoned*, zeroing the second *phoned* (yielding *John phoned, and Mary*) permits (and virtually requires) *and Mary* (the residue of the second sentence) to move to a point where all that follows it is the antecedent for the zeroing (namely, *phoned*): hence, *John and Mary phoned*. This condition accounts for the other *and*-residues, e.g. *John composes, and Mary plays, light piano music*; and it explains why there is no permutation in *John plays piano and Mary violin* (because the above stated point, which has to be before the antecedent *plays* but not before the non-antecedent *piano*, does not exist).

A slightly different length-permutation permits a final verb to move to before the interruption (3.4.4) on its subject: *My friend whom I told you about is coming* to *My friend is coming, whom I told you about*. And under some higher verbs, a verb can move to before its subject: *He let the book fall* to *He let fall the book*. Calling these permutations is a descriptive convenience; their histories can be different. And some apparent permutations are not that at all (e.g. the passive, 1.1.2).

3.4.4. *Interruption: Modifiers*

There is one last freedom of linearization which is of great importance because it creates the semantic relation of being a modifier (adjunct). This is the possibility of interrupting one sentence by a secondary sentence bearing the 'aside' intonation, usually one referring to the first. It is directly observable in, for example, *Science fiction—science fiction I won't touch—is his favorite*, or in *You're in danger—and I'm speaking seriously—of failing math* (or *You—and I'm speaking seriously—are in danger of failing math*). In terms of the partial order, the aside-intonation (which can be written as semicolon) is the highest operator, an O_{oo} whose first argument is the preceding primary sentence and whose second argument is the following secondary sentence: *Science fiction is his favorite; science fiction I won't touch*. In terms of linearization, the semicolon conjunction plus its secondary sentence can interrupt the primary, instead of coming after it.

In the case that the first word, or word-sequence, of the interruption is the same as the primary-sentence word or sequence (with minor adjustments) that immediately precedes the interruption, a new situation is created, for the interruption is then placing a secondary sentence after specially relevant words in the primary. The words in the primary (here *science fiction*) become the antecedent; and the repetition of those words at the beginning of the secondary is reduced to a 'relative pronoun'—*who, which, where*, etc. The whole interruption is then called a relative clause, and its semicolon aside-intonation is generally reduced to the weaker intonation between commas: *Science fiction, which I won't touch, is his favorite*. In this situation it is semantically felt to be a secondary sentence 'about' the particular preceding word or words which are the antecedent of the relative pronoun. The relative clause is then a modifier, called adjunct, on the antecedent, which is its host. This is the source of the adjunct–host relation, the major syntactic relation other than operator–argument.

There are several considerations which favor the above explanation of the development and nature of the relative clause. First, if one tries to state the structure and actual word-content of all relative clauses, one finds that they have no unique composition, but simply are identical with the set of all sentences, each lacking one word or word-sequence which is such as could be referred to by the *wh*-word.[7] It follows that the relative clause must have been a sentence that repeated the antecedent words of the primary sentence. Second, if one asks what words in the secondary sentences can be replaced by initial *wh*-words, one finds, surprisingly, that they are precisely those words in the secondary sentence which can appear at its front either normally or by the fronting linearization (3.4.2, G109, 121). It must be, then, that the secondary sentences in which certain words were reduced to *wh*-words (relative pronouns) had those words already in the front position (rather than that the *wh*-word was formed wherever the words would normally have been and then brought to the front). This is not hard to understand: the front position of a word gives it the status of being the topic of the sentence, which as noted above is indeed apt for a secondary sentence

---

[7] Z. Harris, 'Co-occurrence and Transformation in Linguistic Structure', 3.6.2, *Language* 33 (1957) 283 ff.

about the antecedent: *Science fiction—science fiction I won't touch—is his favorite* exhibits the relevance of the interruption more directly than *Science fiction—I won't touch science fiction—is his favorite*.

Finally, the interruptions: one might think that the secondary sentence which is the source of the relative clause came after the primary, in ordinary sentential succession, and was thereafter moved to the antecedent. However, the fact that the relative-pronouned material was precisely that which could be at the front of its sentence before it was pronouned means that the fronted secondary sentence already had special relevance to particular words in the primary—those which its front words repeat, and after which it indeed appears in the primary sentence. This makes it desirable to assume that the secondary sentence was not successive but rather an interruption in the primary, coming immediately after the antecedent host words. More importantly, taking it as an interruption removes the need for an addressing apparatus—which is lacking in language—in terms of which we could locate the host. We would need this addressing apparatus in order to know to which primary words the secondary words were referring, e.g. if *The big boy hit the little boy* were to be derived from successive sentences *A boy hit a boy; the boy was big; the boy was little*. And we would need it in order to move the secondary sentence up to the antecedent in the primary. Languages do not have an addressing apparatus which can be attached to a secondary sentence to move it precisely to those words in the primary to which its front words refer. However, if the secondary sentence interrupts the primary we do not need to locate (or address) the antecedent host. We merely say that when a secondary beginning with certain words interrupts a primary right after the same words, relative pronouning can result. In other situations, the secondary sentence simply does not develop into a relative clause.

Indeed, in English grammar, interruption by a front-repetition secondary sentence removes the need for a full addressing apparatus for all cross-reference. The reason is that it brings two referentially related items (having various partial-order statuses) into a fixed distance in the linearization. Specifically, first, having the repeated words come at the front of the secondary sentence and, second, having the whole secondary sentence come right after the antecedent means that the two occurrences of the given words are next to each other in the sentence. To locate in the linearized form the antecedent of the words to be pronouned requires no more than the word *next* or *preceding* (in respect to the linearization), and the reduction to pronoun requires a metasentential statement that the second occurrence of these neighboring (next) words refers to the same objects (or in certain situations to members of the same class of objects or events) as the antecedent occurrence (5.3.3). With no more than this apparatus, it is possible to derive the relative pronouns and relative clauses. In English, it is then found that with the aid of the relative clauses all other referential situations can be derived (5.3). There are also other avoidances of the need for a full addressing system to locate antecedents: most repetitional zeroings depend on definable relative positions in the partial order, which can be stated a priori without locating the antecedent in the individual sentence: e.g. zeroing of words in a second sentence under *and* if the same partial-order position in the first sentence is filled by the same words. And many other pronouns do not depend on any particular relative position in the partial or linear order (e.g. *he, they*). The addressing requirement is thus circumvented.

In summary, we find that the partial ordering of words which creates sentences has a normal linear representation in the word-sequence of a sentence, but that certain items in the partial order have optional alternative positions, of particular types, in the linear representation. One of these types, the front position, makes a particular word or sequence the topic of the sentence. Another, the interruption by a secondary sentence, makes the secondary sentence a modifier of the primary sentence, or of a particular word in the primary. The alternative linearizations are almost entirely optional, and do not otherwise change the meaning of the sentence or the grammatical relations (i.e. the partial order) within them. They do, however, have an effect on the constraints on equiprobability of the words making sentences; but this effect is of a second order, consisting of a few kinds of alteration in the relative position.

Language Structure:
The System Created by the Constraints

4.0. *Introduction*

The methods of 3.1–2,4 yield a base set of sentences from which the other sentences of the language are derived. The dependence of 3.1, together with the linearization of 3.4.1–2,4, characterizes the set of possible sentences; it eliminates what cannot occur in the language. The likelihood gradation of 3.2 creates not-well-defined sets of 'normal' and marginal sentences, on which the reductions of 3.3 operate. Separating out the different constraints that contribute to the departures from equiprobability in the word-sequences of all sentences distinguishes the base subset of sentences, in which reductions have not taken place. This subset is structurally recognizable, because the operator–argument relation in it has not been obscured by zeroing, reduction to affix, length permutation, and other transformations. It is also purely syntactic, without having morphology carry part of the syntactic functions, since all affixes that can be derived from free words have been separated off, leaving the vocabulary to consist of syntactically indivisible words. That is not to say that one may not see internal construction, as in the *con-*, *per-*, *-ceive*, *-sist* of *conceive*, *perceive*, *consist*, *persist* or even the *-le* of *dazzle*, *settle*, etc., but that such constructions cannot be usefully derived from separately functioning morphemes or words, joining in the operator–argument relation.

Since reductions are optional, except as in 3.3.3 (where the missing sources in the base can be created), the unreduced sentences remain in the language by the side of their reduced forms: *I prefer for me to go first* and *I prefer to go first*. And since reduction is paraphrastic, changing only the shape of a word but not its existence or its operator–argument status in the sentence, the sentence that results from it does not have different substantive information than the sentence on which it acted. Hence the information of reduced sentences is present in the unreduced ones from which they have been formed. It follows that the base set of syntactically transparent unreduced sentences carries all the information that is carried by the whole set of sentences of the language.

4.1. *The base set of sentences*

In the base, the syntactic classification of vocabulary is into the N, O_n, O_{no}, etc. categories of 3.1. The syntactic operations in it are the operator–argument partial ordering, with optional interruption by secondary sentences and optional front-positioning of words. In English, operators carry operator-indicators (the 'present tense' -*s*), which are replaced or overruled by argument-indicators (-*ing*, *that*, *for . . . to*) when they become the argument of a further operator. This holds for all sentences. Similarly, all sentences can carry the metasentential *I say* with certain appended material on it such as *not*, *and*, *or*, *in query*, *in command*, and also the 'aside' intonation which attaches a sentence as secondary to its primary one. When an operator is defined as acting on an operator (as its argument), it acts on the whole sentence that the lower operator had created, since all operators in a sentence carry their arguments with them (by 3.1), so that each creates a sentence of its own. Given *destroy* on the pair *bomb*, *atoll* (yielding (1) *Bombs destroyed the atoll*), then *think* on the pair *we*, *destroy*, yields *We think bombs destroyed the atoll* (for plural, tense, *the*, cf. 8.2). Further operators and transformations on the higher operator (here, *think*) do not enter into the lower sentence, which remains as an unchangeable component sentence in the larger one. Thus in *Secretly we think the bombs destroyed the small atoll completely*, the *completely*, operating on *destroy*, can move only in the contained sentence (1), and *secretly*, on *think*, can move only outside the contained sentence. The *completely* cannot move into *we think* (no *We completely think the bombs . . .*), for any moving had to be done right after *completely* entered its sentence (the component one), and *we think* had not yet acted upon that sentence then.

In the base, few words are multiply classified, i.e. belong to more than one of the N, O_n, etc. classes; most of the apparent multiple classification arises in the derived sentences, and is due to zeroing. And few base sentences are ambiguous, other than for homonymy and for unclarities in the meaning-ranges of words; the great bulk of ambiguous sentences are due to degeneracies in reduction—to two different sentences being reduced to the same form.

It follows from the above that this relatively simple grammatical apparatus is adequate for all the substantive information which is carried in language. This would not be so if the meaning of further operations depended on whether the operand had been reduced, for then the meaning that required a reduced form would not be available in the (unreduced) base. But this apparently does not happen.

Certain situations present difficulties for the base, but can be resolved. For example: as has been seen, the auxiliaries in English differ from all other operators in that they act only on a subject-less verb: *John can go first*, but no

John can for him to go first, and no *John can for her to go first*, in contrast to *John prefers to go first* by the side of *John prefers for her to go first*. The description of *can* in the base was regularized in 3.2.4 by saying that like other O_{no} operators it operates on a whole sentence, but one in which the subject must be the same as that of *can*, and must then be zeroed: a non-extant but well-formed *John can for him to go first* is then necessarily reduced to *John can go first*. The cost of this derivation is that the source sentence does not remain in the language, hence is not in the base, since its reduction is required; the information of the reduced sentence cannot then be expressed in the base. A resolution of this difficulty can be reached by finding some word or sequence which is approximately synonymous to *can*, and selectionally close to it, which we can consider a variant of *can* in this form, and under which the above zeroing is not required. We can consider that some such sentence as (1) *John is able for him to go first*, which is in the base, occupies there the place of the non-extant (or no longer extant) (2) *John can for him to go first*, while the reduced forms of both, *John is able to go first* and (3) *John can go first* both remained in use. Such a resolution is supported by the fact that historically some auxiliaries indeed were able to have a repeated subject under them. A more general support is that change in word use consists in principle not in a word being dropped or replaced within a sentence, but in sentences which contain that word competing with approximately paraphrastic sentences containing another word, and losing out to them. Given (3) we have to posit a source (2), and if (2) does not exist it must have been replaced by an informationally roughly equivalent sentence such as (1).

This situation is particularly common in the case of affixes, which are analyzed, to the extent possible, as being reductions from free words, mostly operators, in the base. Even when such a reduction is historically evidenced, the free word may have meanwhile been lost; affixes are less amenable to historical replacement than free words. Thus, as noted, the *-hood* of *childhood* was once a compound form of a free word *had* 'state, condition' which was lost from the vocabulary, whereas the affix occurrences of it were not lost. But *had* was not simply lost. Sentences containing some other word, such as *state*, came to be used more than corresponding sentences containing *had*. When free *had* was no longer said, some word such as *state* remained in its place, like a suppletive variant of it, so that the unreplaced *-hood* is now the affixal reduction of *state* rather than of the old *had*. Such established historical cases are rare. For most cases in English, saying here that a given affix is a reduction of a particular free word in a particular position is not a claim of history but a statement that the operator–argument relation of that affix to its host is the same as that which the given free word would have in the stated position, and that the selection (relative likelihoods) of the affix, as to host and to further operators on it (and so its meaning), is approximately the same as that of the given free word in the stated position.

With the above statement as criterion, we can reconstruct a less restricted variant for every word or sequence whose occurrence is restricted. This does not place in the base something that is appreciably different in information from that which is outside the base. Of course, such methods are acceptable only if they apply to a few special situations (even if many words are involved in one or another of these situations), so that the methods are used only to regularize the aberrant situations to make them conform to the main character of the language.

Such reconstructions could also be used in any rare situation in which a reduction in a given word admits a class of further operators that would not otherwise come into the sentence, or admits a selection of further operators different from the selection that the given word has when not reduced. In each case we would say—artificially—that the further operator indeed requires the reduction, but that we can assign to it in the base some synonymous variant that does not require the reduction.

The three difficulties above, which detract from the informational adequacy of the base (and potentially detract from the applicability of the theory), arise only in encapsulated situations in a language. One can more readily find cases of reductions and transformations which block a class of further operators, but this does not interfere with the base carrying all the information which is carried in the language. One example is that the fronting linearization of a sentence (3.4.2) blocks the operator-indicator -*ing* (but not *that*) which appears when the given sentence comes under a further operator: by the side of *I like this* we have *This I like*; but there is no *I reported this my liking* by the side of *I reported my liking this*, while there is *I reported that this I like* as well as *I reported that I like this*.

The base words are generally uni-morphemic, i.e. affixless, since affixes are analyzed, as far as is possible, as reductions of free words in the given position. The only morphemes in English which are not operators or arguments (or reduced forms of such) are: indicators of argumenthood (e.g. -*ing*, *that*, G40) or of operatorhood ('present tense' -*s*, G37), carriers for these when these are suffixes (*be*, 8.2, G54), and words which can be analyzed as constructed from regular word sequences (e.g. *the*, G237).

Within N and the various O classes, the similarities and gross levels of likelihood grading create a few fuzzy subsets (3.2.5). Such are for example the argument-repeating O_{no} operators which have a high likelihood that the subject of the O under them be the same as their own subject (*John tried to phone*, rarely *tried for Mary to phone*), or that the object be the same as their subject (*John suffered defeat*), and so on (3.2.3). Some second-level (i.e. $O_{\ldots o \ldots}$) operators have the operator under them almost always future in respect to themselves; under some of these higher operators the future tense on the lower operator is omitted (e.g. the 'subjunctives': e.g. *order* in *The commander ordered that they be shot*, G283). Some arguments (the indefinites) occur freely under virtually all operators; others (the classifiers) under the same operators as occur on any of particular other arguments (their classificands). Such likelihoods common to sets of words yield in the base many subsets, such as indefinites, quantifiers, prepositions, reciprocal verbs; these are of major or minor importance in the grammar of a language (3.2.2). They may have few or many members, and one or several characterizing properties. In almost every such subset there is a common element of meaning, related to its special syntactic constraint (or lack of constraint), though it is doubtful that one could guess a priori which shared elements of meaning will be represented in a likelihood-similarity subset (cf. 3.2.3, end). Some words may occur with the subset property only irregularly or with marginal acceptance: how acceptably can one say *John tried for Mary to phone*? (The equivalent in French is quite grammatical; in English one would have to say *John tried to have Mary phone*.) And in cases where a subset is characterized by more than one property, one can usually find words which have some but not others of the properties (3.2.5).

In addition to the subsets in respect to likelihood, there are some individual words which have peculiarities of likelihood of such importance as to have a syntactic character worthy of remark. For example, one does not ordinarily find a sentence such as *Mary is here and Mary is not here* except as part of a joke or a metalinguistic discourse; but one can say *Well, he's a politician and he isn't*. One also does not find a sentence such as *Mary is here and John knows that she is not here* or *John knows that she is here and I know that she is not* (but the sentence occurs with *just*, in *John just knows that she is here and I know that she is not here*). If the vocabulary is factored on co-occurrence grounds into maximally independent vocabulary-elements, *know* may be replaced by a combination of words that would reduce the syntactic (combinatory) property above to the property of *not* above; but such factoring is beyond the present confines of syntax.

4.2. *Sentences derived from the base*

From the base sentences, the reductions (3.3) create the other sentences of the language, which can thus be considered to be derived from the base. These reductions are changes in particular word-occurrences, not recastings of the whole sentence. As noted, they are made on the basis of the likelihood of the word in respect to its argument or operator in the sentence, and they are made as soon (in constructing the sentence) as these conditions are satisfied. Reductions are thus located at specified points inserted in the semilattice of the source sentence which they contain (8.3). Since reductions are only changes in shape, the unreduced sentence is not merely a source, but is actually contained in the reduced sentence. Although the types of reduction are similar in many languages, the specific reductions differ; and it is here that we find the main difference among individual language structures.

When a reduction is optional, each sentence it creates is an image of a source in the base. When a reduction is required, each sentence claimed to result from it (e.g. one containing *can*) has a source in the base only if we can assign to the reduced word (e.g. *can*) a suppletive word in the base (e.g. *is able to*). There are also cases where the reduced form and its source both exist, but intermediate reduced forms do not. Thus *He rotated the globe on its axis* (by the side of *The globe rotated on its axis*) can be derived regularly from a source *He caused the globe to rotate on its axis*, going through length-permutation to *He caused-to-rotate the globe on its axis* and compounding-permutation (3.4.3) to the non-extant *He rotate-caused the globe on its axis*, where zeroing of the broad-selection last element of the compound (*cause*) yields *He rotated the globe on its axis*. In such cases we need only say that intermediate steps in the succession of transformations are required, in this case the zeroing after the compounding; the compounding itself was not required. That is, what is optional here is the product of steps, not each step alone.

The reductions leave room for degeneracy, producing ambiguities: two different base sentences, each undergoing different reductions, may end up with identical visible word-sequences. For example, the ambiguous *John has shined shoes* is obtained from two sources: (1) *John has the state of John's shining shoes*, via repetitional zeroing of *John's* and reduction of *the state . . . ing* to *-ed*, this being roughly the historical source of the 'perfect' tense (G290); (2) *John has shoes; these shoes are in the state of one's shining them*, via zeroing of the indefinite *one's* and of the repetitional *them*, and reduction of *the state of . . . ing* to *-ed* (this is offered as the source of the passive, 1.1.2), and reduction of semicolon plus *these shoes* to *which* (yielding *John has shoes which are shined*) followed by zeroing of *which are* and permutation of the residual *shined* to before the host. All these are widespread and well attested English reductions. The two derivations admit different reductions of further operators: for example, (1) appears in *John has shined his shoes*; (2) in *John has well-shined shoes*.

If a given sentential word-sequence has more than one source from which it can be derived by known reductions, then for each derivation the given sentence has the meaning of the corresponding source. The ambiguity of such a sentence is not a matter of vagueness, nor of a broad range of meaning, but of a choice between two or more specific meanings, those of the specific source sentences.

There may be other disturbing situations in deriving sentences from the base. One is that the reduced sentences may lose their semantic relation to their source. This can happen if a reduced word, for example a word with an affix, has come to be used with different likelihoods for the operators on it, no longer the likelihoods (selection) of its component words: e.g. *romance* 'story' from 'in Roman (language)'. To the extent that this happens, the reduced sentences can carry information that is unavailable in the base sentences, although one can usually replace the semantically changed derived word by a definition given in base vocabulary (largely uni-morphemic). In any case, for the great bulk of sentences their derivation from base sentences preserves meaning: this is because the reductions do not change the partial ordering, and the likelihoods of further operators are usually not materially changed.

How to analyze each sentence follows from the above. Given the constraints which characterize sentences, we know that base sentences can only contain words in operator–argument relations, specifically N, O_n, etc. In each sentence, we know that there can only be as many N and O as are required by the various operator classes in the sentence, and that for each O_n there must be one N (possibly reduced) that is not being used as the argument of any other operator, and for each O_{nn} there must be two N, for each O_o there must be an O, of any kind, not otherwise used as argument, and so on. In each base sentence the words are all visible, and their few possible positions relative to their operator are given by the linearizations, including the possibility of interrupting sentences. In a derived sentence, on the other hand, one has to look for the traces of reductions—evidence of zeroing, pronouns, affixes, and coalesced words like *the* (G237). Given the domains and conditions for the reductions, it is thus possible to construct an effective procedure, complex though it be, for computing the derivation and base composition of a sentence. The procedure would of course have to allow for degeneracies as mentioned above.[1] An effective procedure might not be assured if sentences were characterized as being transforms of whole other sentences rather than as being due primarily to reductions on words as the words enter the partial order (as they are here). Nor would it be assured if for some sentences the only characterization was that they were due to analogical processes on some other sentence—which they may certainly be—rather than that they also, even if analogical, must have the form of operator–argument and reduction relations among their words.

The reductions in a language might be such that in many cases a derived sentence would be reachable from the same base sentence by more than one path of different successions of different reductions. In such a situation we may have a grid of reductions; to be a sentence a word-sequence might then need only be reachable along one or more paths of the grid from some other sentence. This would be a grammar of elementary differences rather than of

---

[1] The sentence-generating system of ch. 3 does not suffice directly for an analysis ('recognition') procedure, which requires an orderly accounting of all similarities among the generated forms. A computer program using a detailed grammar of English, based on the contiguity of grammatical constructions, is in operation, analyzing scientific texts in English and some other European languages, and transforming them into informational data bases: N. Sager, *Natural Language Information Processing*, Addison-Wesley, Reading, 1981. An algorithm, with the dictionary written in the form of the algorithm, has been produced by D. Hiż. An algorithm for the operator–argument recognition of sentence structure has been developed and implemented by S. B. Johnson.

elementary reductions. However, no case of such a situation has been established.

In addition, obtaining sentences by the constraints of Chapter 3 makes it possible to characterize each sentence not only by its operator–argument and reduction orderings but also by the sentence-subsets and the sub-language to which it belongs. This will be discussed in 8.1 and in 10.1.

The reductions are not always the grammatical events that historically produced the sentence, for, as noted above, some sentences are formed on the analogy of others, or borrowed, and the like. And even when the reductions are historical, they are not necessarily recognized as such by the speakers of the language, who may have constructed their own perceptions of how certain sentences relate to each other. But in any case, establishing these derivations for every sentence structure, and so presumably for every sentence, shows that a single kind of relation, namely reduction based on high frequency or low information as between operators and arguments, suffices to derive the forms of all sentences, and in general to preserve the information, from the base sentence.

The consistency in the whole set of derivations provides a single measure of complexity in terms of which we can find the 'cost' of each derivation. We may judge that some sentences can be derived directly by the stated reductions even if in historical fact, or in popular perception, they have a different origin. A case in point is the Modern English status of -ing as an argument indicator on virtually any sentence, and the status of the 'progressive tense' is . . . -ing as derived from the above -ing under on or the like; the historical sources of these two were phonemically different, and not etymologically related (G41, 295, 373). We may also judge that some forms require so complex a derivation, passing through such strange intermediate sentences, that we may prefer to sacrifice the reductional derivation—especially in computer processing—and say that at the given point the language is using a form or process outside the operator–argument universe (e.g. for tense, cf. G265); this, even if in fact the sentence went through this derivation historically, but with some of the intermediate sentences having gone out of use. But if we want an apparatus for organizing language-borne information, then the operator–argument and modifier relations which are reconstructed, based on this one method of derivation, from the sentences of a language will suffice in all cases; the reconstruction may also reveal subtleties of meaning and use.

Several of the reductions create new subsets of words: those which accept a particular reduction. Some of these are large and important, and may even be sharply bounded: e.g. the distinction between verb and other operators (adjectives, prepositions, and even predicate nouns). This distinction arises because those operators which are less stable in respect to tense and time-adverbs (e.g. have high likelihood of changing their tense within a discourse) receive the operator indicator -s (which serves as the present tense, and can be replaced by the past tense *-ed*) directly on them: *walks, sleeps*. In contrast, the other operators receive it as a separate word, the indicator *-s* being then carried by the word *be: is ill, is here, is an uncle*. The difference in durativity which is involved here is graded over all operators; but the structural difference of carrying the *-s* directly or, alternatively, on *be* presents only two choices. Thus certain operators are marked off from the others as being more actional and receiving the *-s* directly: the verbs. This division in English is strongly conventionalized: few words have both forms except by one being derived from the other.

Another example of a reductional subset is the prepositions. These O_{on} operators have already been recognized as a subset on likelihood grounds, because they occur between very many O and N: *John fell from the stage, John sat near the tree*, etc. In addition, their high frequency and the broad selection of their widely applicable originally spatial and temporal meanings enabled them to spread by zeroing or metaphor into having many kinds of argument (G80): between two N as in *Mary is in the house* from *Mary is present in the house*; after one N as in *Mary is in* from *Mary is in the house*; after one O as in *Mary came in* from *Mary came in the house, The boss is in* from *The boss is in the office*. As in many subsets, membership is not sharply defined: some prepositions have developed from a preposition plus a noun (*before*); others have developed from other classes (*during* from a verb). And not all members have all the subset properties: *before, during* do not occur with one N (as subject: no *John is before*, normally), and *during* also not with one O (as subject: no *He wrote this during*, but one can say *He wrote this before*).

Similarly, consider the auxiliary O_{no} verbs can, *may, will, ought*, etc., under which the subject of the lower operator, and the *to* on it, must be zeroed, and which do not carry the present-tense *-s* (3.2.4). These are found to have neither a closed and definite membership nor fully common properties for all members. Typically, we have *John can go* by requiring zeroing from a non-extant *John can for him to go*, and the set of verbs that have this required zeroing is commonly thought of as being well defined. But *will* occurs also as an ordinary operator, with a related meaning 'to desire'. *Ought* is erratic about zeroing *to: John ought to go* but *Ought John go?* as well as *Ought John to go?*, and *John ought not go, John ought not to go. Need* has *John need not go*, and in special environments *John need go only occasionally*; but *need* is also an ordinary verb in *John needs to go, John doesn't need to go, John needs Mary to go*, etc. And *let, make* zero *to* while admitting different lower subjects: *John let it go*.

One important successive application of reductions can create not a subset but an apparent escape route for words from the N class to the O class and vice versa, and less dramatically can create non-coherent jumps in the selection and meaning of words—all these being changes that would not result from ordinary extension of selection or from analogical processes.

This is metaphor, which can be derived from two sentences under the conjunction *as*, provided everything beyond what is needed for one sentence is zeroable (indefinites or classifiers). It is seen for example in *I can see his wanting to go* (where *see* appears as O_{no} instead of its base status as O_{nn}) from (roughly) *I can treat his wanting to go as one's seeing things* (via the artificial *I can treat-as-seeing his wanting to go*, G405); in the source, *see* is correctly O_{nn}. Such a derivation explains for example how one can say *The car smashed into the wall without getting smashed* (or: *without smashing*): we obtain *The car smashed into the wall* from *The car moved into the wall as a thing's smashing* (via the non-extant *The car moved-as-smashing into the wall*), to which *without (the car) smashing* can be adjoined. It also explains how one may say *He peppered his lecture with asides* but not *He peppered his lecture*, though one can say *He peppered his soup* (from *He put pepper into his soup* similarly to the zero causative, 3.3.2). The source is *He put asides into his lecture as one's putting pepper into things*, which is reducible to the non-extant intermediate *He put-as-putting-pepper asides into his lecture*, thence into *He peppered asides into his lecture*, whence *He peppered his lecture with asides* (G369). (All these are elsewhere attested zeroings in English.) The fact that *as* is involved in the derivation can be seen in such forms as the French *Le juge a ravacholisé X*, where one would not say (except perhaps as a joke) (1) *Le juge a ravacholisé Ravachol*: we see that *ravacholiser* is 'to treat as one treated (the executed anarchist) Ravachol' (which cannot be said of Ravachol himself as in (1)); it is not 'to do what was done to Ravachol' (which could be said to Ravachol, and which therefore would permit (1)). The source here would be in English *The judge treated X as one's treating* (i.e. *as in the treatment of*) *Ravachol* (G406).

4.3. *Systemic properties following from the constraints*

From the operator–argument dependence of 3.1 above, it follows that there must exist a set of words (zero-level arguments, N) whose argument is null (or which are defined in respect to their own operators), for otherwise there would be an infinite regress of word classes defined by their argument classes. And there must exist operators (first-level, $O_{n \ldots n}$) all of whose arguments are zero-level, since no other arguments have been defined at this point. The first-level operators create elementary sentences. If thereafter we do not define second-level operators ($O_{o \ldots o \ldots}$), at least one of whose arguments is itself an operator, the system would not provide for any sentences beyond the elementary. It further follows that every sentence must have at least one zero-level word, in order to start forming, and no more first-level operators than it has zero-level words to satisfy their argument requirements; if it is more than an elementary sentence, it contains second- level operators in addition. The base words of the language (excluding whole-sentence words such as *Hello*) fall into a three-level partial order in respect to what they require in a sentence. And the words of base sentences constitute a semilattice in respect to their arguments in their sentence (6.3, 8.3). The reductions in a sentence are locatable as additional points in that semilattice.

Since operators appear in sentences only given the presence of their arguments (before reduction), it follows that when a second-level operator appears with its argument, which is itself a first- or second-level operator, the arguments of that first- or second-level operator are also present. Hence the sentence which is made by the first- or second-level operator is imbedded in the sentence made by the higher second-level operator. Thus every sentence consists ultimately of one or more elementary sentences, possibly with second-level operators on them. And since reduction and transformation only alter the phonemic shape or the position of words, a reduced sentence contains its (usually longer) unreduced source. The contained sentence is the same as it is in its independent occurrences, since the further operators on the words and thus their meaning is largely unchanged—aside from the fact that the meaning of certain words, and of certain reduced word-forms, differs under different operators (11.1). Various word-sequences may remain as idioms: irregular reduction (*the more the merrier*), specialization of meaning (*kick the bucket*), etc.

It should be understood that saying that every non-base sentence contains, or is derived from, a base sentence does not mean that such a base sentence was in use first and then reduced. *I like tennis* is derivable from *I like for me to play tennis* or from *I like people's playing tennis* (by zeroing *play* as high-frequency operator on *tennis*, and *people* as indefinite argument, or *me* as repeated argument). This is so not in the sense that the base forms were said first, or that the speaker thought first of the base form, but in the sense that *I like tennis* says only what the two base forms say more regularly (unless the environment supports other zeroings). And to say *the four-color problem* does not imply that anyone first said *the problem of four colors*, only that the latter is sayable in the language with the same meaning, and that if the former was not indeed reduced from the latter then it was built on the model of compound words which are reduced in this manner (*a four-door sedan* from *a sedan with four doors*).

All sentences created by the partial ordering and reductions can in principle be said in the language. Some are extremely rare, either because of very low likelihood of a word combination, or because a reduction is heavily favored as against its source (e.g. *I want to win* as against *I want for me to win*). Since the constraint of likelihood (3.2) is defined on the resultant of 3.1, it follows that no matter how low is the likelihood of a particular word in respect to a given operator on it, the likelihood is presumed never to reach zero;[2] for if it did, we could not say that the occurrence of words in sentences depends only on the dependence properties of other word-occurrences there. As was argued in 3.2.1, this accords with the data about language, since even those members of an argument class which are least likely to occur under a particular argument (say, *vacuum* under *cough*) cannot be entirely excluded from what is possible. (But outside the base, required reductions may preclude the visibility of words, as under the auxiliaries, 3.3.3.)

Certain other properties of language are due to the reductions. First, the base, i.e. unreduced, sentences have the least constraints on word combinations. This arises from the fact that each reduction, defined as operating on some operator–argument or sentential condition, has as domain either all sentences containing that condition or else a proper subset of these, where stating a proper subset constitutes a restriction on word-occurrences. Hence the set of sentences resulting from a reduction cannot be less restricted than the set existing before the reduction is defined.

Second, the phonemic possibilities in reduction leave room for the creation of morphology, as a structure (and hence a restriction) beyond what is found in the base. In many but not all languages, certain reduced words are attached as affixes (mostly operators attached to their arguments), thus creating pluri-morphemic 'derived' words, which do not exist in the base.

Third, reduction of certain operators to zero makes their arguments seem to take over the operator's status (and vice versa): e.g. the apparent noun *take* in *today's take*, derived from (roughly) *the amount of today's taking things*, or the apparent verb *piece* in *to piece together*, reduced from (roughly) *to put the pieces together*. Such reductions also cause an apparent 'lexical' shift in the meaning of the words (as between *the take* and *to take*, or between *to piece* and *a piece*), although the word can be considered to have the same range of meanings in both environments, with the difference being merely the meaning of the zeroed word (*amount*, *put*)—zeroed in its phonemics but still present in grammar and meaning.

Lastly, many reductions create degeneracies (ambiguity) in sentences, when two reductions, acting on two sentences which differ in form and meaning, produce the same sequence of words.

There are also certain properties which follow from linearization. The alternative linearizations create the syntactic relation of modifier. As noted in 3.4.4, if we interrupt a sentence after a particular ('host') word, and if the interrupting sentence begins with the same word, the two occurrences of the word are successive (neighbors) so that the second can be pronouned

---

[2] As noted in 3.2.4 and 4.1, there may be cases of an argument which is excluded under a given operator (e.g. *can*). The generality of 3.1 may then be salvaged by finding synonymous words in the base which are free of this exclusion and which serve as variants of the restricted word.

without a general addressing system, leaving the rest of the interrupting sentence as a modifier on the host to which it is adjoined.

From all the constraints together there follows the major property of language: that not all combinations of words occur as sentences of a language (2.1). This is a universal and necessary property. In every language not all sequences of the phonemes, or the morphemes (words), in the language occur as sentences, or as discourses.

Various features of the constraints, and especially the fact that each is defined on the resultants of others, have the effect of creating a system (12.4.2). The sentences all have a family resemblance, not only because all elementary sentences are made in one way, but also because the second-level operators act on their arguments (ultimately, elementary sentences) in the same way that the first-level ones act on the zero-level words to make the elementary sentences. Both the deformation of sentences (by argument indicators) to make them into arguments, and also the reductions, create the situation of sentences having somewhat varied paraphrastic forms. The existence of a system into which sentences fit has made it possible to recast sentences paraphrastically in ways that enable almost every part of a base sentence to become an argument or operator in a paraphrastic transformation of it: e.g. *John varnished the violin*; *The varnishing of the violin was by John*; *This violin is John's varnishing* (doubtful, but compare *This dish is my cooking*); *Varnishing was what John did to the violin*, *The varnishing was John's (doing)*.

It should also be noted that the system provides for certain kinds of classification. There are classes of sentences in respect to their creation by operators (elementary and not), and in respect to reductions and other transformations (8.1). There are many classes of words and sentence-segments established as domains of various reductions. And there are the never fully statable and slowly changing similarities and gradings of words in respect to their selection of individual operators over them and arguments under them: examples of such classes are the indefinite pronouns, quantifiers, classifiers (3.2.2).

In all, it is seen that the few constraints of chapter 3 determine a considerable amount of structure.

4.4. *Properties following from additional conditions*

In addition to the properties which follow directly from constraints on equiprobability, there are various ones which are due to other conditions of language, some of these essential and universal, but others non-essential.

One condition that is universal is that no space is definable for speech other than its sequence of phonemes and words. The various junctures that may be found between phonemes or between morphemes are sound differences or pauses which establish boundaries and reduce distances phonetically, but do not create different base distances between neighboring phonemes or words (such as are available for example in the beat of music, or in the meter and rhyme of poetry). Hence, in linearization, an operator has to be contiguous to its operand: it can be before, between, or after its arguments, but no pre-existing distance is definable such that it could be at a distance from them (2.5.4) when it acted.

Another universal condition is the finiteness of the users of language and of their public experience with language. Because of it, everything that they have to learn or to agree on in order to be able to speak to each other has to be finite, even if recursively applied, and the stock of entities and processes in language has to be finite or at most recursive. Hence, the set of independent items of vocabulary has to be finite, though changeable. As to the word classes: if they are defined by their relation—e.g. zero-level argument (N), first-level operators (e.g. O_n),second-level operators (e.g. O_o)—then we have a recursive construction for word classes. This also means that the grammar is finite, and that for every language it is possible to formulate a constructive and not merely descriptive grammar—that is, a grammar which lists entities and processes sufficient to form and decode precisely the word-sequences which are sentences of the language (2.5.1).

A third condition met in language is that in many cases the grammatical constructions are the least that are necessary within the constraints (2.2–3). One instance is that many, perhaps all, languages do not have clear cases of third-level operators, i.e. operators which can be used only given the presence of second-level operators. A second-level operator, such as *cause*, carries information that cannot be carried by first-level operators such as *angry*, *heckle*: e.g. *Their anger caused them to heckle*, but not *Their anger heckled them* or *Their anger heckled them to cause*. But it appears that any information that could be carried by a word operating on a second-level operator could be carried (though less explicitly) by a word which is itself of second level, namely a word that can also operate on first-level operators. For example, *heighten* or *increase* can operate on *correlate* (*The emergency increased the refugee flow's correlation with local repression*), but these can also operate on *flow* or *repress* directly (*The refugee flow increased*).

A similar case is that many languages have few operators with three-part arguments. The information carried by such an operator can in many cases be carried by an operator on two operators each with two-part arguments: e.g. *John put the book on the table* (where *put* has a three-part argument, there being no *John put the book*) can be roughly paraphrased by *John placed the book so that the book was on the table*).

Yet another instance of least structure is the fact that to the extent possible second-level operators have the same morphology (due especially to tense and person reductions), and the same linearization relative to their operands, as do first-level operators in respect to their own operands. For example, first-level operators appear immediately before, between, or after their arguments because no other (distant) position is definable for them. In contrast, second-level operators have one other possible position: they could appear between the arguments of their own operand (e.g. *sufficed* on *He wrote a letter* could have produced *His writing sufficed a letter* instead of *His writing a letter sufficed*). They do not do so; rather they in general appear in the same positions relative to their operand (with its attached arguments) that this operand has to its operands (arguments) in turn. This creates the well-known imbedding of whole sentences inside larger sentences, as against inserting higher operators into the lower sentence (except as interruptions, 3.4.4). Similarly, just as a secondary operator (adverb, preposition, or conjunction) on an imbedded operator cannot appear out of contact with the imbedded sentence, so an adverb on the higher operator does not in general appear inside the imbedded sentence. As noted in 4.1 for English, if *later* operates on the higher *explain*, while *immediately* operates on the imbedded *come*, we can find *later* only in positions 2, and *immediately* only in positions 1 of the sentence:

2 *He* 2 *explained* 2 *that* 1 *he* 1 *had* 1 *come* 1 2

(i.e. *Later he explained that he immediately had come*, or *He explained later that he had come immediately*, etc., but not *He explained immediately that he later had come*, or *He explained that he had come later immediately*, etc.). Thus the adverb on each sentence is always contiguous to that sentence, and the adverb on the higher operator is always outside the lower sentence. This creates the well-known nesting of adverbs, and so for other nestings in language. Together with the basic contiguity properties noted above, this creates the possibility of analyzing sentences in a constructive and computable manner into contiguous strings of contiguous strings of words, with each string referring informationally to its imbedded or imbedding string (ch. 4 n. 1, and 7.6).

Another instance of least structure is the limitation of each word, by and large, to a single kind of argument. If a word had one kind of argument in some cases (say, *cause* requiring a zero-level first argument in *John caused our car to swerve*) and another kind of argument in other cases (say, *cause* requiring a first-level first argument in *The truck's skidding caused our car to swerve*), the resulting multiple classification of words, by their various argument-requirements, would be very complicated. It turns out that in such cases we can generally obtain one occurrence as a reduction from the other kind of occurrence (e.g. *John caused . . .* from *John's action caused . . .* or *John's presence caused . . .*).

A result related to least structure is the fact that each language has only one or a few sentence structures. In English the sentence structure of elementary sentences (operator after its first argument) is preserved in more complicated sentences: (1) the morphology (e.g. tense, plural) and relative position of higher operators are generally the same as for lower operators, as noted above (so that, for example, the subject precedes the verb both in *The refugees crossed the border* and *The refugees' crossing of the border angered the government*); and (2) a secondary sentence, given that it contains an argument that has already appeared in the primary, can interrupt the primary right after the repeated word, so that it creates a modifier (relative clause) that is contiguous to the antecedent (host) word (so that, for example, the phrase *refugees from El Salvador* occupies the same position as the single word *refugees* in *The refugees crossed the border*, *The refugees from El Salvador crossed the border*). In contrast, the interruptions that do not form relative clauses do not merge into the previously existing sentence structure, as in *Tomorrow—so I hear—will be cold*, *Tomorrow will—so I hear—be cold*.[3]

A fourth condition met in languages, which is related to the least-structure condition, is that in various respects the structural possibilities of a language are not fully utilized. For example, both the partial ordering and the linearization make it possible to state an address for each word in a sentence. However, languages do not use complete addressing systems, and the only cases in which the location of a word is cited are those where the location can be identified with the aid of a few simple addressing words such as 'prior' (for relative pronouns), 'co-argument' (for *-self*, as in *John washed himself*), 'same partial-order status' (for corresponding-position zeroing, as in *John plays piano and Mary violin*). Another example is the fact that every language admits only a small choice of reductions, leaving unreduced many a word that could have been reduced, as having little information in respect to a given operator: thus *to be here* or the like is zeroable under *expect* (*I expect John to be here*), and *books* or the like is zeroable under *read* (*John reads a lot*), but *clothes* is not zeroable under *wear* (no *The natives wear*).

A fifth condition, which may be due to the independence of selection from phonemic form (5.4), is the possibility of various similarities among words, in their meaning-range, phonemic form, or reduced form, sufficient to create structural details in various languages, beyond the major structure created by the fundamental constraints. There are frequently used reductions (or combinations of reductions) in particular kinds of word-sequences. Such situations create specialized sentence types: e.g. those that result in the passive form, or in metaphoric uses of words (4.2, end), or in new conjunctions such as *because*, *provided*, *providing* (below). Similarities in treatment of *I*, *you*, *he*, etc., or of secondary sentences that indicate time relations, can create in many languages 'person' (subject, object) and 'tense' affixes which get attached to operators. Such constructions as tenses and subject–object affixes in verb conjugations (or possessive affixes on nouns) can become

---

[3] A widespread condition is the saying of incomplete sentences. These fragments generally turn out to be not merely arbitrary combinings of words, but specifiable segments of sentence-construction.

required combinations (on all verbs, or on all nouns) and thereby take on great importance in the grammar of a language.

The high likelihood of particular operators to occur with particular time modifiers can lead to subclassification of operators into verb, adjective, preposition (4.2), and within verbs into perfective (durative) and imperfective (momentaneous) verbs (3.2.5)—morphologically distinguished in some languages but not in others. Similarly, prepositions, quantity words, and other frequent additions can be reduced to such common affixes as cases and plural on elementary arguments, and even on higher-level arguments, creating nouns and 'nominalized sentences' (sentences appearing as arguments). Here too differences of likelihood in respect to plural, etc., can make a distinction in some languages between 'mass' and 'count' nouns.

There are also similarities and differences in the (morphophonemic) form of these important affixes on various verbs, or nouns, yielding different conjugational and declensional subclasses of verbs and nouns; some may be associated with meaning-differences (e.g. declensions for gender), others may have no semantic association (e.g. conjugations). In many languages, these morphophonemic similarities do not cover all verbs and nouns, leaving various exceptions ('strong' verbs, etc.). Such syntactic constructions can have great weight in the structure of an individual language or language family.

Finally, two processes are found in language which are quite different from those involved in chapter 3. One is analogy: the saying of a word in an operator–argument combination on the basis of another word, which has some correspondences to the first, appearing in such a combination. This is a widespread process, restrained by its own conditions, and almost always dependent on the prior existence of a grammatical construction on the model of which the new form is made. Because of this dependence, many forms which historically have resulted from analogy can be analyzed as though derived by regular reductions from existing sentences: the only cost may be extending the selection of a word, or extending the domain of a reduction.

The other process is change through time. All languages change. These changes are primarily of certain types. One type is a slow shift in the basis of articulation, which changes certain sounds, and in particular situations changes their phonemic relations. Another type is of a word or affix extending its selection, sometimes into that of another, so that it competes with the other for the same selectional and semantic niche, and may replace it. Yet another type is extension of the domain of a reduction (i.e. which words are subject to it), or of its conditions or frequency of application. And there are changes due to the borrowing or invention of new words. These changes are largely related to the structure of the existing language, so that major alterations in the structure of language are rare indeed.

The change in selection, i.e. in which words have higher than average frequency as operators or arguments of a given word, is widespread, and important both for meaning and grammar. Many words change their meaning in this way, especially when with particular other words. When a word combination, including word plus affix, changes its selection considerably it may have to be considered a new word, no longer a combination of its component words (11.1). This is not necessary so long as the meaning difference is regular. For example, a great many verbs have both intransitive and transitive uses and meanings (*The disc rotated, He rotated the disc*), but one form can be derived from the other: e.g. by zeroing in the causative (*He caused the disc to rotate*), or by the 'middle' (*The glass broke, The book sold well*, G368). More widespread are the many metaphoric uses of words, as in *I see what you say*, but these can all be derived from the conjunction (O_{oo}) *as* plus indefinites and classifiers (roughly, *I react to what you say as one's seeing things*, 4.2, end). However, selection changes in individual words or in small sets of them can change the grammatical status of the words or introduce the possibility of new classes into the grammar. We see the first case in *provided, providing*, where a great extension in their arguments took them from a secondary-sentence verb, as in *You can make the trip, provided with enough money* to a conjunction as in *You can make the trip provided* (or *providing*) *there is no objection*. Possibility of a new grammatical construction is seen in the extension of *take, give, have*, etc. from their original use in *take an apple, give an apple*, etc., to *take a look, give a look, have a look*, etc., where most of their original meaning is lost and where they could be analyzed as a new class of 'preverbs' with a modifier-like relation to the verb that follows them. (Within the uniform theory presented here this analysis is avoided, 8.2.)

The net effect of all these conditions is that languages have various detailed structures over and above what the constraints of Chapter 3 create, going beyond the corollaries of those constraints as surveyed in 4.3. This further structure includes subsets of words within the basic argument and operator classes, and various grammatical paradigms and standardized sentence types such as the passive, the interrogative, the imperative. These forms can, however, always be derived from the fundamental constraints plus one or another of the conditions of 4.4. And the information can always be found in the base structure, in the base sentences corresponding to the derivations.

5

Metalinguistic Apparatus
within Language

5.0. *Introduction*

Certain features of language structure create a family of metalinguistic devices within language, i.e. of sentences talking about sentences and their parts. These metalinguistic devices include: metalinguistic sentences (5.2, 9.2), and a sublanguage that serves as metalanguage of the whole language (10.2); cross-reference (pronouns, 5.3); the use-mention distinction (5.4); the self-actualizing (performative) properties of *I say* (5.5); and metascience operators in science languages (10.4).

5.1. *The metalinguistic capacity*

Two features underlie this capability of language. One is the fact that phonemes are types and not merely tokens; that is, the set of word-sounds is effectively partitioned (by distinctions, which create types) into phoneme-sequences (7.1–2). The other is the lack of any relation, in respect to meaning, between phoneme and words; that is, the meaning of words is not referable to any meaning of their phonemes.

As to the first: the status of phonemes as types means that every occurrence of a word-sound can be named, as being a case of one known phoneme or another. We should not take this property for granted, as something which holds for every system that could seek to represent (as language does) the information of the world. The objects and events of the world could be referred to not only by words but also by, say, pointing at them, or drawing a picture of them. However, the latter kinds of representation cannot be reused, to represent the representations themselves. True, we can point at (someone) pointing, or draw (someone) drawing; but we do not have a general way of uniquely pointing at all pointings which are specifically at a given set of things, or drawing all drawings of a specific set of things. In contrast, we can take all word-occurrences that name a given set of things (say, the word *chair* for certain seats or the word *bench* for others, or *walk* for certain perambulations or *stroll* for others) and we can uniquely identify these occurrences by the name of their phoneme-sequence, e.g. the names of the successive letters c, h, a, i, r. Since phonemes (and phonemic letters) are characterized by their distinction from other phonemes, every occurrence of a sound satisfying the properties of a particular phoneme distinction can, as a token, be called by the name of its phoneme, as a type. The name of a phoneme—say, *vee* or *doubleyou*—is thus itself a word, with its own phoneme or letter composition (e.g. v, e, e), and with a set of objects (word-sounds) to which it refers; and that name can be used in sentences, e.g. in sentences that give the composition of each word. This makes it possible to use words—that is, to talk—about phonemes, word-occurrences, and about words as sets of word-occurrences. Such talking is called metalinguistic. Once phoneme-sequences are associated with experiences of the world (and so can be used to talk about them), nothing prevents their being associated with a particular phoneme-sequence (and so being used to talk about it).

As to the second feature above, the lack of semantic relation between phonemes and words: that the phonemic components of a word are not its semantic components is a necessary property of language (5.4). Because of this property, the operators which are selected by a word when that word is said as a phoneme-sequence differ in general from the operators that are selected when the same word is said as a carrier of meaning, of a dictionary definition. It will be seen that this property underlies the metalinguistic constructions. More explicitly, it gives rise to the distinction between use and mention, and also to the distinction between direct and indirect discourse, and in particular the effect of *I say* operating on a sentence.

The fact that the metalanguage sentences are sentences of the language makes it possible to talk in the language about words and occurrences and structures of language (speech), and also about the meaning of words and their relation to the world observed by the speakers.

5.2. *Metalinguistic sentences*

The basic metalinguistic sentences, from which all others can be constructed, are made of the names of phonemes with operators on them. The operators specify the sound-differences among the phonemes, their location relative to each other in utterances of the language, and which phoneme-sequences constitute words in the language. These metalinguistic sentences are not about sounds as such, for individual sounds (tokens) are in general evanescent events which cannot be named—that is, cannot appear as words—unless they have been classified into fixed types. But the names of the phonemes and their classes (e.g. the word *phoneme* itself), and the names of those phoneme-sequences which satisfy the conditions of being a word, are all words or word-sequences, of zero level (i.e. primitive arguments), and they occur under a particular selection of operators. We then have sentences such as: *Speech sounds with property A (or differing by A from other speech sounds) are called occurrences of em*; *Em is a phoneme, written m*; *Some phoneme-sequences constitute words*; *Mary is a word*; *The letter-sequence m, a, r, y constitutes the word Mary*. The metalinguistic sentences are thus sentences of the language itself: *C precedes a in the word cat* (or: *Cee precedes ay in the word composed of the letters named cee, ay, tee*) is not structurally different from *The band precedes the marchers in the parade*. The metalinguistic sentences are a separate subset, and a part of a metalanguage, only in that their elementary arguments refer to segments of utterances, ultimately to phoneme-sequences. Thus the arguments used in metalinguistic sentences are mentions (i.e. names) of the words (and phonemes) used in the sentences of the language. Other than this the metalinguistic sentences are themselves sentences of the language.

Given sentences which identify particular phoneme-sequences as constituting words, we can have sentences whose arguments are the words thus constituted. Such sentences can state (morphophonemically) that different phoneme-sequences constitute the same word, as alternative forms of it (e.g. *is* and *am*, or initial *e* and initial *iy* in *economics*), or that a single phoneme-sequence can, in different occurrences, constitute different words (homonyms). Such sentences can also state the location of particular words in respect to each other in utterances of the language, and also their meaning, in terms of other words, either in isolation or in particular combinations.

A phoneme is defined for the utterances of a given language (or possibly a specified set of languages), but—crucially—not for sounds outside language or in other languages. This restriction of domain holds necessarily for metalinguistic sentences, which have been described above as sentences whose arguments ultimately are phonemes, or phonemic distinctions. When the domain of utterances is unspecified, the metalinguistic statements can only assert the existence of phonemes, words, and their combinations. When the domain is all utterances of a language—all that are observed or all that are predicted—or a specified subset of them, we can state regularities of form and of occurrences within the domain. These regularities may be of phonemic shape, as in similarities or transformations among word-occurrences; or they may be of word combination, as in defining operators and arguments. We thus reach statements about phoneme classes, word classes, grammatical structures, and types of meaning for words and sentences.

A new situation arises when the metalinguistic statement is made about a particular utterance. Such statements can be used freely, as in *John is brash—and that last word is an understatement*. The utterance in question has to be identified—that is, cited—within the metalinguistic statement; the utterance cannot be referred to unless it (or a pronoun for it) is grammatically included in the metalinguistic statement itself. This means that the utterance in question and the metalinguistic statement about it form together a larger operator–argument sentence: e.g. *The utterance which consists of John bought a book has the word John as first argument of the word bought*, or *The word John is the first argument of the word bought in the utterance which consists of John bought a book*. In particular, the instruction to carry out a reduction (3.3) can be a metalinguistic statement attached to the sentence in which the reduction is to be made: (1) *Sheep are voracious and sheep eat all day, in which sentence the second occurrence of the word sheep is zeroable*.

The same holds for the sameness statements of 5.3.3, which are sentences conjoined to the sentence of which they speak, whether or not they are followed by an instruction to reduce the repeated word: as in *John lost some time and Mary lost some time*, to which one can conjoin (roughly) *where the second argument of the first operator consists of the same words as the second argument of the second operator*; also in *John deserves John's being appointed as chairman*, to which one can conjoin (roughly) *where the first argument of the first (higher) operator has the same referent as the first argument of the second (lower) operator*.

This situation makes it possible to consider reductions in the form of a word, and other transformational variants in the form of a sentence, to be simply morphophonemic variants of the attached instruction to carry out that reduction. In an exaggerated way, one can consider that in (2) *Sheep are voracious and eat all day*, the lack of the required first argument for *eat all day* (i.e. the absence of the second *sheep*) is simply an alternative form for (3) *in which sentence the second occurrence of the word sheep is zeroable*. If we said that (2) was a variant form of *Sheep are voracious and sheep eat all day*, we would still need a grammatical statement (in the metalanguage) saying that which is said in (3); and so for all other reductions and transformations. But if (2) is taken as a variant form of (1), then all we need for the language is a single grammatical statement, namely that the phonemic addition which comprises a transformational instruction adjoined to a given primary sentence has as phonemic variant the particular phonemic change in the primary sentence which is stated in the adjoined instruction. To regard the absence of the second *sheep* in (2) as a phonemic variant of (3) in (1) is not different in principle from regarding the initial sound *e* of *economics* as a variant of the morphophonemic instruction in *In the word economics (with initial sound iy), the initial iy is replaceable by e*.

The effect of sentences carrying metalinguistic information about themselves is even more far-reaching than the case above. In 9.1 it will be seen that conjoined sentences have the structural property of being typically connected by word-repetition; that is, there is likely to be some word that occurs in both sentences. (1) *Mary went home because Mary (or: she) was tired* is more likely to occur without question than (2) *Mary went home because Jane was tired*. And (2) itself raises no question if we add (3) *Jane is Mary's daughter*, which supplies a repeated word for each of the component sentences in (2). And if (2) without (3) raises no question, it is because the hearer knows (3) or the like, so that for him (3) can be zeroed. Indeed, any two sentences can be acceptably conjoined if we add a repetition-supplying sentence. We can now consider a similar treatment for the structural properties of single sentences. Clearly, all the properties of a given sentence, including definitions of the words, can be stated in metalinguistic statements about that sentence which are adjoined to it: e.g. (1) *John bought a book, where the word John is first argument of the O_{nn} word bought and the word book is second, with John being the name of a man, bought meaning purchased, and book meaning a printed volume.* (An n-fold ambiguous sentence would have n alternative sets of metalinguistic statements adjoinable to it.) When the grammatical status and the definition of each word is understood by the hearer, the conjoined metalinguistic sentences which state them are zeroable. Evidence for their presence and their zeroing may be seen in the following observation: virtually any arbitrary phonemic sequence satisfying English syllabic constraints becomes an English sentence, acceptable and understandable to the hearer, if it carries adjoined metalinguistic statements identifying its parts as words satisfying their argument-requirement: e.g. stating that its initial sub-sequence is the name of some foreign scientists, its final sub-sequence is the name of some new gas, and the sequence between these (which would have to end in s, or ed, or t) is a scientific term for some laboratory operation perhaps derived from the name of the scientist who developed the operation. In this last case, differently from (1), the grammatical and definitional statements are presumed not known to the hearer and hence not zeroed, as they can be in (1).

It follows that all sentences can be thought of as originally carrying metalinguistic adjunctions which state all the structural relations and word meanings necessary for understanding the sentence, these being zeroed if presumed known to the hearer. Every sentence, such as in (1) and (2) above, is thus a self-sufficient instance of what is permitted by the metalanguage (the grammar and dictionary definitions of the language). We can thus append to a sentence in a language all the metalinguistic statements necessary for accepting and understanding it, with the whole still being a sentence of that language. This is possible because the grammatical and dictionary descriptions of a language are stated in its metalinguistic sublanguage (10.2), somewhat as in computer programs from von Neumann's original plan and on, the instructions have the same computer structure as the data on which they operate.

One might ask where these metalinguistic sentences about a sentence are adjoined, since many of them disappear in the course of constructing the sentence. The observations of 3.3.4 and the considerations of 8.3 suggest that they enter into the making of a sentence as soon as the words they are speaking about have entered the sentence (since the reductions which some of these engender are carried out then). This clearly holds for those metalinguistic adjunctions which are made known to the hearer only by their being carried out, namely permutations and reductions, including those that are based on a sameness statement (e.g. pronouning): these adjunctions are replaced morphophonemically by the act of carrying them out. In the case of the relevant grammar and dictionary statements, which are in most cases known to both speaker and hearer, the issue of their location is moot since these are zeroed on grounds of being known.

5.3. *Reference*

5.3.0. *Introduction*

All referentials (e.g. pronouns) can be derived from cross-reference, i.e. from having one word in a discourse refer to another occurrence of a word in the same discourse. The capacity for cross-reference arises from three properties: having repetition, having words for stating the relative location of words in the discourse, and having words for stating sameness of the repeated words.

We have seen that metalinguistic statements can refer to the relative location of elements in utterances of a language. This applies to the easily perceived linear order of phonemes or of words (especially to being before, after, or next), and to the partially ordered 'grammatical relations' of words (e.g. subject of, i.e. first argument of). When an utterance is cited in a metalinguistic sentence, the utterance segments can be located linearly or grammatically in respect to the metalinguistic sentence which contains it and refers to it.

5.3.1. *Repetition*

Under all two-argument operators, there is a possibility of word-repetition in the arguments, i.e. of a word-occurrence in the second argument being a case of the same word (and possibly referring to the same referent) as an occurrence in the first argument. The grammatical importance of such repetition (in high-likelihood situations) is seen in its reduction to pronouns or zero. Under O_{nn}, we see this in *The boy washed himself*, which is derivable (G134) from a reconstructed *The boy washed the same boy* (where we have the same referent). Repetition is frequent under certain O_{nn}, as in *I prefer to go now*, derivable from *I prefer for me to go now* (*me* having the same referent as *I*); and under O_{oo}, as in *John plays violin and Mary piano* from . . . *and Mary plays piano* (where *plays* is the same word with the same meaning). Repetition has high likelihood also in successive sentences of a discourse (9.3).

Repetition may occur in a position where it has high likelihood but where the first occurrence (the antecedent) is not in a fixed favored position in respect to the repetition. In such cases, the repetition can be reduced to a pronoun rather than to zero. For example, given O_{oo} under which repetition is expected, but without relative positions being specified (9.1), we find e.g. *The mail did not come though everyone was awaiting the mail* reducible to . . . *though everyone was awaiting it* (but not to *everyone was awaiting*). And under *and*, where repetition has highest expectancy in the corresponding grammatical position, any repetition in other positions is reducible only to pronoun: *The play was good and the audience liked it*. But when repetition occurs in the fixed high-expectancy position in respect to the antecedent, it can (and under some operators must) be reduced to zero. Under *and* this happens when the grammatical position of the repetition is the same in the second sentence as that of the antecedent in the first: *The play was good and the play was applauded* is reducible to *The play was good and it was applauded* and also to *The play was good and was applauded*. Under O_{oo}, everything after the tense is zeroed in the second sentence if it is the same as in the first sentence: *I will support John for the job if you will support John for the job* is reducible to *I will support John for the job if you will*. Under O_{nn} (as above), if the second argument has the same referent as the first, it is reducible to pronoun plus *-self*: *The man heard himself on the tape* (from *heard the same man*). But under particular O_{nn} where sameness of the two arguments may be especially frequent, it is also reducible to zero: *The man washed himself* and *The man washed*. In the case of the O_{no}, each operator has a particular relative position for zeroing repetition (3.2.3). For example, *I promised him to go* is from *for me to go*, but *I ordered him to go* is from *him to go*, while *I offered him to go* can be from either. But under O_{on}, nothing is reduced to zero: the second argument of *surprise* is not zeroed either in *His meeting her surprised him* or in *His meeting her surprised her*.[1]

5.3.2. *Locating the antecedent*

When a pronoun such as *he* or *it* appears in a discourse, or in most positions under O_{oo}, the hearer does not know what preceding word it refers to (unless there happens to be only one appropriate word, e.g. masculine

---

[1] The analysis of pronouns as repetition (mostly within a sentence) may explain why referential pro-verbs are hard to come by. By the semilattice property, only one word-occurrence in a sentence can be a free operator (the 'main verb', the highest operator in the semilattice). Hence if an operator word occurs twice (as repetition) within a sentence, at least one occurrence, and in many cases both, would be as argument of a higher operator; there would never be two occurrences as free operators, one pronounced as repetition of the other. In contrast, the pronounings that are met with are between two argument occurrences, either of N, e.g. in *John washed himself*, or of nominalized operators, as in *Mary's happiness is more important than John's*, or in *I won, which he had hoped to do* (where *which . . . do* is as though a pronouning of *do winning*, from *He had hoped to do the winning*).

singular for *he*). However, in particular positions under certain operators, a pronoun or zero is recognized by the hearer as referring to a word in a particular antecedent position in the sentence: in *John phoned Mary and wrote me*, the zero before *wrote* refers only to *John*. To state the information required for this recognition, it is necessary to have an addressing apparatus which is communicated to the hearer within the sentence itself. In the hearer's learning of his language, this recognition arises without addresses being said: the hearer, in hearing whole sentences and gathering their meaning, recognizes that in particular positions the pronouns or the absences of an expected word (e.g. of the subject of a following verb) always refer to an antecedent in a particular other position. Nevertheless, it is important for us to know that the information needed by the hearer in order to recognize the sentence and to locate the antecedent can be stated in metalinguistic sentences (of ordinary grammatical construction) adjoined to the given sentence. To state the relative location of the antecedent, these sentences need just the existing vocabulary of the language.

Languages have such words as *latter*, *former*, *earlier*, *before*, *after*, *next*, *following*, *inverse*, *first*, *second*, etc. In the metalinguistic sentences which talk about utterances, these predicates can be used about word- and phoneme-occurrences: e.g. *He was influenced by Stravinsky and Prokofiev, more by the former than by the latter* (meaning 'former said' and 'later said'), *The first three across the finish line were Smith, Jones, and Bannister, in that order* (i.e. the runners were in the order of saying their names here; also . . . *in reverse order*). Other locational words have arisen which refer explicitly to word-occurrences, e.g. in English *aforementioned*, *respectively*, vice versa, and the common use of *latter*.

The utterances or word-occurrences that are being talked about must be in the same sentence as the metalinguistic statement about them, or at a stated distance from it in the same discourse, and this for two reasons. One reason is that, as in the examples above, the material cannot be talked about unless it is presented to the hearer: we would not say *c precedes a* unless we add *in the word cat, etc.* The other reason is that it cannot be identified unless it is at a stated grammatical distance from the metalinguistic statement or at a stated linear distance from the beginning of that statement. If we want to express within the metalinguistic statement the hearer's recognition of the position of the antecedent, we can do it by having the metalinguistic statement interrupt right after the word-occurrence that it is talking about. Thus (1) *A small boy disappeared*, which is equivalent to (2) *A boy who is small disappeared*, is composed of (3) *A boy disappeared* interrupted by (4) *A boy is small* plus (5) the information that the two occurrences of *boy* refer to the same entity. For the hearer, this information is carried here by the *who*. To express this information by existing language material in an adjoined metalinguistic sentence, we need only to consider *wh-* as an O_{oo} (conjunction) introducing a secondary sentence (i.e. a sentence under the secondary intonation marked by dash or semicolon, as in *A boy disappeared; the boy was small*) and to consider *-o*, *-ich*, etc. as reductions of a repetition of the preceding argument. Then in *A boy—a boy is small—disappeared* we need a further interruption after the second *boy* stating (5) that the two successive occurrences of *boy* have the same referent. Thereupon *-o* replaces the second *a boy* plus the sameness statement, and *wh-* replaces the dash, all of which yields (2) which is reducible to (1).

Of course, neither the history of language nor the recognition process of the hearer goes through this reconstruction. But the reconstruction shows what apparatus is needed for cross-reference. And it shows that language developed in a way that always permitted an as-though reductional derivation from a regular predicational source, and that all the information carried by language—in this case, that contained in pronouns—can be expressed in a purely predicational—in this case metalinguistic—source.

Aside from the relative pronouns in *wh-*, there are certain other cases of reduced repetition with fixed antecedent. Here too, given the reduced item, we have to say that it is the same as a word at a fixed position relative to it. In English, the main cases of this are zeroings under O_{oo}, O_{no}, and O_{nno}. Under O_{no} and O_{nno}, it is when the first or second argument of the second (lower) operator is the same as the first or second noun argument of the first (higher) operator, depending on what is the highest-likelihood position for repetition under the particular higher operator. As has been seen, under *promise* we have *I promised him to go* from *I promised him for me to go*; under *leave* we have *I left him feeling sad* from either *I left him, I feeling sad*, or *I left him, he feeling sad*; under *order* we have *I ordered him to go* from *I ordered him for him to go*; and so on (3.2.3, G148).

We see that in all cases where the antecedent of pronouning or zeroing is in a fixed relative position, that position is easy to state, either linearly or in the partial order. In the case of the relative clause, if the secondary sentence interrupts the first immediately after the antecedent, and the repeated word is at the beginning of the secondary, then the two same words are next to each other so that the antecedent can be identified simply as predecessor. To this end, it is relevant that in English the words that can be reduced to a relative pronoun are precisely those words in the secondary sentence that can be said at its beginning even when that sentence is said by itself: words that are never said (even for stress) at the beginning of a sentence are never replaced by a *wh-* pronoun (when that sentence appears as a secondary one). This means that the presence of the relative pronouns (*who*, etc.) at the beginning of the relative clause is due not to permuting a pronoun but to the repeated word (before being pronouned) being already at the beginning of the secondary sentence before the secondary sentence was brought in as an interruption of the primary: e.g. *John, whom I asked, refused* is from *John— John I asked—refused* (originally from *John—I asked John—refused*). Since, then, the relative pronouning does not take place except when the antecedent and its repetition are already next to each other, we must judge that what makes it possible to reduce the repeated word to a *wh-* pronoun is its nextness to its antecedent, where the fact of repetition is most obvious and the naming of the antecedent location is easiest: the preceding word.

Somewhat similarly, in the zeroings, under O_{oo}, O_{no}, and O_{nno} (above), which have a fixed position for their antecedent, that position is easy to recognize and state in terms of the basic grammatical relations (the partial order of dependence): the corresponding position under O_{oo}, and a fixed semilattice-relation of lower to higher argument under O_{no} and O_{nno}.

Thus the only referentials whose antecedent is always at a fixed relative position (*-self*, relative pronouns, and zeroed repetitions) are ones for which the antecedent position is easily indicated, linearly or in the partial order. This suggests that the ability to reduce the repetition in these cases was based on this ease of locating the antecedent. That means that in the formation of referentials, i.e. in the reducing of repeated material, the development of the language and also the recognition by the hearer do not use the general partial-order and linear addressing systems that are inherently possible in language. If either or both of these addressings were usable over their full range by the speaker and the hearer, every phoneme- and word-occurrence could be identified and referred to, including the position of any antecedent relative to the referential. There would then be no reason for referentials with fixed antecedent to occur only in situations where the location of the antecedent relative to the referential is especially easy to state and to recognize.

Conversely, we must judge that the other referentials, whose antecedent is not limited to a single position, indeed do not refer to any particular position (which, as seen, could not be addressed or named within the use of language as we see it). For example, in *John prefers that he be appointed now*, the *he* can be a repetition of or refer to the preceding *John* or any other human masculine singular noun in a nearby preceding sentence. One might think that these free pronouns refer to a disjunction of all preceding appropriate nouns: either to *John*, or else to the masculine singular noun before it, or to the one before it, and so on. But each one of these alternatives (except the first *John*) would have to be identified in a complex address utilizing the apparatus of a general addressing system. Since no fixed antecedent uses the (in any case, non-existent) large addressing apparatus, we have to judge that the non-fixed antecedents are not simply a disjunction of fixed antecedents. All this means that we have to understand such free pronouns as *he* and *it* to be not a reduced repetition of one or another preceding noun but a word meaning 'aforementioned (human) male' (or: 'male mentioned nearby'), 'nearby-mentioned thing', etc. This, even though the speaker knows which word is the antecedent, and the hearer may guess it on some grounds.

5.3.3. *Sameness*

We have seen here that the fixed-antecedent referentials are reduced repetitions (and thereby cross-references), whereas the other pronouns are words with their own cross-referential meanings. The information which triggers the reduction of repeated words is that a given word is the same as the word at the fixed antecedent position, and when the hearer comes upon the referential this is the information it gives him (5.2). It remains to be seen what 'same' means in this case. When the antecedent carries a 'determiner' (e.g. demonstratives, article, quantifiers) the *same* usually means 'has the same individual referent', as in *I saw a bird, which flew off immediately* or *Someone, who shall remain unnamed, goofed*. Note that *A man came and a man left* may refer to a different man, but *A man came and left* refers to the same man, and hence would have to be derived from *A man came and the same man left*. But compare *I brought him a cup of tea, which he needed badly*, where *-ich* indicates the words *cup of tea* but not its individual referent. Otherwise, *same* means 'is a case of the same word' (which includes having the same selection, hence meaning) as in: *All day, men came and left* (not necessarily all of them the same men); *He is educated, which they are not*; *John plays violin and Mary piano* (but hardly *John plays chess and Mary piano*, where different selections of *play* are involved, except as a joke); *I will go today because he did yesterday* (not the same act of going); *This one grew more than the others* (with each doing its own, possibly different, growing).

Also in *he, she, it, they*, there is a distinction between 'same word' and 'same individual referent', except that here these sameness properties are part of the meaning of the pronominal words, rather than being part of the metalinguistic instructions about the antecedent. *He* and *she* almost always refer to the same individual referent as in some appropriate nearby word. But *it*, *that*, and *they* may have merely the same meaning as some nearby word: *Eat less bread! Do it now!* (Wartime governmental billboards in England); *He left early, and I did that too*; *People may succeed, or they may fail.*

5.3.4. *Demonstratives: Reference to the external situation*

Up to this point, 5.3 has analyzed the cross-reference of words and zeros (word-absences) referring to word-occurrences in the same discourse. There remain a few pronouns which are demonstrative or deictic, i.e. refer overtly to an object or situation not mentioned in the text but present to the speaker or hearer: e.g. *That is a beautiful tree, This tree is beautiful*. These can be derived from the already-established intra-text cross-reference pronouns by assuming a zeroable indefinite noun attached to the zeroable metalinguistic *I say* (5.5.1): e.g. *I say of something here that this is a beautiful tree* (where *this* is a cross-reference to *something here*), or *I say of something at which we are looking* (or *of which we are speaking*) *that that is a beautiful tree*.

Such words as *here* and *now* can be derived from *at the place in which I am speaking* and *at the time at which I am speaking*. And *I*, *you* can be considered reductions of *speaker*, *hearer* when these words occur in a discourse as repetitions from a zeroable *The speaker is saying to the hearer*, which one might presume to introduce every discourse (5.5.1): *I will phone you* as though derived from a reconstructed *The speaker is (hereby) saying to the hearer that the speaker will phone the hearer*.

Before analysis, the kind of meanings of all these words differ from the kinds of meanings of the other words of the language. In the other words, the meanings or referents may vary with the environing words (the meaning of *child* in *a child of his time* as against *a child prodigy*), but they do not vary with the time or place in which the sentence is said. But in the words discussed in 5.3.4, the referents vary with the particular bit of speaking: which tree is *this* and who is meant by *I* depends on the speaking situation. The information of demonstratives, including the tenses (below), is for this reason unique also in not being transmissible beyond face-to-face communication. The derivation from an *I say* component underscores, and does not eliminate, this property. Nevertheless, it is a gain for the compactness and simplicity of grammatical description if we can eliminate the new kind of meaning required for 5.3.4; and this can be done by assuming a zeroable *I say of some thing that that* . . . (as source of demonstrative *that*, *this*), or *The speaker is saying to the hearer at the place and time of speaking that* . . . → *here and now*. It will be seen in 5.5.1 that there are many grammatical reasons for assuming such an introduction to be implicit in each discourse or sentence that is said (not merely composed). Note that in sentences that are composed but not really said (e.g. in grammar exercises), *I*, *now*, and *that tree*, and the like do not have specific referents, as they do in each said sentence.

The derivability of tenses from *before*, *after* connecting a said sentence to the *I say* operator on it (G265) means that the tenses are equivalent to demonstrative time-adverbs.

5.3.5. *Summary: Reference from cross-reference*

In 5.3.4, it was seen that words which are deictically referential in an utterance can be derived from cross-reference by reconstructing a source that contains a zeroable *The speaker says*, i.e. that refers to the speaking (or writing) of the utterance. Once this is done, we do not need to know any extra-linguistic context, i.e. anything outside the sentence itself (with its adjoined zeroed metalinguistic statements, 5.2), in order to understand it as a sentence.

As to cross-reference, this was analyzed as a reduction of a second occurrence of a word stated to be the same word (or to have the same referent) as another occurrence (the antecedent) within the utterance. Here a distinction was found between two types: first, there are pronouns whose antecedent was not at a fixed relative position (e.g. *he*); here the antecedent is characterized merely as having occurred nearby in the discourse. Second, there are pronouns and zeroings whose antecedent is at a fixed (and simple) position relative to the pronoun or zero, in the linear order of words or in the partial order that creates a sentence (e.g. *who*); here the position of the antecedent and its sameness is given in a metalinguistic adjoined statement.

5.4. *Use/mention*

We now consider the second feature that underlies the internal metalinguistic capacity of language (5.1), namely, the fact that the meaning of a word is not referable to any meaning of its phonemes. The independence of word meaning from its phonemes is necessary, for language, for if the meanings of words were composed regularly out of the meanings (or the effects upon meaning) of their phonemes, the number of distinguishable word meanings would be limited by the combinations and permutations that one could construct, in reasonably short sayable sequences, out of the meaning contributions of the twenty-odd phonemes that the language has. Given this necessary independence, what we ·say about a word (i.e. its selection) when it is taken as indicating its referent differs from the selection of operators which appears on the phoneme-sequence of that word: e.g. *Mary phoned me*, but *Mary consists of four phonemes*. Note that one cannot pronoun a word's occurrence under the one selection from its occurrence under the other (5.3.3): we do not pronoun *John and Tom* to make *I saw John and Tom whose second phonemes are identical*, but we do in *I saw John and Tom, in whose names the second phonemes are identical*.

The sameness statements do not support pronouning when the selections and meanings of a repeated word are sufficiently different, as in *play* of 5.3.3 or *heart* below. Going beyond that, when certain occurrences of a word have a coherent selection totally different from that of other occurrences of that word, we say that those two sets of occurrences belong to different homonymous words: e.g. the occurrence of *sound* in the senses of 'noise' as against 'strait' and 'healthy'; but no homonym for the occurrences of *heart* in the senses of 'organ' and of 'center (as of lettuce)' (which we consider to be different meanings of a single word). In the case of the two sets of operators on *Mary*, we hesitate to say that, for example, (1) *Mary* in *Mary has* (or: *consists of*) *four letters*, which is not pronounable from (2) *Mary* in *Mary phoned me*, is simply a homonym of (2) in the way that applies to the various occurrences of *sound*, for then every word would be a homonymous pair.

That is, we do not want to say that *Mary is written with a capital m* and *Mary phoned* have two homonymous words *Mary*. A more special treatment is clearly required here.

Aside from the difference in selection between the phoneme-sequence of a word, and a word as bearer of a referent, there are occasional other differences. On the one hand, there are cases of different occurrences of the same phoneme-sequence constituting different words as referent-bearers: the homonyms, above. On the other hand, there are cases of different phoneme-sequences constituting the same word. This happens not only in what are recognized as suppletive variants of a word (e.g. *is*, *am* as variant forms of *be*) but also in what are considered alternative pronunciations of a word. For example, the semantically single word written *spigot* consists of two alternative phoneme-sequences, one with medial phoneme *g* and one with medial phoneme *k* (and with older spelling *spicket*). Both this problem and that of the homonymous pairs can be avoided if the underlying sentences are taken as, for example, *The phoneme-sequence that constitutes the word Mary consists of* (or *has*) *four letters*, reducible to *Mary has four letters*, as against *The object referred to by the word Mary phoned me* (or, for that matter, *has four letters to mail*), reducible to *Mary phoned me*. (The evidence that what was reduced was indeed *the phoneme-sequence* is that the predicate on it, *consists of four letters*, is in the selection of *phoneme-sequence*.) In the reduced forms, the distinction is expressed by saying, in Quine's terminology, that in the latter case the word *Mary* is 'used' in the sentence, while in the former case it is only 'mentioned': 'mentioning' a word is 'using' its phonemes. These useful terms isolate and name the distinction; to derive or explain the distinction grammatically, we need underlying sources as above, while noting that the zeroing of *the phoneme-sequence that constitutes* leaves quotation marks as its trace (in writing: '*Mary*' *has four letters*). This derivation explains why a mentioned plural noun is singular: *Bookworms is on p. 137 in this dictionary* (from *The phoneme-sequence that constitutes the word bookworms is on . . .*); contrast *Bookworms are all over in this dictionary* (the latter from *The objects referred to by the word bookworms are . . .*). The derivation above also explains why one does not say *Mary has a little lamb and four letters*, which would be a possible zeroing from *Mary has a little lamb and Mary has four letters*, but not from *The object referred to by the word Mary has a little lamb and the phoneme-sequence constituting the word Mary has four letters*. The importance of this explanation is that it remains within the universe of word-occurrences and their reductions, without appealing to such considerations as whether Mary is 'taken' as a meaning or as a spelling. When the source sentences are reconstructed, the use/mention distinction, and the Bourbaki 'abus de langage', becomes redundant.

As to the failures in one-to-one correspondence between phoneme-sequence and word-referent, they can be covered by appropriate underlying statements. Thus if a word has alternative phonemic forms, the sentence underlying its mention would be something like *One of the (alternative) phoneme-sequences that constitute the word economics begins with the phoneme e*.

In the metalinguistic sentences of 5.2, the operators speak about the occurrences of a word, hence about the occurrences of its phonemes. This applies to such a sentence as *The phoneme-sequence t, e, n constitutes a word*, or *Ten means the successor of nine*. Thus the metalinguistic sentences speak about a word as 'mentioned' even when they give a definition of the word or indicate its referents.

5.5. *Performatives*

5.5.0. *Introduction*

A distinction related to use/mention appears under the word *say*, which has a unique status in language structure. In various languages, *say* appears with two forms of second argument (object): *He said* (or: *I say*) *'Mary is here'*; *He said* (or: *I say*) *that Mary is here*. There are syntactic differences between the two forms. Under the bi-sentential operator *but*, which requires in principle two contrasts between its arguments, we can find *I said 'Franklin Roosevelt died in 1945' but he said 'President Roosevelt died in 1945'*; but we would not normally find (even with contrastive stress on *President*) *I said that Franklin Roosevelt died in 1945 but he said that President Roosevelt died in 1945*. We ask if these two forms of the argument of *say* can be derived in a way that would explain why the two forms differ in respect to *but*. With an eye to the use/mention sources of 5.4, the obvious sources to propose are: for the first, *I said the words* (or *word-sequence*) *Franklin Roosevelt died in 1945 but he said the words President Roosevelt died in 1945* (with *the words* replaced by intonation of quotation marks); and for the unlikely second, *I said words which mean that Franklin Roosevelt died in 1945 but he said words which mean that President Roosevelt died in 1945* (with zeroings of *words which mean*), where any form of that name and indeed any synonym of it or of *died* could have been what was actually said. In the second source, *but* is clearly less likely than in the first. In both of the proposed sources, the second argument of *say* is *word*, thus eliminating the problem of having two forms of the argument (above).

The relation between a quoted sentence and *that* plus the sentence is also seen in the difference between *Yesterday in Boston the candidate said to the strikers 'I have come to support you here and now'* and *'Yesterday in Boston the candidate said to the strikers that he had come to support them there and then*. Such differences as *here/there* hold not only under *say*, but also under *tell, state, repeat, assert, claim, insist, promise, think, agree*, and other words.

Among all these there is one, *I say* or *One says*, in the present tense, that has a peculiar property which serves to explain the relations noted above, and also several other exceptional forms in language.

This property is a metalinguistic analog to the performatives discussed in logic, as when a judge says *I (hereby) sentence you to* . . ., in which his saying that he is sentencing constitutes the act of sentencing. For a speaker to compose the sentence *I say 'John left'* or *I say that John left* (or possibly *One hereby says that John left*) is the same as for him to actually say *John left*. Hence, the *I say* or *One hereby says* contributes no information, and is zeroable. Indeed, the sentence *John left*, and any sentence that is said (aloud or to oneself) or written (except e.g. as an example of a sentence) can be considered a reduction from: *I say*, plus *that* or quotes (or equivalent marks such as a colon), plus the sentence in question. The *I say* is not zeroed when it is not used performatively, as in *I say it's spinach and I say to hell with it*; or when it is used contrastively, as in *I say John left but Mary denies it*, or *I say that he left, but I don't mean it*. In such cases, it may even be repeated, as in *I say that I say that he left, in order to make it clear*. But a sentence cannot be suspected of carrying an infinite regress of zeroings of *I say*, from an infinite repeating of the performative *I say*, because that zeroing takes place upon the occasion of the saying, and once the sentence has been said, it cannot repeat the performative *I say* and its zeroing.

5.5.1. I say

There is varied evidence of the original presence of *I say* or *One hereby says* and their subsequent zeroing. To any sentence one can add such implicitly metalinguistic phrases as *to tell the truth*, or *to be specific*, or *not to make too fine a point of it*. The zeroing of the subject in these phrases means that there must have been an earlier occurrence of that missing subject (of *tell*, etc.) which had served as repetitional antecedent for the zeroing of that subject. This other occurrence of the subject could not have been present except as an argument of some operator; and the combination of that earlier subject with its operator had to have been zeroed on some grounds, since they do not appear. The only candidate for such zeroing that would have been possible in every sentence and that would have left no trace (except for the very saying of the sentence) is the performative *I say* or *One hereby says*; no other subject and tense of *say* would make it performative, nor would any other verb with different meaning.

In many cases there is direct evidence of a metalinguistic *I*, as when the insertable metalinguistic phrase is *not to repeat myself* (where *myself* could arise only from *for me not to repeat me*), or *to make myself clear*; also in *I hasten to add* or *at least I think so* or *and I say again*. Here, the *repeat, say again*, *add*, and *at least* each implies in different ways an *I say* or *One (hereby) says* on the primary sentence: *John left and, I hasten to add, not secretly* from *I say John left and I hasten to add . . .*; *John left, at least I think so* from *I say John left, at least I think so*; *He's wrong. I say again: He's absolutely wrong* from *I say he's wrong, I say again . . . .*

The widespread zeroing of *I say* explains certain forms where the analysis shows that *I say* has been zeroed by mistake. For example, sentences such as *He's back, because his car is in the garage*, which do not make sense as said, clearly have been reduced from *I say he's back, because his car is in the garage*, where the *I say* should not have been zeroed, because it operated only on the first sentence and is in turn operated on (as first argument) by the *because*. Another example is in sentences such as *The Times says that our stupid mayor will be speaking tonight*; since the newspaper said neither *stupid* nor *our*, the source here has to be *I say that the Times says that the mayor who I say is our stupid one . . .* Here, zeroing the second *I say* left the *who is stupid* to appear as modifier on *mayor*, which is incorrect since *mayor* is under *the Times says*. Consider also the sentence in which a reporter writes, about a woman speaking to her husband, *She said Mary, their daughter, would go to the police station*. Had the woman said to her husband *Mary, our daughter, . . .*, this would have been reported as *She said Mary, their daughter, . . .*; but of course she did not say *our daughter*. She must have said *Mary will go* in which the reporter inserted something like *who I say is their daughter*, where *I say* should not have been zeroed because (as with *stupid* above) it is one speaker's insert in reporting another person's remark.

5.5.2. *Imperative and interrogative*

There is grammatical evidence that the imperative and interrogative are constructed like the performative *say*. In the first place, even languages such as English, in which the imperative has no subject, carry evidence of a zeroed subject. In *Look!* we find no direct evidence, but in *Wash yourself!* the *-self* indicates that the second argument *you* is a repetition of a first argument *you* under *wash*. This second-person first argument is visible in such forms as *You stop that!*, (which may well be different from *You (there)! Stop that!*) and in the older *Go thou!* As to third-person subject, there is no *Wash himself!*; but *Let him wash himself!* is a way of getting a pseudo third-person imperative on *wash* from what is grammatically a second-person imperative on *let*; note also *Let's go* as a second-person imperative form for a first-person imperative meaning. A rare third-person imperative may be claimed in *Somebody call the police!*

The invisible existence of a higher metalinguistic operator is indicated by the rare English *That he go!* (common in French: *Qu'il parte!*). Here we can reconstruct *I ask* (or *request*) *that he go*, with the subjunctive of *go* under *ask*; in effect, for modern English, this consists in zeroing the *will* which has high likelihood in the last argument of such operators as *request*, *beg*, *demand*, *command*, *desire*. Given the high likelihood of repeated *you* here, as in *I ask that you go*, *I beg of you that you go*, the *you* of *you go* is zeroed in the direct-quotes form *I beg of you: 'Go!'*, where the imperative intonation (the exclamation mark) is a phonemic extension of the *beg*, *request*, operators. Given the imperative intonation, the *I ask you* or *I beg of you*, etc., can now be zeroed, as having probability of 1, and hence contributing no information, in the presence of that intonation: this yields *Go!* (In *Would that he went*, usually without an imperative-like intonation, only the *I* subject of *would* has been zeroed.) Furthermore, given the zeroability of *I say* above, it is possible to think of *I request* as composed of *I* say plus an adverbial *imperatively*, *as request* (as in 5.5.3), where the adverbial portion imposes the imperative intonation and is then zeroable, while the *I say* portion is zeroable performatively: this admits of the existing intermediate form *I say (to you): Go!*[2] (G351).

The interrogative is far more complex, and its metalinguistic structure is in English more evident. It is generally considered that grammatically there are two kinds of questions: (*a*) the yes–no questions (*Will John leave? Yes.*) and (*b*) the *wh-* questions (*Who will leave? John.*) However, the yes–no questions can be readily derived from (*c*) disjunctions of questions—or, for that matter, a querying of a disjunction of possible answers: *Will John leave?* from *Will John leave or will John not leave?* It is derived by optionally zeroing any material that is repeated under not (if one of the disjuncts contained repeated material under *not*), and then optionally zeroing the *or not*. Hence *Will John leave or not leave?*, *Will John leave or not?*, *Will John leave?*

It is for this reason that *Will you?*, which is reduced from *Will you or won't you?*, means (aside from nuance) the same as *Won't you?*, which is reduced from the equivalent *Won't you or will you?* (*or* and *and* being semantically commutative). The extended form (*c*) is a source, and an always-possible constituent, of the short (reduced) one, and is a special case of the fact that every question offers a disjunction, and any disjunction of questions constitutes a question: *Will John leave or will Mary leave? Will John leave or will Mary return or will everybody stay put?*

When we now consider (*b*) the *wh*-question, we find that, e.g., *Which one will leave?* is derivable from a distributive *For each one, will this one leave or not?* This last form, in turn, is derivable from *I ask, for each one, whether this one will leave or this one will not leave*. Note that the source disjunction is not of questions but of possible answers. Even in such a question as *What*

---

[2] The sentence on which *I request* operates can be a disjunction (or conjunction) of sentences, as in *Take it or leave it!* from *I request that either you take it or you leave it* (and not from *or* operating on two occurrences of *I request*). However, this differs from the essential occurrence of a disjunction of sentences as the only and complete operand of *I ask*, *I query*, in the interrogative form below.

numbers are prime?, where no finite disjunction of answers seems possible, this analysis offers a finite source: *I ask, for each number, whether that number is prime or that number is not prime*. The same source is then found grammatically adequate for (*a*) the yes–no question too: in the example above, *I ask whether John will leave or John will not leave*. We thus arrive at a single source from which all interrogatives can be derived by established reductions: namely, *I ask* (or: *enquire, wonder, query each answer*, etc.) plus *whether* (in English *wh-* plus *either*), all operating on two or more non-interrogative sentences. In this structure, the disjunctive *either* imposes *or* between the operand sentences (like what *and*, *or* do, 5.5.3). Like the imperative, the *ask* can be considered as composed of *say* plus an adverbial *querying* or *in request to choose either*, where this adverbial generates a question intonation on the operand sentence-disjunction and is thereafter zeroable, and the residual *I say* is thereafter zeroable.

This analysis is not merely a matter of finding a source that has operator–argument form. It produces and explains many peculiarities of the interrogative. In English, it yields the question form out of the ordinary form of the operand sentences (*This one will leave* under (*b*), and *John will leave* under (*a*), above). For English, it takes the *wh-* of *whether* to be the same morpheme as the *wh-* of the relative clause. Also, conclusively, it is able to indicate which words in an operand sentence can be asked about—namely, those that can be pronouned with *wh-* in the relative clause (in effect, those words that can be brought to the head of their sentence). It provides for multiple questions (*Who saw whom?*) and various special forms such as *You'll go, won't you?* (G331).

5.5.3. *Adverbs on* say*:* And, or, not, any, *tense*

There is a final utilization of performative *say*, which provides explanatory derivations for important but peculiar grammatical constructions, but presents problems—or as yet presents problems—for an orderly analysis of complex sentences.

There is grammatical evidence that the performative *I say* carries certain adverbial modifiers (secondary operators on it) which produce various special words whose grammatical properties are peculiar and not otherwise explicable. Such are the logical *not, and, or*, and certain exceptional English forms such as *any*; cf. also the source of the demonstratives (5.3.4). In many languages, the words for *not, and, or* have peculiar grammatical properties (G387). For example, in English they, alone of all operators, cannot take the form of verbs, and cannot carry tense either directly or on an associated *be*: for *He phoned and she wrote*, there is no *His phoning was and she wrote* comparable to *He phoned because she wrote*, *His phoning was because she wrote*. Hence they are not ordinary operators, but neither do they fit any other word category. They are also unique in language as being basic operators of set theory and logic. In addition, *not* and the interrogative do not preserve the selection of the words under them, whereas operators in general preserve such selection: thus *the stone spoke* and *I want the stone to speak* are rare and peculiar combinations, but *The stone didn't speak* and *Did the stone speak?* are much less so. Instead of disregarding these peculiarities and also the logic status of these words, we can to some extent explain them by deriving the words from modifiers on the *I* say (e.g. *for each one* in the source of the interrogative, 5.5.2, and in the source of *any*, below). This can be expressed in artificial restructured sources, such as *I say negatively* (or *I deny, gainsay, contradict, disclaim, claim as false*) *that John is here*, reduced upon zeroing of *I say* (or *-dict* or *claim*) to *John is not here*; *I say both* (or *jointly*, or *co-state*) *Mary is here, John is here*, where *I say both* operates on a pair of sentences and is reduced upon zeroing of *I say* or *I state* to *Both Mary is here and John is here* with *both* then zeroable; also *I say either* (or *alternatively*) *Mary is here, John is here*, reduced to *Either Mary is here or John is here* with *either* then zeroable. (Possibly, also, the intonation that marks a secondary sentence, written as semicolon or dash, could be obtained from *I say with an aside* operating on a sentence-pair: e.g. *Mary is coming, She phoned*, yielding *Mary is coming; she phoned* or *Mary—she phoned—is coming*, ultimately *Mary, who phoned, is coming*.)

As to *any*: in *Anyone can go* the word means roughly 'every', whereas in *I didn't see anyone go* or in *Will anyone go?* it means 'even one'. Also, there is the peculiarity that *any* rarely occurs with only a single operator. In spite of the popularity of phrases like *Anything goes*, the more customary form has two operators, as in *Anything small will do*, or *Anyone who can goes there*. All these forms and distinctions can be obtained within the existing grammar if we take *any* to be an adverbial form of *one*, in particular *(one) by one* (which is roughly its etymological source) operating specifically on the *say*. Then *Anyone can do it* is from, roughly, *I say for each one that this one can do it* (hence 'everyone can'); and *I didn't see anyone do it* is from *I deny for each one that I saw this one do it* (hence 'not even one did'); and *Will anyone do it?* is from *I ask for each one whether this one will do it* (hence 'will even one do it?'). The selectional preference of *any* for two operators is thus in order to relate the two predicates in respect to the one-by-one matching: *For each one if he can then he goes there*, above (G327).[3]

Finally, one of the most important grammatical advantages of recognizing the metalinguistic *say* operator is the explanation which it offers for the tenses. If we consider a sentence such as *A person who registered as late as*

---

[3] There are also less clear cases of words whose meanings or positions can be explained by saying that they operate on the performative *say*. One such is seen in *hardly, scarcely*, whose occurrence in many sentences such as *John hardly deserves it, John scarcely deserves it* would be more understandable if the *hardly, scarcely* referred to the saying of the sentence: that is, it would be hard for the speaker to say that John deserves it.

tomorrow will still be able to vote on Tuesday, we see a known phenomenon, in which the secondary operator is tensed to express its relation to the primary operator as taking place before or after it; the source is *A person who registers as late as tomorrow before that* (i.e. before voting) *will still be able to vote on Tuesday*, with *before that* reduced to *-ed* on *register* (G265). Similarly, if a verb operates on another, the lower is tensed as being before or after the higher: *Yesterday I heard that Mary will* (or: *would*) *come later that day*, from *Yesterday I heard that Mary comes later that day, after my hearing it*, with *my hearing it* zeroed as repetition, and *after* reduced to *will* (G269); in the now less usual form *would*, the lower operator adds to its tense the tense of the higher operator. (If, under a particular higher verb, the lower verb is usually future in respect to it, the future tense on that lower verb can be zeroed: *I request that he go* from *I request that he will go*.)

We can now say that in the same way, the highest operator of a sentence, i.e. the operator of a primary sentence, can be seen as tensed in respect to the *I say* that operates on it (G267): *I say: 'I will go'* or *I say that I will go* from *I say (that) I go; my going is after my saying it*, reduced to *I say (that) I go which is after my saying it*, then to *I say (that) I go after*, to *I say (that) I will go*, to *I will go* (G268). In English, in many situations, when a lower verb is tensed in respect to a higher one, then any tense that the higher verb gets, in its turn, from *say* is added to the lower tense: e.g. in the *would* above. The source was *I state my hearing yesterday, which is before my saying it* → *I state my hearing yesterday, before* → *I state I heard yesterday* → *I heard yesterday*. (Note that one may prefer to have *before* and *after* as adverbs on the higher verb (including *say*) rather than *after* and *before*, respectively, as modifiers on the lower verb: *I say beforehand that I go* → *I say that I will go*.)

In this way, the 'absolute' tense of the highest operator (*I will go, I heard*, above) and the 'relative' tense of those below it (e.g. *come*)[4] are seen to be the same: the highest operator's tense is relative to the *I say* above it. Thus tense in general can be derived, paraphrastically and by attested reductions, from sentences with time-order conjunctions but no tense. Specifically, it can be obtained as a reduction of an adjoined statement giving the time relation of two operators in the sentence, and finally the relation between the time of speaking about an event and the time of the event spoken of. This, rather than an external semantic instruction about tense in sentences (G265).[5]

[4] Note that since a relative clause is a secondary sentence, the lower operator *registered* above receives *-ed* from being 'before' in respect to the primary operator *able to vote*. The higher *able to vote* receives *will* from being 'after' in respect to the time of saying the sentence, i.e. in respect to the *say* on the sentence.

[5] There are many subtle grammatical considerations in the syntax of tense and of *before* and *after* which support the convoluted derivation offered here. In addition, taking tense not as an *ad hoc* primitive construction (outside the operator–argument framework) but as a derivation from time or order conjunctions between a sentence and the statement that it is being said leaves room for various non-temporal uses of tense, deriving from non-temporal uses of those

In sum: *say* and the like can carry certain adverbial modifiers (*as false*, *both*, *either*, *demand*, *querying*, *of something* (5.3.4), *one by one*, *before*, *after*, etc.) which can impose respectively *not*, *and*, *or*, imperative and interrogative forms, deictic pronouns, *any*, tense, etc., on the operand sentence. Of these, *both* and *either* (the latter includes the interrogative *whether*) require the operand to be a sequence of sentences rather than a single sentence. Uniquely, the *and* and *or* differ from the two-sentence O_{oo} operators of grammar in that they are not restricted to being binary, and in that they are semantically associative and commutative: *He left because she phoned him because it was late* means different things in different binary groupings, but *He went and she phoned and I stayed* does not; also *She stayed because he left* and *He left because she stayed* mean different things, but *She will stay or he will leave* and *He will leave or she will stay* mean substantially the same, though not in nuance. Also the disjuncts of the interrogative are commutative. All of these adverbial modifiers relate the operand sentence to the performative operators on them. In particular, *not*, *both*, *either* are relevant to logic and set-theory because they deal with the saying of sentences.

Most generally, some of the operators on sentences are metalinguistic in that they can operate on the words of the sentence as phoneme-sequences and not only on those words as indicators of particular referents. Of these, one operator, *say*, or a small set (*say*, *ask*, etc.), with *I* or *one* as subject, has performative effect and is therefore zeroable upon the saying of the operand sentence.

As indicated at the beginning of 5.5.3, the adverbs on *say* require more investigation before they can be adequately formulated in a theory of language structure. The least of the problems is that the reconstructed sources are clearly non-historical. Whereas the imperative and interrogative can be seen to be related to their proposed sources, the sources for tenses and *and/or/not* are convoluted beyond what would be said in natural language. What remains relevant, however, is, first, that all of these (and such words as *any* in English) have combinatorial properties and levels of meaning that differ from the rest of language, and, second, that these grammatical and semantic properties can be obtained, as suggested above, from entities and relations of a syntactic theory that suffices for the rest of the language. The serious problems arise in accounting for all the environments in which these proposed adverbs on *say* occur and do not occur, and above all in tracing how these derivations function together in the kinds of complex sentences that many languages have.

ordering-conjunctions: e.g. using the past for 'contrary to fact', as in *If he were here I would know it*; or using the future for reducing the speaker's responsibility for what he is saying, as in *I think what you want will be in the fourth aisle*. These considerations are surveyed in Z. Harris, *Notes du cours de syntaxe*, ed. M. Gross, Editions du Seuil, Paris, 1976, p. 158.

6

On the Mathematics of Language

6.0. *Why a mathematical approach?*

As was noted in 1.2, there are reasons to expect language to have a mathematical structure. One is that if new words entering language are to fit into the existing system of word combinations, the existing words have to be characterized by their combining-relation to each other rather than by any individual properties: the new word may have a kind of meaning or of phonemic composition different from the existing words, but it will have to fit into one or another of the kinds of word combination which make sentences. Another reason is that in the absence of an external meta-language (2.1), the elements of language can only be identified by their departures from equiprobability, which is the kind of property that can be studied mathematically (differently from meaning or phonemic composition). A third reason is that the speakers of a language, who have limited (though generalizable) public experiences of recognizing speech entities and their combination, can understand an unbounded number of sentences (more than are ever actually said), since there is no structurally relevant specifiable upper bound to the length of a sentence: hence there must be some constructive or recursive characterization of sentences.[1]

Ultimately, the question is whether language can be defined as one or another set of entities (elements, or particular relations among elements) closed under stated relations or operations. As noted in 1.2, this does not

---

[1] The expectation of useful mathematical description of the data of language stems from developments in logic and the foundations of mathematics during the first half of the twentieth century. One main source was the growth of syntactic methods to analyze the structure of formulas, as in Skolem normal form and Löwenheim's theorem, and in the Polish School of logic (as in the treatment of sentential calculus in J. Łukasiewicz, and the categorial grammar of S. Leśniewski and later K. Ajdukiewicz, cf. ch. 1 n. 6), and in W. V. O. Quine's *Mathematical Logic* (Norton, New York) of 1940. Another source is in the post-Cantor-paradoxes constructivist views of L. E. J. Brouwer and the Intuitionist mathematicians, and in the specific constructivist techniques of Emil Post and Kurt Gödel, in recursive function theory, and from a somewhat different direction in the Turing machine and automata theory. Cf. S. C. Kleene, *Introduction to Metamathematics*, van Nostrand, New York, 1952, and Alonzo Church, *Introduction to Mathematical Logic*, Princeton Univ. Press, Princeton, 1956. In linguistics, the 'distributional' (combinatorial) methods of Edward Sapir and Leonard Bloomfield were hospitable to this approach. Cf. also Nelson Goodman, *The Structure of Appearance*, Bobbs Merrill, New York, 1966.

mean the set of phoneme-sequences or word-sequences under concatenation, for language is only an unstated subset of the set of all the phoneme-sequences or of all word-sequences. Rather, the search has to be conducted on the basis of whatever characterizes the particular combinations that occur in language. It has been found here that this can best be done not by listing the combinations themselves (1.0) but by stating constraints on combination, and that one can define new less-constrained higher-order elements either as classes of more-constrained alternant elements or else as sequences of more-constrained elements. As one does this, the listing of elements is increasingly replaced by stating relations among (the newer) elements. All this brings the description to where mathematical formulation is more readily usable: a hierarchy of (newer) elements with simpler and more regular relations both among the higher classificatory elements and among the lower elements which make up each class. It was found that almost everything in grammar could be characterized in terms of three such relations. Two of these are complex: the different phoneme-sequences which constitute the same words (by morphophonemic variants or by syntactic reduction); and the particular probabilities for those word co-occurrences whose probability is higher than zero. The third is simpler: the partial ordering of words whose co-occurrence probability in sentences is higher than zero, i.e. whose combinations produce sentences. It is here that a mathematical formulation was most immediately indicated.

Closer scrutiny of this partial ordering of words in a sentence showed that it did not start with some particular list of words, but was based entirely upon a single relation among words in respect to their co-occurrence.[2] The ability of words to enter into all sentence-making combinations depends on nothing except this relation among them, so that the set of combinations (sentences) constitutes a mathematical object. The importance of this relation is that not only does it create the set of possible sentences, but also that all further grammatical events and mathematical structures in language (6.3) are defined on resultants of this relation. When this linguistic structure is studied as a case of applied mathematics, we have here not an example of applying known formulas to calculations relevant to a given science, but an example of finding particular mathematical structures holding for particular aspects of the real world—again, holding precisely and not by over-generating and then whittling down.

More generally, Chapters 3 and 4 show that the whole structure of syntax is determined by a system of constraints each of which states a departure from equiprobability in word co-occurrence. Syntax and the information it expresses (or distinguishes) and engenders is thus characterized by a partially

---

[2] This is called the autonomy of syntax, i.e. that no phonetic or semantic property of words is considered in determining what constraints on word combination create sentences. The syntactic structure nevertheless conforms to meaning-differences, but the phonemic structure of words does not in general.

ordered set of departures from equiprobability, something which is usefully subject to mathematical investigation.

6.1 *Language properties having mathematical relevance*

There are various more or less directly observable universal properties of language which make the data amenable to particular mathematical treatment.

(a) **Discreteness**. In the analysis of language, it turns out that the only entities which have usable grammatical regularities are discrete. In particular, phonemes are discrete elements, and words and sentences are specifiable sequences of them. Although phonetically there may be a continuum of difference among sounds, what assigns particular ranges of sound to particular phonemes is a system of relative distinctions, which can be represented by discrete marks or names (7.1). And although the successive sounds of an utterance are in general a continuously changing sound, the sequence of phonemes which represent them is identified by successive cuts in the continuous utterance sound.[3]

Measurement of length does not have to be considered a property of phonemes. The sound between successive cuts may be of different lengths, and different phonemes may differ in length, but these lengths do not distinguish or characterize the phonemes. In some languages some phonemes have distinct long and short sound segments: the long ones can be considered double-length occurrences of the phoneme, or simply two occurrences of the phoneme. There are also certain long phonemic intonations which extend over a sequence of words, e.g. in the interrogative type of intonation and the imperative type; but each of these can be represented by a single discrete entity, since the grammar does not distinguish continuously differing grades of each of these intonations. Those linguistic phenomena which are not representable by discrete elements may indeed carry information (e.g. intonations of hesitation, or of exaggerated matter-of-factness), but they do not enter in any regular way into grammatical relations to other elements (and are generally restricted to face-to-face situations, being not precisely repeatable), and are thus not included in grammar.

(b) **Pre-setting**. The phonemic distinctions of a language, and the sequences of phonemes that constitute words, have to be learned by users of the language, and are hence pre-set in advance of use. When a person hears a phoneme or a word, he need only hear enough to distinguish it from all others that could occur in the given environment, and to know that it is the 'same' as a repetition of it.

---

[3] Z. Harris, *[Methods in] Structural Linguistics*, Univ. of Chicago Press, Chicago, 1951, pp. 26–33, 60–4, 70–1.

(c) **Repetition**. If the hearer thereafter transmits what he has heard to someone else, he pronounces his own rendition of the phonemes. It is this that makes the transmission of utterances a repetition, somewhat as in the reforming of pulses at gates in a computer. In contrast, if the hearer tries to transmit to a third person a sound which has not been pre-set to him, he can only approximate it in an imitation. [4]

(d) **Error compounding**. The discreteness and pre-setting of the elements, with the result that transmission is by repetition rather than imitation, reduces by far the danger of error compounding in transmission. If hearers transmit continuous data which they have received (e.g. intonations of anger, or gestures), especially if those are not categorized in respect to pre-set elements, there is a considerable likelihood of errors being compounded by succesive imprecisions in one direction, in imitating the signals at successive transmissions. In contrast, when the signal is characterized by discrete and pre-set elements, it can tolerate a considerable amount of noise in transmission from one hearer to the next without producing error.

(e) **Arbitrary signals**. The fact that the grammatical elements must be pre-set makes possible what has been called the arbitrary character of linguistic signals, i.e. the fact that the sounds out of which words are composed do not in general indicate the meanings of the words, and differ arbitrarily as among languages. A relation between word sounds and word meanings would be useful primarily if it eliminated the need for language users to learn a fixed and jointly accepted (pre-set) stock of elements. But the fixed stock, with phonemic distinctions, is precisely what is required for transmissibility without error compounding.

Also, a regular relation between sounds and meanings would limit the possible word meanings to what could be obtained out of short sequences of 20 or so phoneme meanings (if such meanings existed), something which would be grievously inadequate for language. Furthermore, if the linguistic signals were not arbitrary, but inherently related to their sound composition or their meanings, they could not in many respects be identically pre-set for all participants. For if the meanings of sound-sequences were not learned, but left for each speaker and hearer to judge on the basis of his individual experience, the speaker and hearer would not necessarily be in agreement, since experience varies in detail for different individuals.

Instead of going by the phonemes individually, the hearer hears certain phoneme-sequences (or broken sequences, or replacements) as words, and then judges from non-language context, and also from the range of combinations in which each word or morpheme appears, what can be the meaning-range associated with each word or morpheme. The discrete and pre-set character of linguistic signals therefore leaves them arbitrary, and indeed

---

[4] For the distinction between such repetition and imitation, cf. Kurt Goldstein, *Language and Language Disturbances*, Grune and Stratton, New York, 1948, pp. 71, 103.

requires that there be no general correlation between sound and meaning. This does not preclude limited cases of onomatopoeia, i.e. of meanings in particular words being attached to particular sounds in those words.

The fact that the grammatical elements (characteristic sounds, and the collecting of these to form morphemes or words) are arbitrary symbols, which are related to meanings only by convention, makes the word-sequences available for study as mathematical sets; for anything which we establish about language as composed of these sequences of elements will hold for any physically different set of elements (e.g. letters) onto which the original set can be mapped.

(f) **Linearity**. Speech, and following it writing, is in a rough way linearly ordered, although in some languages certain phonemes may be pronounced simultaneously (e.g. vowels and nasalizations); and intonations with phonemic or morphemic status are in general simultaneous with words (e.g. contrastive stress) or with sentences (e.g. the interrogative intonation). Certain elements or sequences are nested within others, but even then we can consider the initial segments of these to be linearly ordered (e.g. the component sentences *So he claimed* and *It was not loaded* in respect to *He believed that . . .* , in *He believed—so he claimed—that it was not loaded*).

The linearity of physical elements is not to be confused with linear orderings in mathematically defined sets for the description of language. In the latter case, we are describing language in terms of a set which is closed under some operation; and we ask whether the set is linearly ordered. This turns out to be not quite the case, although we can come close to this result. In the former case, of the physical elements, we can only say that language events are a proper part of the set A of linear orderings of words (or morphemes, or phonemes). Going beyond the phonemic composition of words: as has been noted, the set A that is closed under the word-concatenation operation is not natural language, and we need another relation between words to characterize what subset of A a natural language is. But it remains true that in each language event the phonemes, words, and sentences are linearly ordered (at least in respect to their initial segments), or can be readily mapped homomorphically onto a linear ordering. In addition, it is seen in 9.3 that repetitive structures within the linear ordering can create a double array.

(g) **Sub-sequences**. The sequences of phonemes and of words that make utterances are segmentable into sub-sequences: the phoneme-sequence of a word into syllables, the word-sequence in a sentence into (transformed) component sentences (e.g. clauses). Given a large set of sequences, it is found that each element—phoneme or word—has a particular frequency of occurring with other elements, not in terms of fixed transitional probabilities but in more complicated dependences within a sub-sequence which can be captured in stochastic processes and in local and global partial orderings on the sequence.

(h) **Contiguity**. It was seen in 2.5.4 that all relations and operations in language are on contiguous entities. This is unavoidable, since language has no prior defined space in which its phoneme- or word-occurrences could be located. This condition of contiguity does not preclude successive segments of an operator from being placed next to different (but already present) parts of its operand (as in the case of discontinuous morphemes, 2.5.4), or between parts of its operand as in the nested sentence example of (f) above.

(i) **Finite length**. There is nothing to indicate that utterances, i.e. occurrences of speaking or writing, must in principle be finite in length. However, the sentences, whether these are defined structurally by some relation among segments of the utterance, or stochastically by sequential regularities of the segments (7.7), are necessarily finite in length; for until a segment of an utterance is finished we cannot be sure that it has satisfied throughout its length the conditions for being a sentence. In actuality, sentences are reasonably short except for *tours de force* of various kinds.

(j) **Number of sentences**. For a given number of phonemes, and also for a given number of words, the set of sequences is recursively denumerable. But for the subset of these sequences that constitutes a language, the question as to their number is more complex, and depends on how we characterize the criteria for their being in the subset.[5] In any case, the set of possible sentences as fixed in the grammar is not known to be finite, since to every sentence one can in principle add without limit some further clause.

(k) **Finitary grammar**. Aside from the non-finiteness and recursivity of language as a defined object, one should also consider the finiteness of the speakers and hearers. The phonemes and vocabulary, and the ways of combining words into sentences, all have to be used by the speakers of a language in a way that enables them to understand each other. The users, and their experiences with the language, are all finite. Hence it must be possible to characterize a language by a finitary grammar: a finite stock of elements and also of rules (constraints) of combination, with at least some of the rules being recursively applicable in making a sentence.

We now consider the number of event types (as against tokens) in a language, i.e. of occurrences which are linguistically distinguishable from each other. The number of items that have to be described is affected by the fact that languages change. In particular, new words and new combinations enter the language, and some go out of use. Over a very long time, with successive generations of speakers, the vocabulary of the changing language may be large to an extent that we cannot judge. However, there are no doubt limits—barely investigated so far—on the growth of the vocabulary, and on the rate of change in the structure of language. The number of elements and

---

[5] Whether a subset of a denumerably infinite set is itself known to be denumerable depends on the availability of methods of enumerating the members of the subset. This is a crucial difficulty in characterizing language.

of rules of combination available at any one time is not very large. It will be seen in 12.4.5 that in respect to an appropriate theory of language one can in principle count everything that a user has to know in using a language.

(l) **Computability**. It follows, from the discreteness, linearity, contiguity, and finitary grammar, that the structure of sentences is computable. That is, finite algorithms can be created that can synthesize, and much more importantly analyze, any sentence of the language. Although constructing such a language, especially for analysis (recognition), is a daunting task, it has not proved impracticable (ch. 4 n. 1).

6.2. *Increasing the amenability to mathematical description*

It was seen in 6.1 that many properties of language data are amenable, with various exceptions, to description in mathematical terms. This makes it desirable to regularize the properties not only for efficiency and elegance of theory, but also in order to reach a description of language in terms of such elements and relations as permit interesting decompositions, mappings, and constructions that are closed over relevant domains. The criterion is that any mathematical structure which is posited should apply in a specified way to observables in the data and to their stated relations, and not to abstract concepts whose precise relation to the regularities of the data has yet to be established.

Such regularization means finding effective ways to treat each 'exception' as a non *ad hoc* special case of the regular relation. This may require a modified formulation of the relation. For example, not all words need to be pre-set, i.e. known to the hearer, in order to be understood: a sentence can contain an unknown word whose boundaries are given by its neighbors and whose meaning can be guessed from the rest of the sentence. Or the regularization may be accomplished by finding special conditions that together create the apparent exception. Thus in linearization of phonemes, the nasalized vowel in the French *don* ('gift')—overtly a single sound—is linearized as vowel plus *n* (in the absence of following vowel), in order to agree with the vowel plus *n* sequence when the word appears with following vowel, as in *donner* ('to give'). Also, in defining words (or morphemes) as being each a sequence of contiguous phonemes, allowance has to be made for cases where the phonemes are not contiguous. An extreme case is in what is called 'agreement' ('government', 'concord') in grammar: e.g. plural-agreement in *The stars rise* as against *The star rises*. Instead of defining here an *ad hoc* grammatical relation between the verb and its subject, additional to the partial-ordering relation which creates sentences, we can say that in English the plural morpheme (and also *and*) on the subject of a sentence has a discontinuous segment located on the verb (consisting of dropping the -*s* on it): *The stars rise*, *The sun and moon rise* (G37, 169, 263). Occurrences of the plural morpheme when not in subject position, or with other tenses, do not have this detached segment (*The star emits photons*, *The star rose*, *The stars rose*). This then becomes a correction on the contiguity of phonemes in a morpheme (the plural morpheme), but avoids any special subject–verb relation.

In addition to cases in which exceptions can be treated as special or marginal cases, there are some properties of language data which present essential difficulties for mathematical treatment. Even after the welter of sentences is made more manageable by recognizing the different likelihoods of word-choices in operator–argument relations and the existence of reduction in sentences, we find that the co-occurrence likelihoods differ for each word in a way that limits classifiability, and we find that the criteria for reducibility are in many cases so complex as to preclude predictability of reduction. As a result, the set of sentences is not well defined: there are a great many sentences which are marginal either in that speakers disagree as to whether they are sentences (e.g. *A man who I met called me*) or in that speakers agree that they are borderline (e.g. *The baby took a crawl to the chair*).

All of these difficulties involve encapsulated, although in some cases major, aspects of language (e.g. likelihoods and reductions), or special cases of general properties. It is therefore possible to deal with them within a precise and regular structure, and this not by *ad hoc* deviations but by arguably justifiable additions to the structural relations or to their domains. A major step is to try to reach independence of the elements. This includes separating out, in a given occurrence, features that appear elsewhere in different combinations. For example, consider pairs of words which have the same selection except in respect to time words. The fact that the difference in some neighboring time-adverbs between *is* and *was* is the same as between *walks* and *walked* (and so in many other verb-pairs) shows that *is* and *was* contain one element in common (as most of the other verb-pairs do), and that *was* contains a variant of the -*ed* (past-tense) morpheme seen in the verb-pairs. And comparison of *He is*, *I am* with *He walks*, *I walk*, etc., shows that *am* is a post-*I* obligatory variant of *is*. Such analyses lead to systemic regularization only if they are based on full consideration of the 'behavior'—i.e. the relative occurrence—of the entities. Thus given the co-occurrence of plural markers on both subject and verb (above), we say that the morpheme occurs on the first argument (the subject) and has a detached segment on the verb, rather than the other way around, because its occurrence on arguments can be independent of its appearance on the verb (e.g. when the argument is second rather than first, as for *photon* in *The star emits photons*), whereas its appearance on the verb is dependent on its occurrence on the subject.

In general, then, expressing the properties of entities by relations among them is a precondition for mathematical description, e.g. in seeking a least grammar with most regular domains for the operations of the system. The regularity of domains is a matter not only of avoiding small *ad hoc* domains as in the examples of 1.1.1–2 but also of large-scale regularizations. One such is the avoidance of any distinction between operators on a sentence and operators on a verb, which requires evidence that what seemed to be operators on a verb are derivable from operators on a sentence (with zeroed arguments in the lower sentence). Another large regularization of domains is achieved by separating (a) the permanent partition of words into argument-requirement classes (N, O_n, O_{no}, O_o, etc.) from (b) the changeable graded likelihoods of operator–argument selection within (a). Isolating (a), which constitutes the mathematical object underlying language, from (b) leads to many advantages in structure-description and in meaning-correlation. Of course, all the regularizations, great and small, have to be explicitly justified in terms of the linguistic data.

A more fundamental step toward mathematical treatment of language is replacing the observable combinations of entities by the constraints that limit the combinations to some generalization of those that are observed, and by the transformations (including reductions) that replace certain combinations by others. These replacements are important because, as noted, mathematical combinatorics on the letters and words as directly observed is inapplicable to language, since the latter essentially falls short of all combinations.

These preparings of language for mathematical treatment have the additional effect of making the structure of language more similar to that of mathematics and logic. Thus, the *not, and, or,* and in a different way the quantifiers, which are special here in that they operate both in language and in mathematics–logic, are seen here to operate on sentences (rather than on words) in a manner closer to mathematics and logic than appears on the surface (G244, 321, 387). And the imperative and the interrogative, which are absent from the syntax of mathematics and logic, are found to be derivable in language from assertions (G331, 351). Furthermore, the structure of the metalanguage (as a science sublanguage) is closer to the syntax of mathematics than is the structure of natural language as a whole (10.2).

6.3. *Mathematical structures in language*

The properties of language surveyed in 6.1, and the processings mentioned in 6.2 and in Chapter 2, make it possible to state various mathematically definable operations and various sets closed under mathematically definable relations, which together characterize almost all the grammatically and semantically important universal structures of language. Beyond this, individual languages and language families have special structures which are less amenable to mathematical characterization: e.g. syllable types, morphological systems, morphophonemic systems such as conjugations and declensions, genders, verb 'aspect' (of duration), etc. The major operations and sets will be noted here; brief descriptions of the most important structures are given in 7.3, 7.6–8, 8.1.5–6, 8.3, 9.3.

(a) First, we consider those operations which do not involve the syntactic analysis presented in Chapter 3. Phonemes are a partition of sounds (after minor adjustments for independence of sound-occurrences) based on a set of distinctions which is constant over the population even if individual pronunciations differ (7.1–2). There are then stochastic procedures on utterances for finding word (and morpheme) boundaries as sub-sequences of phonemes (7.3), and sentence boundaries as sub-sequences of words (7.7). These arise because there are far fewer distinct words than possible short sequences of phonemes, and far fewer n-word sentences than possible sequences of n words. It is possible to define a hierarchy of rewrite rules in such a way that an elementary sentence ((g) below) is found inside each given sentence by a system of recursive equations that expand the parts of the elementary sentence into corresponding parts of the given sentence (7.8). And it is possible to define a recursive expansion of the elementary sentence (as a single entity) into the given sentence, by defining the structure of a type of 'center' word-sequence and of a number of types of 'adjunct' word-sequences, such that the adjoining of any sequence, as indicated by its definition, to any other sequence A creates a new sequence of the same type as A (7.6). To the extent that each word can be marked by what are its possible neighbors in the segments of a sentence, it is possible to use a cycling free group cancellation to test whether a word-sequence is a sentence of the language (7.8).

(b) Second, the set of word-sequences under dependence on dependence constitutes a mathematical object. We start with the fact that not all combinations of phonemes make words or sentences, as follows from the effectiveness of the stochastic procedures above. We now consider for each given word the set of words which have frequency zero of occurring with it, in a regular way, in any sentence (or utterance) of the language, including the shortest, in any corpus of utterances no matter how large.[6] The relation of the given word to every word in the complement to the above set, i.e. of every word that has positive frequency no matter how small of occurring with it in a regular way, is called dependence (of operator word on argument words). For mathematical interest, the crucial step is that the dependence is not on a fixed list of neighboring words but on the dependence properties of the neighboring words in turn (3.1): *eat* can be the ('first-level') operator on

---

[6] 'In a regular way' here means in the same relative position in all sentences, including the shortest, after correcting for any reductions (3.3) and linearization differences (3.4.2–4).

(and only on) any pair of words which in themselves do not depend on any words (equivalently, that depend circularly only on the first-level operators), namely two 'zero-level arguments' N; 'second-level' operators (e.g. *because*) act on (and only on) any pair of words that in turn depend on other words. Every word which depends on any stated dependence properties can form a sentence, by combining with words whose dependence properties satisfy it. Every sentence is a sequence which satisfies the dependence-on-dependence of every word in it; and every such sequence is grammatically a possible sentence, even if nonsensical (e.g. *Vacuum chides infants*) or artificially complex (e.g. the proposed restructured source for tense, as in *I state his leaving; his leaving is before my stating it* as source for *He left*).

As noted previously, the dependence relation creates the meaning 'predication' (11.2), or carries that interpretation, which holds for every system having this dependence, such as mathematics, even if the system lacks the other features of language (e.g. likelihood differences). The predicational meaning does not require any of the other features of language, nor does it require the particular physical content (sound, ink marks) or meanings of the words involved. The dependence on dependence is defined on arbitrary objects, not necessarily words at all; and the abstract system it creates is thus a system of categories.

(c) This mathematical object is not the whole of language, but it underlies language in that the other sets and relations which are essential to language are defined on its products. To obtain the structure of natural language we need two additional relations: the differences in frequencies of co-occurrence, and reductions in phonemic content.

The distinction between zero and non-zero frequency of combination, in terms of which the dependence was defined, leaves open the question of randomness, or stable or variable differences, in frequencies of each word relative to the various words on whose set it is dependent. What gives language its unique informational capacity is that there are indeed such differences and that they are roughly stable, although marginally changeable over time; and they correlate with—and to some extent create—the individual meanings of each word involved (3.2). These stable differences can be expressed as inequalities, or even a grading of likelihood (or probability), of each operator in respect to its possible arguments (and vice versa).

The reductions are constraints whereby word-occurrences in a given sentence which have highest likelihood (or otherwise no informational contribution) in respect to their argument or operator there can be reduced in phonemic content (3.3). The reductions are relations between sentences because every occurrence of an operator on an argument constitutes a sentence, and the reduction changes that into a physically (phonemically) different sentence. There may also be some stated other grammatical transformations of sentences, not based on these considerations. And there may be morphophonemic variants of particular words, by which the word has different phonemic content in different operator–argument or linear environments (7.3). In meaning, all of these changes are paraphrastic; they leave the substantive information of the sentence unaltered.

(d) The dependence-on-dependence relation imposes a partial ordering (of just three levels, in most cases) on the words of a language in respect to their co-occurrences in the set of all sentences; some (second-level) words depend on at least one (first-level or second-level) word that depends on something not null, with other (first-level) words depending on one or more (zero-level) words that depend only on null (equivalently, only on the first-level words). This ordering induces an equivalence relation on the words of the language, which partitions them into second-level, first-level, and zero-level. The dependence also imposes on the word-occurrences in each sentence a tree structure, specifically a join (union or cup) semilattice:[7] an ordering in which for any two word-occurrences there is a word-occurrence in the sentence which is their least upper bound (l.u.b., union). For all a, b, the greatest lower bound of a, b, if defined, is either a or b. This means that no one word-occurrence can be an argument of two different operator-occurrences.[8] Thus *John phoned and came* is derived from *John phoned and John came* (*John* in the first example is not simultaneously the argument of both *phoned* and *came*); *I know John came* is not from *I know John* and *John came* but from *I know* on *John came*; *I asked John to come* is from *I asked John for John to come* (with zeroing of *for John*); *The man I wrote answered* is from *The man whom I wrote answered* from *The man answered; I wrote the man*.

In the semilattice, for word-occurrences a, b, c, with $b > a$, in the sentence, b covers a (i.e. is the immediate operator on a) if there is no c ($c \neq a$, $c \neq b$) such that $b > c > a$; we call a the argument of b. Two or more word-occurrences which are covered by the same word-occurrence as operator are co-arguments (of each other) under that operator. Each set of co-arguments is linearly ordered; and this ordering is semantically relevant, distinguishing the subject and object of an operator. The word semilattices are the semilattices of the base sentences.

Equivalently to the semilattice, the word-occurrences of a sentence constitute a set with one operation: to each a, b in the set there is associated an element $a \lor b$ (corresponding to the l.u.b. of a, b). The operation is

[7] The common uses of the term 'tree' include cases that are not applicable to language. For example, in genealogical trees, the relation of two cousins to their six instead of eight grandparents does not occur in the dependence relations among words of a sentence: as noted below, a word-occurrence cannot be the argument of two different other word-occurrences.
[8] The sameness statement (5.3.3) presents a problem here, because it refers to two words which are already the argument of other operators. The problem is avoided by a solution proposed by Danuta Hiż, which excepts the metalinguistic statements from the lattice of the sentence (8.3).

idempotent: $a \vee a = a$; associative: for all c, $(a \vee b) \vee c = a \vee (b \vee c)$; and commutative: $a \vee b = b \vee a$. This last means that the semilattice is unchanged by different linearizations of the base partial order, in particular by permutation (3.4.3).

We now consider the reduced sentences which are derived from the base by reducing particular words in the operator–argument structure that makes those sentences. Here a related, 'word-and-reduction', semilattice of a sentence is formed by ordering the reduction instructions as points in respect to the word semilattice.[9] This ordering is available because the description of reductions holds that they are made as soon as the conditions for them are satisfied in the constructing of the sentence. For example, in *I like teaching*, from *I like my teaching something*, the zeroing of *my* is carried out right after *like* operated on the pair *I*, *teach*: this, since *teach* brought with it the pair *I* (*my*), *something* (the latter having been zeroed as being an indefinite), and *like* permits the zeroing of the subject of the operator under it if it is the same as its own subject. This semilattice of a sentence (e.g. in 8.2.2, 8.3) is a precise description of it as an ordering of the entry of words and reductions into the making of the sentence—and of each component sentence of it (as sub-semilattices).

(e) The denumerability of the sequences of phonemes of a language, and of the words in a language, cannot be applied to the particular subset of them that constitutes the utterances of that language. However, given the partial ordering in sentences, one can say which parts are iterable without specifiable limit, thus creating for the theoretical structure the seemingly unbounded set of possible sentences out of a finite vocabulary.

(f) In the word-sequence of a sentence, concatenation is semantically non-associative and ambiguous: *The yellow and green cards* can be derived both from *The cards which are yellow and green* and *The cards which are yellow and the cards which are green*. In contrast, the successive entries of operators, which (together with reductions) describe the same sentences as concatenation would, are associative and non-ambiguous. Mappings and operations on sets of sentences can therefore be more conveniently carried out on the entries, and in particular on the operators, in these sentences, than on the overt word-sequence of the sentences.

We consider first the unary second-level operators, i.e. those whose argument-requirement dependence includes precisely one operator (with possible zero-level arguments too). The set of these is a semigroup with successive application (i.e. next later entry in the making of the sentence) as

[9] There may be one or more simultaneous (independent) reductions located at a particular point, i.e. immediately above a particular point in the word semilattice. In case more than one reduction is located above the highest operator (the maximum, or universal, point) of the word semilattice, we obtain correspondingly more than one maximal point. It is therefore necessary to define an artificial new universal point, the sentence intonation of that sentence, which provides the l.u.b. for the more than one maximal reduction points if such exist.

operation; one can consider the null operator as identity. The object here is not the operator itself, but the application of it to its argument. For present purposes, given A applied to (operating on) an argument X, and then B applied to A (with its argument), we consider B applied to A without argument as a complex operator BA applied to the argument X. A product of two operator-successions in the semilattice is itself an operator-succession; the multiplication is associative.

The binary second-level operators, i.e. those whose arguments include precisely two operators, form a set of binary compositions on the set of sentences. In the base set, each binary second-level operator can act on every pair of sentences, although its likelihood of occurrence is higher on sentence-pairs which contain in their base form a word in common. In the whole set of sentences, certain reduced forms (e.g. question) do not appear under certain binary operators: not in *I am late because will you go?*; this may be resolved by defining here some required transformations which would replace these sentences by their non-prohibited source, in this case *I am late because I am asking you if you will go*. Products of these binaries are in general not associative in their semantic effect, and correspondingly in the history of making the resultant sentence. Here the structure of the base semilattice, which is contained in the word-and-reduction semilattice of the resultant sentence, is crucial.

(g) The most important algebraic structures in the set S of sentences are those which arise from equivalence relations in S in respect to the particular base (operator–argument) semilattice in each sentence, and in respect to the highest operator (the upper bound of all words in the semilattice), or to the word-and-reduction semilattice. These equivalence relations identify the informational sublanguage (the base) and the grammatical transformations, as will be seen below.

We note first that the resultant of every operator is a sentence. Every unary second-level operator acts on a sentence (and possibly also an N) to make a further sentence; and every binary second-level operator acts on two sentences (and possibly an N) to make a sentence. Every reduction acts on a sentence to make a (changed) sentence. All of these, in acting on a sentence, preserve the inequalities of operator–argument likelihoods in the operand sentences (except in statable circumstances). The unary second-level operators are a set of transformations on the set of sentences: each maps the whole set S of sentences into itself (specifically, onto a subset of sentences which have that second-level operator as their latest entry, i.e. their universal point); and the binaries map $S \times S$ into S. The reductions and other transformations are a set of partial transformations on S, each mapping a subset of S (sentences whose last entries are a particular low-information operator–argument pair) onto another subset of S (sentences containing the reduction on a member of that pair).

The preservation of inequalities of likelihood under transformations of the set of sentences, i.e. under the second-level operators and under the reductions, is of great importance. Without it, there would be no semantic connection between a sentence and its occurrences under further operators or reductions. The second-level operators and transformations preserve the likelihood inequalities and the meanings in their operand sentences, although with a reasonable number of specified exceptions. The second-level operators also add their own meanings and likelihoods in respect to their argument. However, some metasentential operators such as *deny* (the source of *not*) and *ask* can change the likelihoods of operator–argument pairs under them; hence the inequalities in the resultant sentences (with their new higher operator) need not be the same as in the corresponding operand sentences. As to the reductions, these preserve with few exceptions the inequalities of likelihood and the meaning of their operand sentences. In some cases, a reduction may raise (or lower) likelihoods of occurrence in the resultant sentences, but for the most part equally on all its operand sentences, or only stylistically and idiomatically.

We consider the set S', where each word-sequence which is grammatically ambiguous in n different ways is considered to be a case of n different sentences, each with its own base semilattice. S' is a semigroup under the binary operator *and*: for any two sentences A, B we have A *and* B as a new sentence C. (This, after adjustments are made for A, B pairs which do not take *and*.)

We present now a structure which isolates the minimal subset of the set of sentences as a residue of the second-level operators and reductional transformations. It has little importance when the great bulk of transformations are products of such simple reductions as have been established for English, and when the minimal (elementary) sentences can be characterized, as in Chapter 3, as the resultants of first-level operators on zero-level arguments. However, in a language in which we do not have so clear a picture of the structure of the set of transformations or of the set of base sentences, a way of identifying the elementary sentences is useful; and it indeed was useful in the early stages of the development of the transformational theory (8.1). To obtain this structure, we form the set S'' of sentences, which consists of all the sentences of S' that contain no O_{oo} (including *and*, *or*). We then take an equivalence relation in S'', whereby two sentences are in the same equivalence class if they exhibit the presence of the same succession of unary operators and the same orderings of particular reductions. In the natural mapping of S' onto its quotient set E of these equivalence classes, the kernel of the mapping, i.e. all the sentences which are mapped onto the identity of E, contains only the elementary sentences.

A different and more important structure is obtained if in the set S' we take an equivalence relation by which two sentences are in the same equivalence class if they have the same ordered word-entries (i.e. the same operator–argument semilattice). Since almost all reductions are optional, each equivalence class contains (with possibly certain adjustments) precisely one reduction-less sentence. The set of these is the base set (which includes the elementary set), from which the other sentences are derived by reductions. The base set is closed under the word-entry operation: any word-sequence satisfying this form is such a sentence, and the composition of two sentences of the base under *and* is again a sentence of the base. Hence the base is a sublanguage (10.1).

(h) The sentences in the language or in the base cannot be listed: they are constructed as sequences of words in O_n, O_o, etc., or of proper subsets of these, the subset being due to extreme likelihood properties or to reductions. Each reduction is defined on a particular domain of words—all the words of O_n, or of O_o, etc., or else a proper part of one of these classes; it cannot be defined on more than a whole argument-requirement class. Hence if there are successive reductions on some word class in a given base structure, the domains of the successive reductions are monotonically decreasing. Hence the base set of unreduced sentences is the most unrestricted in S, i.e. is least limited in respect to particular subsets of words.

The distinction of dependence, likelihood, and reduction in the construction of sentences makes possible precise characterization of several other grammatical entities or relations. For example, one can obtain in an orderly way all the component sentences of a given sentence. One can present the unambiguous forms of each ambiguous sentence: either the base form, or the given form associated with distinguished reductions or component sentences, sufficient to separate one meaning of the ambiguous sentence from another. One can design a procedure for translating between two languages by decomposing a sentence (via its reductions) into operators and their arguments, thereafter translating these (a far simpler task than translating the original sentence), and then carrying out equivalent reductions, if desired, in the new language. Alternatively, one can create a common grammar of all that is identical or similar in the two languages, and use it as an intermediate translating or grammar-describing station.[10]

---

[10] Such constructions are presented in Z. Harris, *Mathematical Structures of Language*, Interscience Tracts in Pure and Applied Mathematics 21, Wiley-Interscience, New York, 1968.

III

Grammatical Analysis

7

Analysis of Sentences

7.0. *Classes and sequences*

The theory of syntax presented in the preceding chapters has words as its primitive elements. Since one might hesitate at a theory whose primitives are undefined or established only circularly, it is necessary to consider whether these words are determinable by objective and effective methods, independently of the syntactic theory. To do so we have to see how linguistic methods have segmented speech to obtain phonemes as classes of sound distinctions, words as sequences of phonemes, and sentences as sequences of words.

Traditional grammar was an informal description of the shape and meaning of words and sentences. In contrast, descriptive and structural linguistics recorded the shape of successively larger segments of sentences. The search for coherent and regular analyses of sentences in terms of their segments was satisfied by successive steps of classification and sequencing. At various stages, the sets of entities obtained were partitioned into equivalence classes. The set of equivalence classes then constituted a new set of 'higher' entities whose behavior in respect to the sentence-segments was far more regular than that of the earlier entities. Thus: the phonemically distinct sounds are collected (as complementary variants, or free variants) into phonemes (7.2); the word-size segments are collected as complementary or free variants into words ('free morphemes') or bound morphemes (7.4); and words are collected, by similarity of environments, into word classes (7.5). Particularly constrained sequences of the entities that have been obtained are defined as constituting new 'higher' entities, which have new regularities in respect to each other. These are phoneme-sequences constituting word-size segments (which have the status of words or bound morphemes in their sentences, 7.3), word-class sequences constituting either constituents or else strings (by two different styles of grammar, 7.6), and string sequences (or constituent sequences) constituting a sentence as a regular segment of utterances (7.8). In certain important cases the sequences can be characterized by stochastic (hereditary) processes that proceed through them, stating for each first n members of a sequence what are the possibilities for the $n+1^{th}$ position in it. Such methods can define certain kinds of sequences as constituting next-higher entities (words, sentences, 7.3, 7).

It will be seen that all the entities can be identified by objective repeatable procedures, and defined on the basis of the relations in which they participate, without appeal to other properties. It is of interest that all the major segmentations are recognizable in writing, in particular in alphabetic writing. Phonemes correspond roughly to letters of the alphabet. Words are marked by space, and bound morphemes can be considered word forms in which the bounding-space has been zeroed. Virtually all punctuation will be seen to mark the boundaries of sentences; the non-sentence segments marked by comma and dash will be seen in 8.1 to be reduced forms of sentences. But many sentences, reduced or not, which are contained in larger sentences are not bounded by punctuation: e.g. *small* in *A small stone fell*, *We sleep* in *He works while we sleep*.

For the methodological considerations of 2.3, it may be noted that at each stage the classifications allow for simpler statements of the sequencing constraints (on the domain of the equivalence classes) than would have applied to the domain of the entities themselves, before the partitions. And in the case of special problems, such as the exceptions and special forms so common in language, the peculiarities are moved down from the domain of the sequencing relations to the domain of the partitions where they can be treated much more easily.

7.1. *Phonemic distinctions*

The use of language takes place not only in speech but also in writing and thinking; but these are in different ways secondary to speech, and in particular writing is relatively recent. A search for the ultimate minimal elements of language, therefore, necessarily turns to the sounds and sound-features that play a role in speech. Phonemic analysis holds that in every language certain sets of sounds, called phonemes, are distinguished from others by users of the language, with each language having a fixed set of phonemes.[1] The existence of hearers' linguistic distinction of sets of sounds, and the actual list of sounds thus distinguished in a language, can be seen in the following simple test with users of the language.

We begin with two speakers of the language, and two utterances, prefer-ably short and preferably similar to each other, e.g. *heart* (an organ) and *hart* (a deer), or *heart* and *hearth* (a fireplace), or *economics* with initial *e* and with initial *iy*. We ask one speaker to pronounce randomly intermixed repetitions of the two words—the organ and the deer, or the organ and the fireplace—and ask the other, as hearer, to guess each time which word (or

---

[1] See chiefly Ferdinand de Saussure, *Cours de linguistique générale*, Paris, 1910; Edward Sapir, 'Sound Patterns in Language', *Language* 1 (1925) 37–51, reprinted in David G. Mandelbaum, *Selected Writings of Edward Sapir*, Univ. of California Press, 1958, pp. 33–45; Leonard Bloomfield, *Language*, Holt, New York, 1933.

pronunciation) was being said. It is found that the hearer's guesses are correct about 50 per cent of the time for some pairs of words, such as *heart–hart*, and about 100 per cent for other pairs such as *heart–hearth*, or for the two pronunciations of *economics*; and the results are much the same for any users of the language. In the cases where the hearer's guesses are correct close to 100 per cent of the time we say there is a phonemic distinction between the two words (*heart, hearth*; or the two pronunciations of *economics*); otherwise not. In some cases the results of this pair test are problematic, and the decision as to whether a phonemic distinction holds is adjusted on the grounds of later grammatical considerations. But the direct results of this pair test provide a starting-point, a first approximation to the set of ultimate elements adequate for characterization of language.

The ability to distinguish utterances in the pair test does not depend on any recognition or agreement by speaker and hearer in respect to the meaning of the utterances, or even just whether the meanings are the same or not[2]—the more so as a single utterance may have more than one meaning, and there are cases of homonyms (e.g. *heart* and *hart*), which are considered different words with different meanings, but are phonemically identical; also, the two pronunciations of a single word, *economics* (with initial *e* or else *iy* sounds), are phonemically different and are distinguished in the pair test. Rather, then, the distinguishability depends upon, and illustrates, the behavioral fact that when a person pronounces a word he is not imitating its sound but repeating it (ch. 3 n. 4), that is, saying his own version of the phonemic distinctions which he has heard. The role of repetition as against imitation will be discussed in Chapter 12, but the clustering of pair-test responses near 50 per cent and 100 per cent shows that utterance-recognition is based on repetition: an utterance either is or is not a repetition of another. The ability to recognize what is a repetition means that the phonemic distinctions of a language are pre-set by convention. They differ from language to language but can be acquired—in many cases perfectly—when a person learns a new language. Even a predisposition to recognize phonemic distinction in the abstract, or repetition as a relation between behaviors, need not be assumed as an inborn or a priori preparation for a person's being able to recognize phonemic distinctions when he meets them. It may be simply that various pronouncings are maintained at such distances or contrasts from other pronouncings that a person learns to maintain the distance or contrast even if his own pronunciation (or that of a new person he hears) is somewhat different.[3] This is an unusual relation in human behavior, so that the search for analogs to phonemic distinction in other human behavior has not been fruitful.

---

[2] The pair test will equally well distinguish *tez* from *taz*, where no meaning or meaning-difference is present.

[3] Cf. E. Sapir in n. 1 above. Phonemic distinctions are distinctions among timbres, i.e. as to which overtones have particularly high amplitudes.

We have yet to connect the phonetic distinctions between words with a phonetic segmentation of words. Given two words or very short utterances which have been shown to be phonemically different, we can locate the points at which they seem to differ; the difference can be measured instrumentally, although present knowledge does not yield an adequate phonetic characterization of phonemic differences. Thus between *write* and *ride* one may notice a difference (of length) in the vowel and a difference (of voicing) in the final consonant; in this case compactness of description favors taking the vowel difference to be dependent on the consonant difference, rather than the converse. It does not matter if the observations mislocate the relevant differences, or note too many or too few, because comparison with differences in other word-pairs will correct such errors. The phonemic difference can then be stated between two phonetic segments rather than between two utterances. If one of the paired words is phonemically different from yet other words at the same point, we can begin forming a distinction between its segment and all other segments that differ from it in the various distinct words, e.g. between the last segment of *heart* and those of *hard*, *harp*, etc.

If we compare the results from various pairings of a few short utterances, we can describe some utterances as being distinct from others at more than one point. Thus we can make a segmentation such that *hart* differs from *hearth* in the last segment, *hart* differs from *dart* in the first segment, *heart* differs from *hurt* and *dart* from *dirt* in the second segment, and *heart* differs from *dirt* in the first two segments, and *heart* differs from *dearth* in the first, second, and last segments (despite the spelling). We can then try to cover the length of each short utterance in some adequate sample by segments that distinguish it from other short utterances. For each utterance, the minimal number of distinctions necessary to distinguish it from every other utterance, in terms of pair tests, determines the number (and location) of segments which express these distinctions. These are the phonemic segments of the utterance.[4]

A phonemically distinct segment (or segment-type) is, then, a set of segment-occurrences (in one position of utterances) which are not distinguishable one from the other (by the pair test), but each of which is distinguishable from all segment-occurrences not in the set.

7.2. *Phonemes and letters*

7.2.0. *Phonemes*

Many phonemically distinct segments of utterances have complicated restrictions on how they combine with others in the utterance. Some of these

---

[4] Cf. Z. Harris, *[Methods in] Structural Linguistics*, Univ. of Chicago Press, Chicago, 1951.

constraints apply to several phonetically similar segments (e.g. to all voice-less stops such as p, t, k); others are unique to one segment type. The use of segmentation in identifying words—which is its major use in grammar—is made far more regular, and simpler to state, if we impose a partition on the set of segments whereby no two segments in one equivalence class (one part in the partition) are phonemically distinct from each other. Each equivalence class is called a phoneme. The main advantage in this step is to collect complementary phonemic segments, i.e. those which occur in different positions of utterances and with different neighboring segments, into one entity: for example, in English the aspirated stops whose phonetic value is *ph, th, kh* which occur in initial position of words (*pin, tin, kin*) and the unaspirated stops p, t, k, which occur only after s (*spin, stint, skin*) can be considered to be cases of the same morphemes respectively, p, t, k. Also, the two vowel segments in *write* and *ride* are complementary variants of the same phoneme (before voiceless and voiced consonants). Some segments have more than one segment complementary to them (but distinct from each other) in a particular position: for example, not only *ph* but also *th* and *kh* in initial position is complementary to p after s. In such cases the assignment as to which complementary segments are in the same phoneme is made on the basis of various considerations of simplicity, including having least phonemic change in the different occurrences of the same word ('morphophonemics'). For example, if we want least change in the phonemes of *connect* or of *position* when they appear after *dis* (in *disconnect, disposition*), we would take initial *kh* as 'complementary variant' of post-s occurrence of k, and similarly for *ph* in respect to p. For each occurrence of a phoneme, say p, we can always tell from its phonemic neighborhood what phonemic segment it represents, e.g. whether *ph* or unaspirated p.

The sounds in the segments are continuous, and vary during their course, and may dissolve in a continuous way into the next segment, and may have various lengths; but the phonemes have none of these properties. Since they are simply the minimal segment-distinctions necessary for utterance-distinctions they are discrete and have no length. For example, it is meaning-less to look between any arbitrary two phonemes for some other phoneme half-way between them (either in the difference scale, or in the sound-sequence of the word).

7.2.1. *The phonemic system*

The phonemes of a language fall into patterns of similarities and relations. There are phonetic similarities, such as among p, t, k and b, d, g, degrees of phonetic closeness to other phonemes, and degrees of audibility (e.g. the difference between unreleased p, t, k may not be very audible), particular positions in words and similarities in these respects (e.g. English h and the *ng* phoneme occur in limited—and complementary—positions). These inter-relations of phonemes help to decide to what phonemes to assign particular phonemically distinct phonetic segments when the complementary environ-ments do not suffice for unique assignments. (Note that phonemically distinct tones, stresses, and intonations are found by further analysis of distinctions among segments.) The final assignments are made in such a way as to obtain least constraint on combination for each phoneme, and maximum regularities and for the phonemic composition of words (7.3). Some phenomena remain marginal: e.g. the phonemic status of grammatical intonations such as those of a sentence (period, question mark, exclamation point), which extend over a phoneme-sequence but can be linearly located at the end of that sequence. In some cases it may be possible to replace the phonemes by elements which are combinatorially less restricted, at the cost of these latter elements having more degrees of freedom, such as being able to occur simultaneously or having multiple lengths. Aside from such solutions, which may not be worth the cost of a new dimension over a whole grammar, phonemes can be considered as having unit length (aside from the intonations) and as being linearly ordered in the composition of an utterance. In those cases where a single sound is assignable to the combination of two phonemes (e.g. assigning the nasal flap in Midwest American *winter* to the successive phonemes *nt* in that position) the phonemes remain linearly ordered.

We thus arrive for each language at a set of phonemes, as the most freely combining set of phonetic-segment distinctions sufficient to distinguish every utterance of the language from all utterances which are not recognized by speakers as repetitions of it. (The speakers, however, need not be aware of phonemic distinctions and phonemes.) The phonemes give a unique spelling for each such utterance (but cannot distinguish homonyms); and they do this in a precise and objective way that is virtually free of substantive differences of judgement. On this depends the discreteness and precision that can be obtained in the analysis of everything else in language, which is ultimately defined in terms of phonemes.

7.2.2. *The phonemic alphabet*

The basic status of phonemics in representing language may be seen when we consider its contribution to the origin of the alphabet, and to the success of the alphabet as a system of writing. The creation of the alphabet was somewhat of an accident in culture history. As far as is known, the various true writing systems which were developed in early civilizations consisted of marks—generally pictographs—for words (not directly for objects of the world). In some cases, a mark came to be used as a phonogram for any similar-sounding word, in addition to the one pictured by it ('rebus' marks). Some systems developed further, with marks that were originally pictographic coming to represent particular syllables (of the originally represented word); these marks were then used for segments of words, in whatever word those syllables occurred. The alphabet, however, was not a further stage, beyond syllable marks, toward which various writing systems developed. Its formation was a single historical set of events—a fact which needs explaining— and all alphabets are either later stages of the original one (in many cases preserving the order of the letters), or else are modelled upon an already existing alphabet.

The single creation of alphabetic marks apparently developed out of Semitic (Canaanite) speakers' contact with Egyptian hieroglyphic writing around the middle of the second millenium BC.[5] In Egyptian as in Semitic, each word began with a consonant; and in the various words which a single Egyptian hieroglyph pictured (all of which, in general, shared the same grammatical root) the consonants were generally constant while the vowels varied with the grammatical status of the word in its sentence. Furthermore, hieroglyphs could be used for other words or segments of words which sounded like those pictured. The pictographs for very short words (having one-consonant roots) came to be used, rebus-style, for the successive syllables of names, especially foreign names, which naturally had no picto-graphs of their own. As a result, the names, or rather their successive syllables, were written with a sequence of these one-syllable pictographs, each of which represented the sound (initial consonant plus vowel) of the pictured word. The root consonant was in general the same for all occurrences of a given hieroglyph (and for the given syllable of the name being written), while the vowel (belonging to different Egyptian case-endings of the pictured word) could be taken as whatever was the vowel at that point of the name which was being written. Early Semitic alphabetic inscriptions in Sinai and nearby suggest that Semitic speakers acquainted with this Egyptian rebus use of one-consonant hieroglyphs saw in it an acrophonic relation. That is, they apparently saw it as pictures being used to represent the first sound of the pictured word—when actually the pictures were representing (non-acrophonically) the full syllable sound of their word, of which the constant part was the first sound alone (the consonant) only in the case of uniconsonantal words. The earliest alphabetic inscriptions show crude acrophonic pictures—new marks used for the initial phonemes of the pictured (not necessarily uniconsonantal) Semitic (not Egyptian)

---

[5] Kurt Sethe, *Vom Bilde zum Buchstaben*, Untersuchungen zur Geschichte und Altertum-skunde Ägyptens 12, Leipzig, 1939; Johannes Friedrich, *Geschichte der Schrift*, Heidelberg, 1966; W. F. Albright, *The Proto-Sinaitic Inscriptions and their Decipherment*, Harvard Theological Studies 22, Cambridge, Mass., 1966. Acrophony, as a new relation between the picture or name of a symbol and the sound it represents (its phonetic value), was reached by a misunderstanding: the people who first used the method for making a set of writing marks apparently thought, erroneously, that that was the basis for the Egyptian hieroglyphs used for names. Acrophony can naturally be alphabetic and phonemic, because it recognizes the initial distinguishing (hence phonemic) sounds of the words of the language, whereas writing systems do not otherwise naturally reach single sound segments.

word—such as a bull-head (*'alpu*) used for the glottal stop ', a house (*betu*) for *b*, a camel (*gamlu*) for *g* (the greek *gamma*), etc.

That the original alphabet was only the application of a technique, rather than a full understanding of sound-segments, is seen in the fact that it did not develop marks for the vowels. The acrophonic reading of the pictographs yielded only the sounds that were word-initial, which were always consonants (in Semitic as in Egyptian). The marks were not thereupon extended to vowels. This was not because of the usual explanation, namely that 'vowels are unimportant in Semitic'—they are essential for distinguishing grammatical forms—but because the acrophonic method which was used to obtain the alphabet could not yield the vowels, since no Semitic or Egyptian word began with a vowel. Thus the acrophonic basis of the original alphabet explains not only the introduction of alphabetic marks but also their original vowellessness. The Greek borrowing of the alphabet, which retained its acrophonic property by keeping the Semitic letter-names (alpha, beta, etc.), yielded the vowels indirectly because Greek disregarded the Semitic laryngeal consonants, leaving their marks for the vowels which followed the unheard laryngeals. (Intentional additions of vowels—and additional consonants and even consonant clusters—took place later in certain extensions of the original alphabet, and at the end of the Greek alphabet.)

To summarize: the alphabet was not made by any phoneme-like decomposition of words into their successive sounds: such decomposition does not appear as a normal final stage of writing development. It was made by utilization of phonemic distinctions in word-initial position. The successive sounds of each word—to the extent that pictographs were available for them—were written with the pictograph for a word whose beginning was that sound; in the original alphabet, this supplied only the consonants of the word.[6] Pictographs became established to the extent that they had distinct beginnings under repetition; and the retention of their names (*alpha*, *beia*, etc.) was essential to their acrophonic use, which is why they were retained in Greek.[7] The vowels, in the words being written, were not written, because there was no picture whose word began with those sounds. Complementary variants of phonemes were not come upon, because they did not occur in the initial position. But as many marks were established as there were phonemically distinct word-initials in the language. The fact that successive segments of a word were marked by the beginnings of various other words was thus not a new recognition of phonetic segmentation: the Sinai writers were doubtless doing only what they thought the Egyptians

[6] We cannot tell whether the early users of the alphabet assumed that the required vowels for a given word could be drawn from the various vowel values of the Egyptian signs (which did not hold for the specific Semitic letter-names *'alpu*, *betu*, *gamlu*, etc.), or whether they simply wrote for a word as many relevant marks as were available to them, which were only those that represented word-initial sounds (consonants).

[7] Later borrowings of the alphabet lost the acrophonic names while keeping much of the original order: e.g. the Latin alphabet and the Arabic.

were doing. And the Egyptians continued with their non-alphabetic system, just writing some long words with the successive marks for short words whose sounds were similar to successive segments of the long words.

The alphabet was thus neither an invention nor a discovery, but a development whose product was novel because existing materials had unused potentialities: the apparent acrophony of marks for one-syllable words. This development has a certain distant similarity to aspects of the evolution of language (12.4.4).

7.3. *Morpheme and word boundaries*

Each utterance is a sequence of phonemes (7.2). However, not every sequence of phonemes of a language constitutes an utterance of the language, and it has been seen that a central problem of language analysis is to find a way of characterizing the sequences which are utterances as against those which are not (2.1). We cannot expect to find an effective procedure that will do this directly by stated constraints on phoneme combination alone. First, it is impossible to describe (or generate) all the phoneme-sequences of the language and only these in a non-hereditary way, for example as a Markov process. For instance, the entry of a new word into the language by borrowing, or by coining technical vocabulary, can create in its intra-sentence environments many new phoneme-sequences. And finite-state transitional probabilities can be disturbed arbitrarily when, for example, a word of limited use, with a rare initial phoneme, is used in a new environment: the phoneme *th* (of *this*) followed by *zzw* may be virtually absent from English, but it is immediately recognized as English if one chooses to say *He bathes Zouaves there*.

Secondly, a single hereditary process cannot be expected to predict directly all the phoneme-sequences in the language and no others. Evidence for this is, for example, that into almost any sufficiently long utterance it is possible to insert, at various points, arbitrarily many segments of utterances, repeated arbitrarily many times, and still have an utterance of the language as a result: e.g. insertion of *very very inadequate* and *in my opinion* into *A performance was given last night*, to yield *A very very inadequate per-formance—in my opinion—was given last night*. In contrast, the charac-terization of the phoneme-sequences constituting utterances can be done in two passes, if we first establish words and bound morphemes as 'word-size' sequences of phonemes, and then utterances as—in a complex way— sequences of words. The word-size segments are those phoneme sequences (or sets of alternant phoneme-sequences) whose combinations in utterances are regular.

The boundaries of morphemes within a word, and more sharply of words within a sentence, can be found by a particular stochastic process on phoneme-sequences in utterances. In a given sample of utterances, we consider all those whose first n phonemes are a particular phoneme-sequence $q(n)$. In the $n+1^{\text{th}}$ position of these utterances, after $q(n)$, the number of different phonemes which are found will be called $v(n)$. The $v(n)$ thus measures the variety of followers after all occurrences of a given n-phoneme sequence in the sample. It is found that in each m-phoneme utterance, if we record $v(n)$ successively for its initial n phonemes, $1 \leqslant n \leqslant m$, the value of $v(n)$ has repeatedly a gradual fall and sharp rise, as n goes from 1 to m. This sawtooth regularity is not found for arbitrary phoneme-sequences which do not constitute utterances of the language. The sharp rises will be seen to mark what we would want to consider (if we know the language) as the ends of morphemes and words. As an example, in Figure 7.1[8] we take the results for a short English sentence, written phonemically.

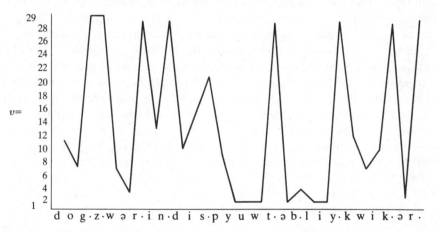

FIG. 7.1. Word boundaries in a sentence. *Dogs were indisputably quicker.* Dots have been inserted between the phonemes, to show where a morphemic segmentation would have been made on syntactic grounds. These dots were, of course, not involved in the test.

Such results have been obtained by this procedure, going both forward and backward, in various disparate languages. Similar results were also obtained for the more difficult case of morpheme boundaries within one word at a time, each initial sequence being matched for its next-neighbor variety, both forward and backward, in a sampling from Webster's unabridged English dictionary, together with various science dictionaries, alphabetized both forward and backwards. In this case, spelling was used unavoidably instead of phonemic sequence, but the results remained in good accord with grammatical segmentation, as can be seen from Figure 7.2.

[8] The figures, and various sections of the text, in 7.3–8 and 8.1 have been taken from Z. Harris, *Mathematical Structures of Language*, Interscience Tracts in Pure and Applied Mathematics 21, Wiley-Interscience, New York, 1968, where more details are given.

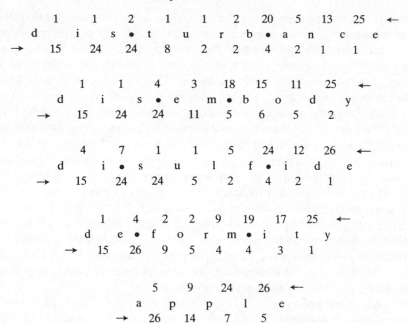

FIG. 7.2. Morpheme boundaries in words. An example of the dependencies in this recurrent stochastic process is given in the list above of individual words, each tested against all the words in English. The probabilities for each outcome, i.e. each possible sucessor, are of no interest here, and are not given. The process states only the number of different outcomes, at each state, which have any positive probability. As in the preceding figure, the dots indicate syntactic boundaries of morphemes, and were not used in the test.

This procedure reveals a difference between the phoneme-sequences that constitute utterances of a language and those that do not. Specifically, all utterances in a language have a constraint that increasingly restricts the possible phonemic successors (hence the gradual decrease of v) up to a point at which the restriction abruptly ceases (the v peaks). This peaking can be interpreted as a segmentation of the utterance into phoneme sub-sequences within each of which the increasing successor-restriction holds. It will be seen that these 'word-size' segments, which constitute words or morphemes, are associated with meanings in a way that phonemes or sequences of phonemes other than word-size are not, and that certain regularities of combination of these word-size segments characterize utterances. The word-size segments between peaks are the ones which have stable selections or other grammatical relations to following or preceding segments. (Note that the initial *sl-*, *gl-*, segments of English do not have such relations, and do not have a peak after their *l*.)

The effect of having this intermediate word-size segmentation between phonemes and utterances (or, rather, as will be seen, sentences) is that the constraints on freedom of combination which we seek as characterizers of utterances, and the meanings which these constraints carry, can be stated on a far larger body of elements—the morphemic segments (including words) —than just the twenty to thirty phonemes that each language has. This makes it possible to expect a much simpler system of constraints than if everything language does had to be stated in terms of the few phonemes it has. By the same token, the segmentation means that phoneme combination short of the peaking constraint (e.g. in syllables), is not directly relatable to the main, and information-bearing, constraints of utterances—those of Chapter 3. That is to say, the phonemic composition of morphemes is arbitrary, as are phonemes themselves, in respect to most of the rest of grammar; and phonemes do not carry meaning. The peaks of v which determine the segmentation result from two facts: that not all phoneme combinations are utilized in making morphemes, and that the number of morphemes in a language is some three orders of magnitude larger than the number of phonemes. The boundaries of a morpheme are therefore points of greater freedom for phoneme combination.

We thus have not only a distinguishing property of those phoneme-sequences which constitute initial segments of sentences, but also evidence that sentences (differently from non-sentences) contain a certain segmentation into 'morphemic' sub-sequences of phonemes; and we have the boundaries of these morphemic segments.

For the user of a language, the morphemes are known not, of course, from this comparison of phoneme-sequence successors, but from being learned, some in isolation and some by their regularity of combination in utterances. They are thus pre-set as the user becomes able to speak or understand the language, except insofar as the meaning of an unknown morphemic segment in an utterance can be guessed from knowledge of the environing ones.

Various adjustments can be made in this procedure, particularly in order to get a morphemic segmentation that has simpler syntactic regularities. One adjustment eliminates the occasional case in which the rise, which marks the end of the morpheme, is gradual rather than abrupt as it should be. For this, we take v not as the number of phonemic followers to each initial sequence $p_1, \ldots, p_m$ (where p_m is some phoneme a and p_{m-1} is a phoneme b) but as the ratio of that number to the number of different phonemes that follow the phoneme a (or the sequence of phonemes ba) in all neighborhoods in which a (or ba) occurs.[9] In this ratio of hereditary to Markov chain dependence, we obtain the hereditary effect more sharply; i.e. the restriction on the followers of $p_m=a$ due to that particular p_m being

---

[9] We may consider the pair ba, rather than just the last phoneme, because many syllable restrictions depend on pairs of neighboring phonemes. The way in which the phonetic property of phonemes restricts their successor does not depend on the predecessors back to the beginning of the utterance.

located in a complete initial sequence $p_1, \ldots, p_{m-1}, p_m$, rather than simply in the phonetic neighborhood $b=p_{m-1}$.[10]

In general, the successor count can show syllable boundaries, because of the greater number of consonants than vowels and because various consonant clusters can be followed only by a vowel. However, these syllabic differences in successor count become far less pronounced as we proceed through the word (e.g. in *body* in Fig. 7.2), because the hereditary effect within the word becomes more prominent.

Certain peaks in some sentences do not correlate with a morphemic segmentation of that sentence. This happens when the first part of a morpheme in one sentence is homonymous with a whole morpheme which can occur in the same initial neighborhoods. Thus in *He was under* . . ., there would unavoidably be a peak after *He was un* because of the prefix *un-* which occurs here with many followers (*He was unreserved, He was uninformed,* etc.). Such wrong segmentations can almost always be corrected by carrying out the corresponding count (of predecessors) backward on the same sentence.

7.4. *Words and morphemes; complementary alternants*

Once we have reached morphemic (i.e. word-size) segments, we can proceed to regularities of their combination in utterances. Before attempting this, it is useful to seek out any special constraints which hold for just one or a few morphemic segments. In particular, there are many cases in which two or more segments are complementary in respect to some feature in the environment, so that the sum of their environments roughly equals the set of environments of many single morphemic segments. Each set of one morphemic segment or of more than one complementary segment is called a morpheme, bound if it occurs only attached to one or a small set of other morphemes, free (or: a word) otherwise.

One kind of complementary variant is seen in *a bet, two bets* (with added *s* phoneme) as against *a bed, two beds* (with added *z* phoneme) and *a child, two children* (with added *-ren*), where after *two* the *-ren* appears only after *child*, *s* only after voiceless phonemes such as *t*, and *z* only after voiced phonemes such as *d* or vowels. Clearly, it is preferable here to define not three plural morphemes (not different in respect to further operators on them), each restricted to different phonetic environments, but one unrestricted plural morpheme with three environmentally complementary phonemic shapes. The spelling of this morpheme would then not be the

---

[10] In the examples above, the two cases in which *v* rises gradually to the peak, 9–14–20 for *dis* and 11–6–9–28 for *kwik*, would thus be replaced by decreasing ratios. Fewer phonemes can follow a consonant, especially a pair of two consonants, than can follow a vowel. Thus 9 is a smaller percentage of the total followers of *i* than 6 is of the total followers of *w*.

phoneme *s*, etc., but a new entity, the morphophoneme *S* defined as a set of several alternants (including a few additional onès, as in *foot, feet* below). Similarly, we find *a knife, two knives, the knife's edge*, where instead of having two restricted morphemes, *knive* before the plural morpheme *S* and *knife* before all other morphemes (including the possessive *'s*), we can define a single morpheme *nayF* where the last phoneme has been replaced by a morphophoneme *F* defined as a set of two phonemes *f, v* with morphemically complementary environments. (In contrast, we see from *a fife, two fifes* that the morpheme *fife* is phonemically and morphophonemically *fayf*, and from *a five, two fives* that the morpheme *five* is *fayv*.)

We see here that cases in which two environmentally complementary morphemic segments have different phonemes at a particular point can be treated by giving them a single morphophoneme, which is a set (a disjunction) of those phonemes: this leaves us with but one, more regular, morpheme, with the complementary alternation being now stated on the phonemic members of the morphophoneme. Morphemes can then be defined as sequences of phonemes and morphophonemes. They could be considered to be sequences of just morphophonemes if the unchanged phonemes in the sequence are taken as morphophonemes with only one, non-alternating, phonemic member. One might then think that we can define morphemes directly as sequences of morphophonemes, each of these defined directly in terms of sounds. But this would be circular, because if we did not have phonemes as utterance-distinctions independently of morphemes, we would not be able to state precisely what are the alternant forms of each morpheme and morphophoneme. More generally, we would miss the importance of phonemes as a rare case of fixed distinctions among behaviors, creating repetition in contrast to imitation, and yielding a system of discrete elements which assures the discreteness of everything further in language.

There are some complementary morphemic segments for which morphophonemes cannot readily be used. In *I am, you are, he is, to be* we have four morphemic segments restricted as no other English segments are. We can form a single morpheme *BE* as a set of complementary suppletive phoneme-sequences, without being able to define successive morphophonemes as its components. (This, too, argues against morphophonemes replacing phonemes as components of morphemes.) Such a solution would be valuable if, as in this case, the set of all environments in which the complementary segments occurred was roughly the same as that in which many single morphemic segments do (*go, come, walk, talk*, etc.), so that *BE* becomes a regular member of a known set of morphemes (the word-class 'verbs', 7.5), whereas each of *am, are*, etc. (as morpheme-alternants) have unique environmental restrictions and lack the regularity of whole morphemes.

Another situation is seen in *They break records, They broke records before too* as against *They brake the car too fast, They braked the car too fast before too*, where we find one morphemic segment *brake* (phonemically *breyk*), having the added segment *-ed* before *before*, while the homonymous other one, *break* (phonemically *breyk*, for the homonymity see below), has a change of phonemes in that position. The change of phonemes from *ey* to *ow* is not a variant of *break*, since it occurs with the same likelihoods ('selection') of further operators as does *-ed* on *brake*; hence it is a post-*break* alternant of *-ed*. Even though we do not obtain it as a morphemic segment in the phoneme-successor procedure of 7.3, we have to accept *ey* → *ow* as an alternant member of the *-ed* morpheme, occurring with the morpheme written *break* but not with *brake* (while *wake* has both *waked* and *woke*). Similarly, in *a foot, two feet*, we have an additional *u* → *iy* alternant of the plural *S* morpheme, occurring with *foot* but not with *boot*.

Once we see that morphemes can have alternant members or forms consisting of phoneme replacements, we can consider cases in which phonemic replacement is the sole form of a morpheme, not just a variant form. A need to recognize such morphemes arises in Semitic word-roots, where words having a common meaning share the same consonants while such grammatical meanings as past, imperative, place of action, etc. are indicated by different intercalated vowels. These non-contiguous phonemic sequences can be fitted into the system of internally connected morphemic segments which are more common in the world's languages, by accepting some basic connected segment, e.g. the root with past tense vowels, as the lexical morpheme, and considering all the other vowel-arrangements—future, nominalization, etc. —as morphemes whose phonemic composition consists mostly in replacing the vowels of the lexical morpheme. In other languages, the meanings and grammatical functions carried out by the Semitic intercalated phoneme-sequences are carried out by ordinary connected phoneme-sequences such as words and affixes.

There are also cases where one has to say that zero is an alternant of a morphemic segment. This is seen in *a sheep, two sheep*, where we either have to say that *sheep* is virtually unique among morphemes in not accepting a plural affix (while accepting a plural verb, as in *The sheep are here*), or else that *S* has zero as its alternant after *sheep*. A larger issue of this kind arises when we compare *The trees fall, They fell the trees*, and *The children sit, They seat the children*, and *We wake at 8, They waken (or wake) us up at 8*, with *The umbrellas stand against the wall, They stand the umbrellas against the wall*. In *fell, seat, waken* we have a morphemic segment or a phoneme replacement, with a causative meaning, which accompanies the moving of the subject (e.g. *trees*) to object position. Given that the unchanged morphemes (*stand, wake*, also *walk, run, rest*, etc.) can by themselves carry this change in subject plus causative meaning, the simplest description is that these words are then carrying a zero alternant of that causative morpheme, an alternant whose phonemic composition is zero.

A somewhat more complicated case is that of *He presents the thesis tomorrow* compared with *his presentation of the thesis tomorrow*, and *He defends the thesis tomorrow* compared with *his defense of the thesis tomorrow*, and *They will write up the thesis* compared with *their write-up of the thesis*, all in contrast to *He will delay in completing the job* compared with *his delay in completing the job*. When a sentence is 'nominalized', i.e. becomes the subject or object of a further verb (8.2), the verbs in the above examples have, respectively, an added morphemic segment, a phoneme replacement, a stress change, or zero, all with a following *of* or the like. These have all to be taken as alternant members of a nominalizing morpheme (an 'argument indicator').

Occasionally one finds two or more morphemic segments which occur together, and which can be taken as parts of a single morpheme even though not contiguous. An important case is what is called agreement in grammar (6.2). This exists, for example, in languages in which the adjective matches its noun as to number and gender: *hortus amplus* 'large garden', *puella pulchra* 'beautiful girl'. Instead of adding to the apparatus of the grammar a special relation of 'agreement' between a noun and its adjective, we need merely say that the *-us*, *-a*, *-um*, and their respective alternants which come at the end of each Latin noun are repeated on each adjective in predicate (including modifier) relation to it. That is, the phonemic shape of *-us* is really *-us*, *-us*, . . ., *-us*, the number of portions in each occurrence being one more than the number of associated adjectives and predicates. This is taken as holding even when no gender suffix appears on the noun, as in *lux aeterna* where *lux*, 'light', 'has feminine gender', i.e. ends in *a* . . . *a* . . . *a* . . . (even though with zero alternant for the initial *a*), because its adjectives receive the repeated portions of the feminine suffix *-a*.

In this survey of major forms of morphemes, primarily such as are found in English, we see phoneme-sequences, with sets of complementary alternants, and also phoneme replacement and zero. We also see that many morphemes are sayable independently, and in many languages carry a separate main stress, and enter into a complex spectrum of combinations: these are words (free morphemes). Other, bound, morphemes lack these properties and are always pronounced, under a main stress, with one or another morpheme of a particular set: these are affixes or phoneme-replacements or zeros, and they form words only together with the morphemes of the set to which they are restricted. In most cases they are attached to words, as in *-ation* on *present*; but there are cases of particular bound morphemes combining with each other to make a word, which then takes on its own stress: *per-*, *con-* and *-ceive*, *-mit* forming *perceive*, *permit*, *conceive*, *commit*.

In summary: we started with morphemic segments, many of which function as independent morphemes. Some morphemes, however, were defined as disjunctions of segments. In other cases, a segment was broken into two morphemes, as in the case of phoneme replacements. These and all the other modifications which we make in the results of the phoneme-sequence successor procedure are always made on the basis of comparing the further constraints on a given morphemic segment with those on other segments having comparable environments, in order to arrive at a simplest set of entities and constraints for all the data. For example, in the case of complementary members of a morpheme, there are three criteria, all simplifying the description: that the members be complementary in respect to some stated near neighbor; that the members be very similar in respect to their farther neighbors (operators) and thus have the same meaning (e.g. *I am*, *you are*, *he is* occur in much the same farther environments); and that the set of environments of all the members together be roughly the same as that of some—preferably many—single morphemic segments (e.g. *be* in respect to verbs).

When the criterion of maximum regularity is applied to the selection of a word, that is, to the set of words with which it occurs with greater than average likelihood (3.2), we can distinguish homonyms, i.e. cases where a single phoneme-sequence occurs in the kind of separate grammatical positions and separate coherent selections that are generally covered by two or more different words. On such grounds, occurrences of *heart* in *a heart attack* and in the *heart of the matter* and even *a hearty meal* would all be considered the same word (the latter uses having the properties of metaphoric and other known extensions of a word's selection); but the same phoneme-sequence in a *hart and a dove* would be a homonym rather than the same word. Conversely, slightly different phoneme-sequences which have the same (synonymous) selection would be considered free alternants of the same word, as in the two pronunciations of *economics*. This assignment might even be made 'suppletively' when there is no phonemic similarity, but there are other special grounds, as between *e.g.* and *for example*. The whole question of marginal cases of being the 'same' word is not crucial and may ordinarily be ignored in many languages: is *case* in *briefcase*, *a small case (of wine)* the same word as *case* in *a small case (taken by a legal-aid lawyer)*? The issue affects zeroing of word-repetition: one cannot say *He got two small cases today, one of wine and one in law*. One can say *He plays violin and she piano*, but can one say except jokingly *He plays violin and she chess*?

The criterion of maximum regularity does not suffice to decide whether certain uniquely restricted segments are morphemic (word-size) at all. One problem is seen in small sets of phoneme-sequences which combine only among themselves. A situation roughly of this type exists in *per-*, *con-*, *-ceive*, *-mit*, etc., in English *perceive*, *permit*, *conceive*, etc. Another situation is that of phoneme-sequences which do not combine freely enough to be considered morphemes, but which have a bit of the meaning-properties and combinatorial properties of morphemes: e.g. the *sl-*, *gl-*, of *slide*, *glide*, *slow*, *glow*, and *slither*, *gleam*, *glimmer*, etc. (11.1). Somewhat more morphemic is the *-le* of *dazzle*, *drizzle*, *jiggle*, *joggle*, etc., which historically was a morpheme, but most of whose attached segments are not recognizable as morphemes now: if the -*le* is to be a morpheme, that to which it is attached must also be one. Another type of complexity is seen in *curb*, meaning 'restrain; a bit; a street border', where British English has the spelling *kerb*, but only for the third meaning. The question whether *kerb* is a different word does not arise in speech; but it arises in writing. Somewhat similarly, *stanch* has two meanings, one of which is the same as one of the meanings of *staunch*, which in American English has two pronunciations, one the same as *stanch*.

The phoneme-sequences identified by the morpheme-boundary procedure have regularities of combination with each other in utterances of the language, but we have seen here that various ones of them have to be adjusted to fit the further regularities. The specific phonemic content of morphemes is not related in any essential way to their further grammatical behavior, and indeed phonemes are a requirement of speech rather than of syntax: the same syntax as in a spoken language, with the same informational interpretation, can be created in a one–one correspondence between morphemes and indecomposable marks such as pictographs. The phonemic composition of morphemes is needed primarily because speech cannot make as many single distinct sounds as the distinct morphemes of a language. It incidentally follows from this that any classification or redefinition of a syntactic problem that reduces the syntactic irregularity to a phonemic-composition irregularity (morphophonemics in a broad sense) is a real descriptive gain, since the syntax is seen to be a system independent of the phonemic composition. An example is moving the dependence of *am*, *are*, *is* on *I*, *you*, *he* from a restriction on verbs in respect to their subjects, to a restriction on the complementary phonemic compositions of the verb *be*.

7.5. *Word classes*

To go from words and bound morphemes of 7.4 to the set of all possible word combinations in utterances is a large step. This is especially so in view of the great number of words, with almost every one being unique in some of its combinations and transformations, in keeping with its unique meaning and its stylistic character. In 3.1–2, it was found that this large step could be decomposed into two crucial successive stages for each word. The first stage determined what word-occurrences the given word never combines with in all base utterances including the shortest (where observation is easier); this creates the argument-requirement word classes. The second stage characterizes the graded likelihood of each word in respect to the various other words in whose environment it has probability >0 in the shortest utterances.

Meanwhile, in traditional grammar, an approximation to these word classes is obtained by related if less global considerations. The traditional classification is based on similarities in combination. It is well known that for almost every word there are many other words whose environments are largely similar to those of the given word. For example, we can collect many words into a class of nouns (*N*), not only because they can all occur in some particular and more or less immediate morphemic environment such as after *the* (or, for most of them, before plural *S*), but even more because they can all occur with one or another of some other collection of words: e.g. every member of *N* can be found before some verb, though not before every verb (*V*). It is then possible to say that the sequence *NV* is a sentence structure, where *NV* indicates some member of *N* followed by some member of *V*, but not all *NV* combinations. The other classes can be characterized comparably (sometimes a bit roughly): each verb occurs with a tense affix and can have one or another preceding *N*; adjectives (*A*) can occur between *the* and *N*; prepositions (*P*) occur before *N*; co-ordinate conjunctions (*C*) can occur between any two like-classified words, and so on. Most of these classes are found in many languages, and serve for the combinations found in constituents or in strings (7.6), on the way to stating the structure of sentences. In English, each of the large classes *N*, *V*, *A*, contains many single morphemes as words (e.g. *man*, *think*, *large*); but each class also contains words which belong to some other class but which carry affixes that make them members of the class in question: *manhood*, *consideration*, *largeness* in *N*; *to man*, *to reconsider*, *to enlarge*, *to up* in *V*; *man-like*, *considerable*, *largeish*, *uppity*, in *A*. These are thus classes of words rather than of morphemes; they are such because the criteria for each class are satisfied not only by its special morphemes but also by the morphemes of other classes when carrying the appropriate affixes.

These classes do not constitute partitions of the set of words of a language. For one thing, some words have the properties of one class with an occasional property of another: e.g. *rich*, in *A*, can occur by itself after *the*, like *N*; and the color and number words in several languages have both *A* and *N* properties. Indeed, some languages have a great quantity of regular or irregular multiple classifications, whereby some occurrences of a word have the properties of one class, say *A*, while other occurrences of it have the properties of another class, say *N*. Furthermore, there are some single words or small sets of words which satisfy none of these criteria: e.g. *yes*, or in English a set containing *only*, *hence*, etc. which have a peculiar bi-sentential environment (1.1.1, G395). In any case, these classes say nothing about the relative likelihood of finding a particular member of one class with a particular member of another. Thus, tensed *V* are stated to have a preceding *N*; but whereas we can find, as a case of *NV*, *The earth slept*, we will be hard put to find *The vacuum slept*, and will not at all find *His departure slept*.

In the final analysis, it will be seen that these word classes, based on observable combinations in the word-sequence, can be replaced by a sharper classification of morphemes based on a more fundamental relation of argument-requirement (3.1, 8.2).

7.6. *Word-class sequences:*
Continuous and contiguous strings

Given the word classes, it is possible to isolate a few types of word-class sequence which suffice as the structural components of utterances. In traditional grammar, what came to be formulated was primarily the near neighborhood of a word, which was the most sharply fixed: above all its affixes, its morphology; and in addition, its immediately neighboring modifiers, constituting such local constructions as noun phrases and verb phrases which were codified by Leonard Bloomfield as immediate constituents.

The structure of the sentence was then described as a combination of immediate constituents, primarily noun phrase (as subject) followed by verb phrase. However, not all modifiers are in the immediate neighborhood of their host. One example of this is *Finally the man arrived whom they had all come to meet* (where the *whom* phrase modifies *man*, and would normally be next to it as in *Finally the man whom they had all come to meet arrived*). Another example, in Latin, is the ability of adjectives to be at various (but not unlimited) distances from the noun they modify.

One can still describe the constituent, even with non-contiguous parts: the subject above is *man* with adjoined *the* on the left and *whom* . . . after the verb on the right. But the difficulty lies in formulating a constructive definition of the sentence. For if we wish to construct the sentence by defining a subject constitutent and then next to it a verb (or predicate) constituent, we are unable to specify the subject if it has a non-contiguous part because we cannot as yet specify the location of the second part (the modifier at a distance). At least, we have no general and complete way of specifying the location of the distant modifier until we have placed the verb constituent in respect to the subject; but we have no general way of placing the verb in respect to the subject as a single entity unless the subject has been fully specified.

A similar constructive difficulty arises with respect to all other non-contiguous phenomena such as discontinuous morphemes, and the special case of them known as grammatical agreement (above, as between subject and verb in respect to the plural: *The name appears below*, *The names appear below*). Arbitrarily many different word-sequences can appear in utterances between the subject and its predicate, and the specification of plural agreement between them is hard to formulate, especially when the predicate precedes the subject as in *Below appear, as John requests, the names of all the participants*. More generally, not every sentence has a consistent analysis into the kind of constituents described above (e.g. what is the status of *that he go* in *It is important that he go?*).

A more adequate and more regular decomposition of sentences into subsequences is to find in each sentence a minimal component sentence (the 'center string': generally the tensed verb with its subject and object) and to show that everything else in the sentence is adjoined as 'adjunct string' to a 'host string' (word or sentence) which is in the sentence (i.e. to the center string or to a previously entering adjunct). In each sentence, the strings consist of sequences of one or more words in the sentence. In a theory of sentence structure, the strings are defined as particular sequences of word classes, each originally continuous, which can adjoin particular other types of strings at stated points—before them, after, or next to particular host words in the string, i.e. in all cases contiguously at the time it adjoins. Each adjunct string is defined as capable of adjoining any host string of a given type, the resultant being again a string of the class of that host string. Thus A is (in English) a left adjunct string of a host N (within a string containing that N). We consider the A as a string adjoined to the string containing the N, rather than as an adjunct to the N alone. D (adverb) is a left adjunct string of A or D (or an adjunct of V in some string) producing DA as an A string (*entirely empty*), DD as a D string (*almost entirely*), as well as DV (*almost fell*). The DD on A is an A string, as is D on DA (both as in *almost entirely empty*); AN and DV are not in themselves strings, since the N and V are merely words within a string.

That the adjunctions are to a string and not to a word in it can be seen in the cases when an adjunct referring to a word is at a distance from the word, but not at a distance from the string of which the word is part: e.g. in *Finally the man arrived whom they had all come to meet*, where the adjunct *whom . . .* depends on the occurrence of *man* (i.e. occurs only in the presence of certain N including *man*); or in Latin, where the adjective (adjunct) that agrees with and refers to a noun can be at a distance from the noun. In both cases, the adjunct is always contiguous to the string of which that noun is a part. Thus adjunction is a relation between strings, even though the grammatical dependence and semantic connection can be between one string and a host word in the other string.

The main string types sufficient for English utterances are found to be:

Center strings (S): chiefly noun plus verb plus tense plus the object required by that verb; also the interrogative and imperative forms of this;

Adjuncts on center strings, entering at right or left or at interior points of the host string and often separated by comma: e.g. D (*perhaps, certainly*), preposition-plus-noun PN (*in truth*), certain idiomatic strings (*no doubt*), and C conjunction followed by a center string (*because they *rioted*) or *P* followed by a 'nominalized' center string (*upon the army's arrival, with defeat certain*);

Right adjuncts on virtually any string or string-segment, consisting of *and*, *or*, *but*, or comma, followed by a string or string-segment of the same type: e.g. *N or N* (*pen or pencil*), *S but S* (*He went but I didn't*).

Left adjunct on host nouns: e.g. *A* (*wooden expression*), *N* (*wood stove*); or on host adjectives, prepositions, adverbs: e.g. *D* on *A* (*very wooden*), *D* on *P* (*completely under*), *D* on *D* (*almost entirely*);

Right adjunct on nouns: e.g. *PN* (*of matches* after *a box*), relative clauses (*whom we met*), *APN* (*yellow in color* after *a box*); on adjectives: e.g. *PN* (*in color* after *yellow*); on verbs: e.g. *PN* (*with vigor* after *oppose*);

Left or right adjuncts on verbs: e.g. *D* (*vigorously* before or after *oppose*). (By right or left adjunct on a noun or a verb is meant an adjunct to the string that contains that noun or verb, the adjunct being placed to the right or left respectively of the host noun or verb in the string.)

These strings are not merely constructional artefacts, but an aspect of the total structure of utterances. This is seen in the fact that many grammatical phenomena are describable in terms of the internal structure of strings or in terms of a host and its adjunct string. This holds for the restrictions as to which word subclass (within one class) occurs with which other word subclass, or which individual word of one class occurs with which individual word of another, or the locating of a grammatical agreement on a pair of classes. Examples are the restrictions between the members of a string (e.g. plural agreement between its N and its V) or beween part of a string and an adjunct inserted into that string (the restriction of a pronoun to its antecedent, or of an adjunct to the particular word in the host string to which the adjunct is directed), or between two related adjuncts of a single string (as in *People who smoke distrust people who don't* where the zeroing of *smoke* in the second adjunct is based on its parallel relation to the first adjunct).

The string property of language arises from the condition that language has no prior-defined space in which its sentences are constructed (2.5.4). Within this condition, there is room for a certain variety. For example, a string b_1b_2 could be inserted at two points of its host a_1a_2 instead of at one, yielding e.g. $a_1b_1a_2b_2$ instead of the usual $b_1b_2a_1a_2$ or $a_1b_1b_2a_2$ or $a_1a_2b_1b_2$; this happens in the rare intercalation *He and she play violin and piano respectively*. But one adjunct c of a host a_1a_2 could not be inserted into another adjunct b_1b_2 of that host: we would not find $a_1b_1cb_2a_2$ or the like (unless c was an adjunct of b_1b_2, or an interruption between dashes). And a part of a string cannot be dependent on an adjunct to that string, or on anything else which is outside that string itself. Any sequence of classes which satisfies the definitional conditions above is called a string; and what is empirical is that it is possible in language to find class-sequences (of rather similar types in the various languages) which indeed satisfy these conditions.

A good approximation[11] to the strings of a language can be obtained by successively excising, from each set of what are tentatively judged to be similarly constructed sentences, the smallest parts that can be excised preserving independent utterancehood (i.e. where the residual utterance does not depend on the excised portion, and retains certain stated formal and semantic similarities to the original utterance), until no more can be excised: each excised segment is an elementary adjunct string, and the residue is a minimal center string. It follows that the successive word classes within a string are required by each other, while each excisable string is an adjunct, permitted but not required by the residue to which it had been adjoined. Thus in *Crowds from the whole countryside demanding their rights ringed the palace*, we can first excise *whole* and *their* and then *from the countryside* and *demanding rights*. In constructing the sentence from elementary strings, we first adjoin the two left adjuncts of *N*, *whole* and *their*, to the two elementary right adjuncts of *N*, *from the countryside* and *demanding rights*; and then the two resultant non-elementary right adjuncts of *N*, *from the whole countryside* and *demanding their rights*, are adjoined to the elementary center string *Crowds ringed the palace* (to the right of a stated *N* in it), yielding the desired sentence.

The result of applying the criteria mentioned here is that we are able to set up a small number of types of strings, with not many specific word-class strings in each type, which suffice for the construction of all sentences, and which satisfy various additional properties. For example, no further operation on the strings can introduce intercalation of them (beyond what may be given in the original insertion conditions for particular strings, as with *respectively*), because any permutation can only be on the word classes within a string, or in moving a whole string from one point to another in the host string (as in permuting adverbs to new positions). No non-contiguity arises, because whereas an adjunct, e.g. of a noun, may be at a distance from that noun, it can never be at a distance from the string which contains that noun. For such reasons, string analysis is convenient for computation.

7.7. *Sentence boundaries*

What string analysis characterizes directly is not the set of all utterances of a language, but the set of all sentences, these being a segmentation of utterances. The reason string analysis produces only sentences is not that we started intentionally with sentences—we started with whatever was a short utterance—but that we did not include sequences beginning with a period among the adjunct strings: such a sequence as *and perhaps she left* is an adjunct string (as in *He left and perhaps she left*) but period + *Perhaps she left*

---

[11] Proposed by Henry Hiż.

(as in *He left. Perhaps she left*) is not. If we look only at the succession of these sequences, we may see little difference, e.g. between *He left and perhaps she left* and *He left. Perhaps she left*. But if we check for restrictions, we find important differences. For example, under *and* there is a rare zeroing to *He left and perhaps she*, but hardly under a period (*He left. Perhaps she.* is dubiously accepted). More importantly, sequences under *and* can be the subject or object of further words, but not sequences under a period: *His leaving and her leaving are clearly related*, but no *His leaving. Her leaving are clearly related*; similarly *That he left and she left surprised us*, but no *That he left. She left surprised us*. For such reasons, the period is not included in the set *C* of conjunctional strings. Hence, although the minimal 'center' strings include any minimal utterance, the result of string adjunction is not the set of all utterances but the set of certain segments of utterances called 'sentences'.

The existence of sentences as a unique grammatical construction is shown, differently for different languages, by various considerations such as above. In a more general way, string analysis can be used for a recurrent stochastic process which locates the sentence boundaries within a discourse. Just as the process proposed in 7.3 shows that a segmentation into morphemes exists (and locates the specific boundaries), so the process proposed below shows that a segmentation into sentences exists (and locates the sentence boundaries).

We consider the word-class sequences which constitute sentences, and we think in terms of distinguishing in them all positive transitional probabilities; from those which are zero. We can now state a recurrent stochastic dependence between successive word classes of each sentence form in respect to the string status of each of those words.[12] In this hereditary dependence of the nth word class, w_n, of a discourse on the initial sequence $w_1, \ldots, w_{n-1}$, sentence boundaries appear as recurrent events in the sequence of word classes. This is seen as follows: if a discourse D is a sequence of word (or morpheme) classes and x, y, are strings (as in 7.6) included in D, then:

a. the first word class of *D* is: (1a) the first word class of a center
                                                string,

                                 or (2a) the first word class of a left
                                                adjunct which is defined as
                                                able to enter before (1a), or
                                                before (2a);

---

[12] Cf. N. Sàger, 'Procedure for Left-to-Right Recognition of Sentence Structure', *Transformations and Discourse Analysis Papers* 27, Univ. of Pennsylvania, 1960. We take the choice of a particular word class w_n in the nth word position of the sentence as the outcome which depends on the selections $w_1 \ldots, w_{n-1}$ in the successive preceding word positions. As before, we disregard the probability weightings of each outcome, and note only which outcomes have a non-negligible probability.

b. if the nth word class of D is:

  the mth word class of a string x containing p word classes, $p>m$

then the $n+1^{th}$ word class
of D is:

  (3) the $m+1^{th}$ word class of x

or (4) the first word class of a right adjunct which is defined as able to enter after the mth word class of x,

or (2b) the first word class of a left adjunct which is defined as able to enter before (3), or before (4), or before (2b);

c. if the nth word class of D is:

  the last word class of a left-adjunct string x, where x is defined as entering before the mth word class of a string y

then the $n+1^{th}$ word class
of D is:

  (5) the mth word class of y,

or (6) the first word class of a left adjunct defined as able to enter before the mth word class of y, and such as is permitted to occur after x,

or (4c) the first word class of a right adjunct defined as able to enter after x,

or (2c) the first word class of a left adjunct defined as able to enter before (6), or before (4c), or before (2c);

d. if the nth word class of D is:

  (7) the last word class of a right-adjunct string x which had entered after the mth word class of a string y which contains p word classes, $p>m$,

or (7') the last word class of a right-adjunct string x which had entered after (7)

then the $n+1^{th}$ word class
of D is:

  (8) the $m+1^{th}$ word class of y,

or (6d) the first word class of a right adjunct defined as able to enter after the mth word class of y, and such as is permitted to occur after x,

or (4d) the first word class of a right adjunct defined as able to enter after x,

or (2d) the first word class of a left adjunct defined as able to enter before (8), or before (6d), or before (4d), or before (2d);

e. if the nth word class of D is: the last word class of a right-adjunct string x which had entered after a string y

then the $n+1^{th}$ word class of D is:

(1e) the first word class of a center string,

or (6e) the first word class of a right adjunct defined as able to enter after y, and such as is permitted to occur after x,

or (4e) the first word class of a right adjunct defined as able to enter after x,

or (2e) the first word class of a left adjunct defined as able to enter before (6e), or before (4e), or before (2e),

or (2e′) the first word class of a left adjunct defined as able to enter before (1e), or before (2e′),

or (9e) null;

f. if the nth word class of D is: the last word class of a center string
then the $n+1^{th}$ word class of D is:

(1f) the first word class of a center string,

or (4f) the first word class of a right adjunct defined as able to occur after a center string,

or (2f) the first word class of a left adjunct defined as able to occur before (4f) or before (2f),

or (2f′) the first word class of a left adjunct defined as able to occur before (1f) or before (2f′)

or (9f) null.

Thus we are given two possibilities for the string standing of the first word class of a discourse; and given the string standing of the nth word class of a discourse, there are a few stated kinds of possibilities for the string standing of the $n+1^{th}$ word class. The possibilities numbered (2) are recursively defined. This necessary relation between the nth and $n+1^{th}$ word classes of a sequence holds for all word-sequences that are in sentences, in contradistinction to all other word-sequences.

We have here an infinite process of a restricted kind. In cases 1e and 1f, 2e' and 2f', 9e and 9f, the nth word class of D is the end of a sentence form,[13] and the $n+1^{th}$ word class of D, if it exists, begins a next putative sentence form. The transitions among successive word classes of D carry hereditary dependencies of string relations. But 1e, f and 2e', f' are identical with 1a and 2a, respectively. The dependency process is therefore begun afresh at all points in D which satisfy 1e, f or 2e', f'. These points therefore have the status of a non-periodic recurrent event for the process.

In this way, sentence boundaries are definable as recurrent events in a dependency process going through the word classes of a discourse; and the existence of recurrent events in this process shows that a sentential segmentation of discourse exists. Given a sequence of phonemes, we can then tell whether the first word is a member of a first word class of some sentence form, and whether each transition to a successor word is a possible string transition. That is, we can tell whether the phoneme-sequence is a sentence —is well formed in respect to sentencehood. In so doing we are also stating the string relation between every two neighboring words, hence the grammatical analysis of the sentence in terms of string theory.

7.8. *Sentence structure*

Traditional grammar referred to sentences, but did not provide a general and precise structure for sentences, because this could not quite be done with the 'immediate constituent' entities of language that were recognized in grammar (e.g. verb phrase). However, one can construct various devices for recognizing the structure of a sentence on the basis of grammatical relations among its successive words, and also for deriving other sentences from elementary ones. For example, it is possible to make an orderly constituent-like expansion of word classes into word-class sequences by a type-theory-like hierarchy of rewrite rules such as (using L here for plural): $AN^1 = N^1$ if one can substitute AN^1 for N^1 (e.g. *dark sky* or *ominous sky*, for *sky*)

---

[13] This holds also if the $n+1^{th}$ word class of D does not satisfy any of the possibilities for cases e, f, since only 6e, 4e, 2e, and 4f, 2f can continue a sentence form beyond a point where it could have ended. In this case the $n+1^{th}$ word class of D is part of no sentence form. Aside from this, if the mth word class does not satisfy the table above, then D is covered by sentence-form sections only up to the sentence-end preceding m.

(obtaining AAN^1 (e.g. *dark ominous sky*) in the class N^1); and $N^1L=N^2$ if one cannot substitute N^1L for N^1 throughout (we cannot put a plural noun N^1L for N^1 before L to obtain NLL, i.e. a noun with double plural). At each point where recursion does not hold, we raise the power of the host. The hierarchical system of equations presents several stages of substitutability: the combining of morphemes to take the place of single words (mostly level 1), the repeatable adjoining of modifiers (remaining at the same level), the non-repeatable expansions (e.g. plural, *the*, and the compound tenses) forming levels 2 and 3; and it keeps track of how a whole sentence is pushed down into being a modifier (e.g. relative clause). For a given sentence, the formula which is obtained, e.g. the sequence N^4V^4 (in English), points to a particular 'central' elementary sentence within it (the level 1 material which is contained in that N^4, V^4). And the successive equations which reach this formula for the sentence point to the way N^1 and V^1 of that elementary sentence were expanded to make the given sentence. In this manner, it is possible to house all structural sub-sequences of a sentence in one relation, and also to show that each sentence contains a central elementary sentence underlying in an orderly way the hierarchies of expansion and replacement.[14]

In string analysis, these expansions are then found to constitute a regular system of contiguous adjunction of originally continuous strings. Each sentence is obtained from its sentential string plus various adjunct strings on it or on previous strings.

It is also possible to construct a cycling automaton with free-group cancellation to determine sentence well-formedness. For this purpose, each word is represented by the string relations into which it enters. We then have an alphabet of symbols indicating string relations, and each word of the language is represented by a sequence of these new symbols. On any sequence of these new symbols we can, by means of a simple cycling automaton, erase each contiguous pair consisting of a symbol and its appropriate inverse, in the manner of free-group cancellation. Then each sequence which can be completely erased represents a sentence.

There are three string relations (i.e. conditions of well-formedness) that hold between words of a sentence: they can be members of the same string; or one word is the head of an adjunct and the other the host word to which the adjunct is adjoined; or they can be the heads (or other corresponding words) of two related adjuncts adjoined to the same string. Furthermore, since sentences are constructed simply by the insertion of whole strings into interior or boundary points of other strings, it follows that the above

---

[14] Z. Harris, 'From Morpheme to Utterance', *Language* 22 (1946) 161–83; reprinted in the writer's *Papers in Structural and Transformational Linguistics*, Reidel, Dordrecht, 1970, where there are fuller treatments of the material presented here throughout Chapters 7 and 8. The hierarchy of rewrite rules is sketched for various languages in the paper cited here, and in the writer's 'Emeneau's Kota Texts', *Language* 21 (1945) 283–9, and 'Structural Restatements, I', *International Journal of American Linguistics* 13 (1947) 47–58; also in pp. 285 ff. of Z. Harris, *[Methods in] Structural Linguistics*, Univ. of Chicago Press, Chicago, 1951.

relations are contiguous or can be reduced to such. For each string consists of contiguous word classes except in so far as it is interrupted by another string; and each adjunct string is contiguous in its first or last word to a distinguished word of its host string (or to the first or last word of another adjunct at the same point of the host).

This makes it possible to devise the cycling automaton. In effect, we check each most-nested string (which contains no other string within it), i.e. each unbroken elementary string in the sentence. If we find it well formed (as to composition and location in host), we cancel it, leaving the string which had contained it as a most-nested, i.e. elementary, string. We repeat, until we check the center string.

In view of the contiguity, the presence of a well-formed (i.e. string-related) word class B on the right (or: left) of a class A can be sensed by adding to the symbol A a left-inverse $b\`$ (or: a right-inverse $\`b$) of B. If the class A is represented by $ab\`$, the sequence $AB$ would be represented by $ab\`b$; and the $b\`b$ would cancel, indicating that the presence of $B$ was well formed. For example, consider the verb $leave$ in the word class $V_{na}$ (i.e. a verb having as object either $NA$, as in $leave$ $him$ $happy$, or $NE$, as in $leave$ $him$ $here$). It would have two representations: $va\`n\`$ and $ve\`n\`$. Then $leave$ $him$ $happy$ could be represented by $va\`n\`.n.a$,[15] which would cancel to v, indicating that the object was well formed.

Specifically: for any two word classes (or subsequences of word classes) X, Y, if (a) the sequence XY occurs as part of a string (i.e. Y is the next string member to the right of X), or (b) Y is the head of a string which can be inserted to the right of X, or (c) Y is the head of a string inserted to the right of the string headed by X—then, for this occurrence of X, Y in a sentence form, we set $X \rightarrow y\`$ (read: $X$ is represented by $y\`$, or: the representation of X includes $y\`$) and $Y \rightarrow y$, or alternatively $X \rightarrow x$, and $Y \rightarrow \`x$.[16] Here $y\`$ is the left inverse of $y$, $\`x$ is the right inverse of x, and the sequences $y\`y$, $x\`x$ (but not, for example, $xx\`$) will be cancelled by the device here proposed.[17]

It should be stressed that what is essential here is the fact that the inverse symbols represent the string relations of the word in whose representation they are included. When a most-nested symbol pair is cancelled, we are eliminating not a segment of the sentence but a string requirement and its satisfaction.

It follows that if a word class Z occurs as a freely occurring string, i.e., without any further members of its own string and without restriction to particular points of insertion in other strings, then its contribution to the well-formedness of the sentence form is that of an identity (i.e. a cancellable

---

[15] The dots separate the representations of successive words, for the reader's convenience, and play no role in the cancellation procedure.

[16] Correspondingly, if Y is the end of a string inserted to the left of X, etc., then $Y \rightarrow x\`$, $X \rightarrow x$; or $Y \rightarrow y$, $X \rightarrow \`y$.

[17] In this notation, it will be understood that $(xy)\`=y\`x\`$, $\`(xy)=\`y\`x$ and $\`(x\`)\`=x=\`(\`x)$.

sequence). The representation would be $Z \rightarrow z`z$. (While such classes are rare in languages, an approximation to this in English is the class of *moreover, however, thus,* etc.)

If, in a given occurrence in a string, a word class has more than one of the string relations listed above, its representation will be the sum (sequence) of all the string relations which it has in that occurrence. For example, if in a given string, N has two adjuncts, A on its left and E on its right (as in *young men here*), its representation there should be $`ane`$; $a.`ane`.e$ would cancel to n.

There are certain conditions which the representations must meet. No two different classes should have the same representation (unless the string relations of one are a subset of the string relations of the other), for then we could cancel sentence forms that have one of these classes instead of the other. No proper part of a whole string should cancel out by itself, for then if only that part (or its residue in the string) occurred instead of the whole string, it would cancel as though it were well formed. If a proper part of a string is itself a string (and so should cancel), then it should be considered a distinct string; otherwise the extra material in the longer string has the properties of a string adjoined to the shorter string. For example, in *NtVN* (*He reads books*) and *NtV* (*He reads*) we have two distinct strings, with appropriate representations for their parts.

Most word classes occur in various positions of various strings. For each of these occurrences there would be a separate representation, and the occurrence would be well formed in a particular sentence if any one of these representations cancelled with its neighbors. As is seen in the next paragraph, for example, N has not one but several representations. We do not know which, if any, of these is the appropriate one for a given occurrence of N in a particular sentence until we see which, if any, cancels one of the representations of the neighboring words in that sentence.

Generally, host words carry the inverses of their adjuncts, as in $`ane`$ above for N; heads (as markers) of strings carry the inverses of the whole string (so that the members of such strings do not have to carry the inverses of their next string member): e.g. in *which he will take* we can represent *which* by $wv`f n`$, *he* by n, *will* by f, and *take* by v, so that after cancellation all that will be left is w to indicate that a well-formed *which* string occurred here. Order of adjuncts can be expressed as follows: in respect to noun adjuncts, N, TN, AN, TAN all occur, but not ATN (e.g. *the star, green star, the green star*, but not *green the star*). We have to give these classes, therefore, the following representations (in addition to others):

T	A	N
t	a	n
	$`ta$	$`tn$
		$`an$

Then the above examples would cancel down to n in the following sequences of representations: n, $t.'tn$ (*the star*), $a.'an$ (*green star*), $t.'ta.'an$ (*the green star*); but no representations could cancel *green the star* (*ATN*).

Permuted elements require separate representation. Thus *saw* as a verb requiring N object is represented $vn\grave{}$; but since the object can be permuted (as in *the man he saw*) the verb also has the representation $'nv$.

Some consequences of language structure require special adjustment of the inverse representation:

Delays. Sequences of the form $x\grave{}x'xx$ will cancel only if scanned from the right. If the scanning is to be from the left, or if the language also yields sequences of the form $yy\grave{}y'y$ which can only be cancelled from the left, then it is necessary to insert a delay, $i\grave{}i$, to the right of every left inverse (and to the left of every right inverse) which can enter linguistically into such combinations. We obtain in such cases $x\grave{}i\grave{}ix'xx$ and $yy\grave{}yi\grave{}i'y$, which cancel from either direction. This occurs in English when a verb which requires a noun as object (and the representation of which is $vn\grave{}$), or certain string heads like *which* (in a representation ending in $n\grave{}$) meet a compound noun (the representation of which is $n'.nn$). The representation $vn\grave{}$, for example, is therefore corrected to $vn\grave{}i\grave{}i$, so that, e.g.

$$\begin{array}{ccc} take & book & shelves \\ vn\grave{}i\grave{}i & n & 'nn \end{array}$$

cancels from the left (as well as from the right) to v.

Conjugates. There are also certain rare linguistic situations (including intercalation) which yield a noncancellable sequence of the form $z'x\grave{}zx$. An $x\grave{}$ which can enter linguistically into such an encapsulated situation has to be representable by its Z conjugate $zx\grave{}z\grave{}$, which enables the $x\grave{}$ and z to permute and the sequence to cancel. If a string AB, represented by $a'a$, encircles X (i.e. X is embedded within AB), once or repeatedly, then the relation of encirclement requires each A to be represented by $xax\grave{}$: then, e.g., $AAXBB$ yields $xax\grave{}.xax\grave{}.'a.'a$, cancelling to x. (Dots are placed to separate the representations of each capital letter.) In some cases, a particular word X (such as markers discussed below) requires the presence of particular other words or classes Y at a distance from it, i.e. with intervening Z, after all cancellable material has been cancelled: YZX occurs, but not ZX without Y. The representation of X then has to include $'z'yz$ so as to reach over the Z and sense the presence of Y. Thus the representation for *than* will check for a preceding *-er, more, less*; the representations for *neither, nor* will check for a preceding *not* (or negative adverb like *hardly*) or a following *nor*: *More people came than I had called*; *He can hardly walk, nor can he talk*.

Exocentrics. Another problem is that of a string XY which occurs in the position of a class Z, i.e. which occurs where a Z was linguistically expected: here the X would be represented by $zy\grave{}$, so that $zy\grave{}.y$ would cancel to z. (This XY includes grammatical idioms.)

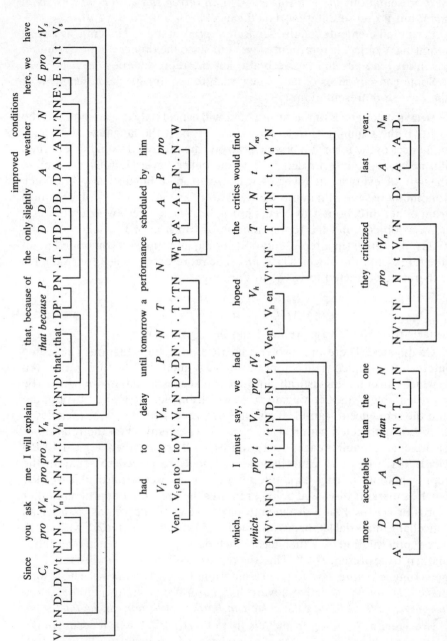

FIG. 7.3. Cycling automaton for an English sentence.

Markers. If, in any linguistic situation requiring a special representation, one of the words occurs only in that situation or in few others, then that word can be treated as a marker for that situation, and all the special representations required to make the sequence regular can be assigned to the marker. This applies to the heads (initial words) of many right adjuncts, e.g. *that, whether (that he came,* etc.). This makes uncouth representations for the markers (e.g. for the *wh-* words), but it saves us from having alternative representations for more widely occurring word categories. Similarly, certain classes of verbs occur as (adverbial) interruptions in sentences: e.g. *Celsius, I think, is Centrigrade.* Rather than put the inverse of these verbs as an alternative representation of every noun (so that the noun plus verb interruption should cancel), we put the inverse of the noun onto the representation of these particular verbs: *think* is not only *V that'* (to cancel in *I think that* . . .) but also *'nd* (so that *I think* is *n.'nd* which cancels into an adverb, as proposed under Exocentrics).

Among the linguistic phenomena, in addition to the above, which were inconvenient for the inverse representation was:

Dependent repetition. The co-ordinate conjunctions permit recurrence of the preceding word class or sequence but not of most others (*N and N, V and V*, but not *N and V*). One or another of the participating words has to carry inverses that restrict the class (or class-sequence) after *and* to being the same as the one before *and*; and we must allow for sequences which may intervene between the class and *and*. For example, we may give each class X a representation $xx'+'$.[18]

Figure 7.3 is an example of one representation of a rather complex English sentence.

The hierarchy of rewrite equations for sentence components, and the cycling automaton for sentence well-formedness, may seem too laborious for use in sentence description. Their chief utility may be in what they show to be possible for linguistic theory, rather than in actual applications. Thus the hierarchical equations show that each sentence has an underlying minimal-sentence structure and a transformational relation to other sentences. And the cycling automaton shows that sentence well-formedness can be formulated computably, independently of the transformational history of the sentence. Nevertheless, the hierarchical equations provide overviews of the syntactic characteristics and differences of various languages. And such devices as the cycling automaton can be developed for special-purpose analyses of sentences and texts.

---

[18] In *XY and XY*, it suffices to provide for *X and X*, since each *X* cancels its *Y*. + represents *and*.

8

Analysis of the Set of Sentences

8.0. *Introduction*

Chapter 7 presented the analysis of occurrences of language as far as it can be done in what is called structural linguistics. The analysis ended up with two general limitations. First, the co-occurrence of individual words were so numerously varied that regularities could be stated only as combinings of classes of words, with the individual word-choice as between the class combinations left unstated. The second limitation was that these regularities held only within sentences but not over whole discourses, i.e. not between the successive sentences of a longer utterance.

Two procedures of analysis point to resolutions of these limitations. The first is the hierarchy of expansion-equations (7.8), which isolates within each larger sentence a central elementary sentence plus a hierarchy of expansions on its parts. This constituted a first step in transformational decomposition of sentences (8.1), by showing that inside every sentence there exists a minimal sentence formed by a short sequence of words of the given sentence. The decomposition into component sentences was completed when it was shown that each of the expansions on the parts of the central sentence is itself a transform of a sentence, ultimately of a minimal sentence (7.6). The complete analysis yielded a decomposition of each sentence into minimal (elementary) ones (8.1.5), the word-choices in the larger sentences being merely those of the component elementary sentences.

The second procedure was the application of the co-occurrence analysis of Chapter 7 within the confines of one discourse at a time, in order to see what restrictions or regularities appear in sentence sequences (ch. 9). It was found that the words which appear in some grammatical relation to each other, in a given sentence, have high likelihood of appearing in later sentences also in a grammatical relation, but not always in the same relation.

Thus, after *The lymphocytes contained antibody* we may see, at one point of the discourse or another, *Antibody was found in lymphocytes*, *the antibody content of lymphocytes*, and *the antibody in the lymphocytes*. It was then found that the set of grammatical relations in which these words were repeated (i.e. the set of linguistic transformations) was not arbitrary (e.g. we would not find here *Antibody contained lymphocytes*). Rather, it had a structural characterization (as preserving the likelihood-grading of word combinations); and it had the semantic property of being paraphrastic. The transformations change the shape and arrangement of the words, but not their choice, so that given a base sentence (4.1) the word-choices of all its transforms are the same and need not be stated separately for each transform. When completed, the transformational relation provided also a structure for the set of sentences (6.3, 8.15).

The process of paraphrastic replacement of grammatical relations is presented in 8.1. In 8.2 it is shown that this process is the result of fundamental operations in sentence formation, primarily reduction. Further-more, it is found that when we undo this process, we reach the base structure that underlies language.

We begin with the question of what causes the paraphrastic variation of grammatical relations between the same word sets in successive sentences. In each discourse particular small sets of words are involved with various but not random grammatical relations among them. For example, in the article analyzed in 9.3, one sentence contains *optical rotatory properties of proteins*, another *rotatory properties of proteins*, and a third *proteins undergo changes in specific rotation*, all exhibiting *proteins* and *rotate* in grammatical relation to each other, but different relations. Also, we find *The rotatory power of proteins is sensitive to experimental conditions*, and we find *under which it is measured* (where the *which* repeats *conditions* and *it* repeats *rotatory power* making it *Their rotatory power is measured under these experimental con-ditions*), and *dependence of the rotatory properties of proteins on wavelength*, and *dependence of optical rotation on experimental conditions*, all exhibiting grammatical relations between *rotate* and *conditions*.

If we want to investigate this variation, we have to see what is common to the variations in many discourses, each with its own small set of words held together by the relations. We find indeed that particular grammatical constructions participate in these variations. (a) There are variations among free-standing sentences: e.g. the transitive active as against the passive, the assertion as against the interrogative. (b) There are also variations between free and bound (component) sentences: the free sentence form ending in a period, as against nominalized sentences (such as *dependence of optical rotation on experimental conditions*, above), and also as against adjectivized sentences such as relative (*wh-*) clauses and subsidiary clauses (e.g. *this being the case*). In each of the (a)-forms one can find approximately the same normally-selected word-choices (i.e. choices with higher-than-average frequency), and similarly in each of the (b)-forms.[1] And just as a word's particular selection, of what are its more-frequent-than-average operators

---

[1] A crucial fact here is that the set of all relative clauses is the same as the set of all sentences (including resultants of 'fronting', 3.4.2), except that an initial argument (or one statably near front position) has been replaced by a pronoun (3.4.4). The same applies, with certain modifications, to some subsidiary clauses. These can therefore all be best analyzed as bound transforms of free sentences.

or arguments, is an indication of its meaning, so also the fact that these different grammatical constructions have the same selection of word-choices indicates that the complexes of grammatical relations within them have the same meaning.

In the inter-sentence transformations, (a) above, we have a system of paraphrastic relations among certain sentential constructions, or among the sets of sentences in these constructions. These relations were called transformations because they constitute a set of partial transformations on the set of sentences, each transformation mapping the subset of sentences which have the properties of one of the constructions above onto the subset of sentences which have the properties of another construction, preserving word-choice and meaning. It is because of this that these transformations provide a decompositional structuring of the set of sentences (6.3).

In addition, transformations between free and bound sentences, (b) above, provide a method of decomposing individual sentences into elementary sentences plus certain expansions. Many of the expansions are 'adjuncts', i.e. transforms of sentences which are being adjoined to another sentence: e.g. the expansion of *article* to *long article*, in *He wrote a long article*, transformed from *He wrote an article which is long*, from *He wrote an article; the article is long*. Other expansions involve 'nominalizations', i.e. transforms of a sentence when it is the subject or object of a further verb: e.g. *destruction of Dresden* in *He saw the destruction of Dresden*, from *He saw people's destruction of Dresden*, *He saw people's destroying of Dresden*, and somewhat differently *He saw people destroying Dresden*, *He saw that people destroyed Dresden*. (The status of these further verbs, e.g. *saw* here, presents certain problems to string analysis and to transformational analysis; the difficulty is resolved in operator analysis, 8.2.)

8.1. *Sentence transformations*

8.1.0. *Word-class grammars and word-choices*

In the sentence analyses reached in Chapter 7, the sentence structure was composed of word-class sequences. In constituent analysis, the components for each sentence were word-class sequences that were not themselves sentence structures. In string analysis, as also in the expansion equations of 7.8, for each sentence there was a central component which was a sentence structure, while the other components were not. As between these two analyses, string analysis (7.6) not only brought in the advantages of contiguity, but also reduced the environment for syntactic events from the whole sequence in a sentence to the internal structure of each string plus the adjunction relation between strings. This simpler framework is a real property of sentences, as can be seen not only from its relation to sentence boundaries (7.7), but also from the fact that transformations, which are a further property of sentences, can be located in respect to the string structure of sentences (as is seen in 8.1.4 in the case of English).

However, all word-class components of sentences fall short in certain respects, as noted in 8.0. For one thing, they do not recognize important similarities in word class, word-choice, and meaning between certain word-class sequences. For example, neither string nor constituent analysis provides in a principled way for the similarities between certain adjunct (modifier) strings and center strings: e.g between AN and N is A (*the long article* and *The article is long*), or between a relative clause and a sentence minus a noun (as in *which he gave to me yesterday* and *This he gave to me yesterday*). For another, word-class sequences as components produce not sentences but only sentence structures: to produce the actual sentence would require characterization of the specific word-choices, which has always been considered a body of data so vast and unstable as to be beyond the powers of grammar. Finally, many problems concerning the meaning of sentences were touched upon only tangentially: for example, paraphrase and ambiguity.

8.1.1. *The transformational method*

Most of these inadequacies are resolved by a further method of analysis, that of transformations. This method utilizes the word-choice similarities between the sentence expansions of 7.8 and corresponding independent sentences. For example, in the expansion equation $AN^1 = N^1$ (as part of some sentence) the normal choice of particular A for each N^1_i was the same as in N^1_i is A. In the great majority of cases, one can say that the word-choice in the independent sentence form N is A is preserved in N which is A and in AN, which are its transforms. The *which is A* and the modifier A are transforms of N_i is A when the latter is adjoined to another sentence containing N_i.

The search for preservation of word-choices was a matter of finding pairs or sets of sentence structures ('frames'), in which the corresponding three or more word-class positions were normally (selectionally) filled by the same n-tuples (pairs, triples, etc.) of actual words. Roughly, one sees this in the active ($N_1 t V N_2$ structure, with t for tense) and passive ($N_2 t$ be Ven by N_1) pair: *John will see the picture* and *The picture will be seen by John*, but not *The picture will see John* nor *John will be seen by the picture*. Similarly, relative clauses have the same word-choices as sentences consisting of the same word-sequences with the relative pronoun being replaced by a preceding noun: *I read a novel which he wrote* and *He wrote a novel*, but not *I read a lamp which he wrote* and correspondingly not *He wrote a lamp*.

There are two problems in making a precise formulation of word-choice preservation. One is that in many sentence structures there seem to be special domains of word n-tuples over which the preservation held. This was resolved by showing that the different domains were due to preservation in different sentence structures. Thus in the passive N_2 *be Ven by* N_1 we find an active $N_1 t V N_2$ for *The picture will be seen by John* but not for (1) *The picture will be seen by candlelight* or . . . *by midnight* where there is no *Candle light will see the picture*; however, we can show that (1) presents the word-choices of (2) *People will see the picture* (or N_1 *will see the picture*) plus (3) *People's seeing the picture will be by candlelight* or . . . *by midnight*. By the side of (2) we have (2') *The picture will be seen by people*, where *by people* can be zeroed—this being itself one of the crucial transformations, discussed below. Then (2') plus (3), i.e. *The picture will be seen by people by candlelight*, can be reduced to (1). In this way we do not have to define subclasses of N, e.g. of place or time or such *ad hoc* categories as (in this case) *light*, for which we would say that the passive *by* N is not preserved from the active subject. Instead, we show that even in these cases the passive preserves the word-choices of the active, but that the second *by* N above comes from a different source.

The other problem in formulating the preservation of word-choice arises from the difficulty of working with word-choice as data. We cannot list all the word triples that occur in the $N t V N$ sentence structure, not only because they are too many but even more because not all speakers agree about all triples and many speakers are uncertain about some triples (as for example in (4) *John wrote a wall*). Furthermore, the list of triples varies with the context—e.g. (4) might be said about wall newspapers or graffiti; and it varies with time as subject matters and customary locutions change. In order to deal with the preservation of word-choice we therefore have to introduce, as a research procedure, a new degree of freedom on grammars: the varied acceptability of sentences. Instead of certain sentence structures having the same word-choices—or not having the same—we will speak of the word-choices having the same acceptability grading in both sentence structures. We can reach this grading if we merely recognize that the n-tuples of words have varied inequalities of unacceptability in each sentence structure of n word-classes.

8.1.2. *Sentence acceptability grading*[2]

To introduce this system of inequalities, we note first that not every word-sequence is either definitely a sentence of the language or definitely not one. Many word-sequences are definitely sentences, e.g. *I am here*, and others are definitely not, e.g. *ate go the*. But there are also many word-sequences which are recognized as metaphoric, or as said only for the nonce, i.e. as

[2] Large sections of 8.1.2–3 and 8.1.5–6 are revisions from Z. Harris, *Mathematical Structures of Language*, Interscience Tracts in Pure and Applied Mathematics 21, Wiley-Interscience, New York, 1968.

extended by the present speaker on the basis of 'real' sentences, e.g. *Pine trees paint well*, perhaps analogized from such sentences as *Spy stories sell well*. Other word-sequences are recognized as containing some grammatical play; and some are pure nonsense sentences (but may be accepted): *The cow jumped over the moon*.

It is also easy to find a word-sequence for which speakers of the language disagree, or readily alter their judgement, or cannot decide, as to whether it is a grammatical sentence of the language or not: e.g. *He is very prepared to go*; *He had a hard crawl over the cliff*. There can be various degrees and types of marginality for various word-choices in a form: e.g. the gradation from *He gave a jump*, *He gave a step* to dubious *He gave a crawl*, *He gave a walk* to objectionable *He gave an escape*.

Alternatively, instead of speaking of different acceptabilities for word-sequences as sentences, we can speak of different sentential neighborhoods. For each one of the marginal sentences can be an acceptable sentence in a suitable discourse, or under *not*; whereas the completely nonsentential word-sequence, i.e. those which do not belong to sentence forms (such as *go wood the*), appear without acceptability ordering, and only as subjects of special predicates such as the metalinguistic ones (e.g. in this case, . . . *is not a sentence*). For extending the grammar, the neighboring sequences in the discourse can supply the source of the extension: *While you had a walk along the valley, he had a crawl over the cliff*. For nonsense, the neighborhood might be nonsensical in the same way, as in a fairy-tale or nursery rhyme. Aside from the obvious cases of marginal sentencehood, many sentences are to be found only in particular types of discourse, i.e. with particular environing sentences; and many of these would indeed be but dubiously acceptable sentences outside of such discourses. Thus *The values approach infinity* is sayable in mathematical discourse but rather nonsensical outside it.

Acceptability does not depend on the truth of the sentence: *The book is here* is a grammatically acceptable sentence even if the book is in fact not here. It does not even depend in any simple way on meaning (as in *The cow jumped over the moon*). While there is a certain connection between meaningfulness, in the colloquial sense of the term, and acceptability, the connection is merely that a sentence which is acceptable in a given neighborhood has meaning in that neighborhood. We cannot assert that it is acceptable because it has meaning (i.e. because the combination of its word meanings makes sense), for in many cases a sentence is acceptable not on the basis of the meanings that its words have elsewhere. If *Meteorites flew down all around us* is acceptable, it is not because this is a meaningful word combination, since *flew* means here not 'flew' in the usual sense but some movement distantly similar to it. Nor can we say that acceptability exists for every word combination which makes sense in any way, for there are word combinations which can make sense but are not acceptable sentences: *Man sleep*; *Took man book*. *The air swam around me* may have low acceptability, although it can have meaning.

The standing of a word-sequence (carrying the required intonation) in respect to membership in the set of sentences is thus expressible not by two values, yes and no, but by a spectrum of values expressing the degree and qualification of acceptability as a sentence, or alternatively by the type of discourse or neighborhood in which the word-sequences would have normal acceptance as a sentence.

The complicated and in part unstable data about acceptability can be made available for grammatical treatment by means of the following pair test for acceptability. Starting for convenience with very short sentence forms, indicated by ABC, we make a particular word choice in each word class except one, in this case A: say p in B and q in C, written as B_p, C_q. Then for every pair of members A_i, A_j of A we ask how the sentence formed with one of the members, A_i in (1) $A_iB_pC_q$, compares as to acceptability with the sentence formed with the other member, A_j in (2) $A_jB_pC_q$. If we can obtain comparative judgements of difference of acceptability, we have for each pair A_i, A_j either (1)>(2) or (1)<(2) or roughly (1)=(2).[3] In this case the relation is transitive, so that if $A_iB_pC_q>A_jB_pC_q$ and $A_jB_pC_q>A_kB_pC_q$ it will always be the case that $A_iB_pC_q>A_kB_pC_q$. We would then have an ordering of the members of A in respect to acceptability in AB_pC_q. If the judgements which we can obtain in this test are not quantitative but rather in terms of grammatical subsets (metaphor, joke, etc.) or language subsets (fairy tales, mathematics, etc.), then some reasonable classification of the subsets could be devised so as to express the results in terms of inequalities (A_i being acceptable in the same subset as A_j, or in a subset which has been placed higher or lower). The results of the test might also take the form of some combination of the above. It is possible that we would obtain a relation in the set A such that for each pair either equality or inequality holds in respect to AB_pC_q, but without the relation being transitive. The system of inequalities, obtained from such a test, among sentences of the set AB_pC_q will be called here a grading on the set AB_pC_q; each sentence $A_iB_pC_q$ which has a grading in respect to the other sentences of the set will be called a (sentential) proposition. As seen in note 3, there may be more than one grading over the A in AB_pC_q, if there is more than one transformational source for ABC.

Other gradings can be obtained for all members of A in respect to every

---

[3] We may obtain more than one answer as to the acceptability difference between $A_iB_pC_q$ and $A_jB_pC_q$. For example, *The clock measured two hours* may be judged more acceptable than *The surveyor measured two hours*, in the sense of measuring off, i.e. ticking off, a period of time, but less acceptable than the latter in the sense of measuring something over a period of two hours. In all such cases it will be found that there are two or more gradings (in the sense defined immediately below) over all members of A in AB_pC_q; each grading is associated with a sharply different meaning of AB_pC_q and, as will be seen later, a different source and transformational analysis of AB_pC_q. Thus the sense in which *surveyor* is more acceptable than *clock* here is the case where *measured two hours* is a transform of *measured for two hours*.

word-choice in *BC*. In the same sentence form *ABC* we can also obtain a grading over all members of *B* for each word-choice in *A* and *C*, and so for all *C* for each word-choice in *AB*. We may therefore speak of a grading over the *n*-tuples of word-choices in an *n*-class sentence form as the collection of all inequalities over the members of each class in respect to each word-choice in the remaining $n-1$ classes.

Given the graded *n*-tuples of words for a particular sentence form, we can find other sentence forms of the same word classes in which the same *n*-tuples of words produce the same grading of sentences. For example, in the passive sentence form N_2t *be Ven by* N_1 (where the subscripts indicate that the *N* are permuted, with respect to N_1tVN_2), whatever grading was found for *A dog bit a cat*, *A dog chewed a bone*, *A book chewed a cup* is found also for *A cat was bitten by a dog*, *A bone was chewed by a dog*, *A cup was chewed by a book*. And if *The document satisfied the consul* may be found in ordinary writing, but *The set satisfies the minimal condition* only in mathematical writing, this difference in neighborhood would hold also for *The consul was satisfied by the document* and *The minimal condition is satisfied by the set*.

This need not mean that an *n*-tuple of words yields the same degree of acceptance in each of these sentence forms. The sentences of one of the forms may have less acceptance than the corresponding ones of the other, for all *n*-tuples of values or for a subset of them. For example, *Infinity is approached by this sequence* may be less acceptable than *This sequence approaches infinity*; but if the second is in mathematical rather than colloquial discourse, so would be the first.

In contrast, note that this grading is not preserved for these same *n*-tuples in N_2tVN_1 (i.e. a sentence structure obtained by interchanging the two *N*): *A cat bit a dog* is normal no less than *a dog bit a cat*; *A cup chewed a book* is nonsensical no less than *A book chewed a cup*; but *A bone chewed a dog* is not normal, whereas *A dog chewed a bone* is.

To summarize:

1. Inequalities. Using any reasonable sense of acceptability as sentence, and any test for determining this acceptability, we will find, in any sentence form $A(X_1 \ldots X_n)$ of the variables (word classes) $X_1 \ldots X_n$, some (and indeed many) *n*-tuples of values for which the sentences of *A* do not have the same acceptability. Alternatively, given this *A*, we will find some *n*-tuples of values for which the sentences of *A* appear in stated different types of discourse.

2. Preservation of inequalities. Often, we can find at least one other sentence form $B(X_1 \ldots X_n)$ of the same variables, in which the same *n*-tuples yield the same inequalities; or alternatively in which the same *n*-tuples show the same difference as to types of discourse.

The degree of acceptance, and sometimes even the kind of acceptance or neighborhood, is not a stable datum. Different speakers of the language, or different checks through discourse, may for some sentences give different evaluations; and the evaluations may change within relatively short periods of time. Even the relative order of certain n-tuples in the grading may vary in different investigations.[4] However, if a set of n-tuples yields sentences of the same acceptability grading in A and in B, then it is generally found that certain of the instabilities and uncertainties of acceptance which these n-tuples have in A they also have in B. The equality of A and B in respect to the grading of these n-tuples of values is definite and stable.

In this way, the problem of unstable word-choices is replaced by the stable fact that the same (perhaps unstable) word-choices which order the acceptability of sentences in one sentence form A do so also in some other one B.

A sentence may be a member of two different gradings, and may indeed have a different acceptability in each (e.g. *The clock measured two hours*, below and n. 3 above). It will then have a distinctly different meaning in each. Some sentences have two or more different accepted meanings; certain ones of these have normal acceptance as members of two or more different graded subsets, in which case they will be called transformationally ambiguous. Consider the ambiguous *Frost reads smoothly* (in the form *NtVD*). We have one grading of a particular subset of sentences, one in which *Frost reads smoothly* and *Six-year-olds read smoothly* are normal, while *This novel reads smoothly* is nonsense; this grading is preserved in the form *N tried to VD*: *Frost tried to read smoothly*, *Six-year-olds tried to read smoothly*, but not *This novel tried to read smoothly*. In another grading, of another subset of sentences, *Frost reads smoothly* shares normalcy with *This novel reads smoothly*, while *Six-year-olds read smoothly* is dubious; this grading is preserved in *One can read Frost smoothly* or *Reading Frost goes smoothly*, *One can read this novel smoothly*, and only dubiously *One can read six-year-olds smoothly*.

The relation between two sentence forms in respect to preservation of acceptability grading of n-tuples is not itself graded. We do not find cases in which various pairs of forms show various degrees of preservation of acceptability grading. The preservation of the gradings is a consequence of this. In most cases, over a set of n-tuples of values, two sentence forms either preserve the acceptability grading almost perfectly or else completely do not. The cases where grading is only partially preserved can be reasonably described in terms of language change, or in terms of different relations between the two forms for two different subsets of n-tuples.

When we attempt to establish the preservation of the grading between two sentence forms, we find in some cases that it does not hold for all the n-tuples of values in both forms, but only for a subset of them. For example, the

---

[4] There are special cases involving language change or style and subject-matter difference. For example, people may accept equally *The students laughed at him*, *The students talked about war*, but may grade *He was laughed at by the students* more acceptable than *The war was talked about by the students*.

N_1tVN_2/N_2t be Ven by N_1 relation holds for *The clock measured two hours/ Two hours were measured by the clock* but not for *The man ran two hours* (no *Two hours were run by the man*). We must then define the relation between the two forms over a particular domain of values of the variables. This raises immediately the danger of trivializing the relation: there are many more word n-tuples than sentence forms, hence for any two sentence forms of the same word classes it is likely that we can find two or more n-tuples of words whose relative grading is the same in the two forms. For example, given only *A dog bit a cat, A hat bit a coat*, and *A cat bit a dog, A coat bit a hat*, we could say that the grading preservation holds between N_1tVN_2 and N_2tVN_1 for this domain. Interest therefore attaches only to the cases where a domain over which two forms have the same grading is otherwise syntactically recognizable in the grammar: e.g. because the domain covers all of a given morphological class of words (all adjectives, or all adjectives which can add *-ly*); or because the domain, or the set of n-tuples excluded from the domain, appears with the same grading also in other sentence forms and is too large a domain for this property to be a result of chance.

One can specify the domain of one grading preservation in terms of the domain of another. For example, the passive is not found for triples which appear with the same relative grading both in N_1tVN_2 and in N_1tV *for* N_2 (checking all N_2, with a given V):

> *The man ran two hours*, and
> *The man ran for two hours*, whereas
> *Two hours were run by the man* is rare or lacking;

as against:

> *The clock measured two hours*, and
> *Two hours were measured by the clock*, whereas
> *The clock measured for two hours* is rare or lacking.

One can also specify the domain in terms of stated subsets of words. Some of the sets of n-tuples whose grading is preserved as above have some particular word, or an inextendable subset of words, or the complement set of these, in one of the n positions. For example, the passive is never formed from n-tuples whose V is *be, become*: no *An author was become by him*.[5]

In other cases, a subset of words in one of the n positions is excluded from the domain of the grading preservation, but only in its normal use; the same words may occur in that position in a different use. In the preceding example, we can say that the passive does not occur if N_2 names a unit of duration or measurement (*mile, hour*) unless V is one of certain verbs (*measure, spend, describe*, etc.). But in many, not all, of these cases, it is not

---

[5] It will be seen later that most of the problems of restricted domains apply to an aberrant case: the analogic transformations. These domains are partly due to the compounding transformation. For the base transformations, the domains are unrestricted, or restricted in simple ways.

satisfactory to list a subset of words which is excluded from the domain of a certain grading preservation between two forms, because these words may be nevertheless found in the domain, but with very low acceptance. For example, *The clock measured for two hours*, excluded in the preceding example, may indeed occur, in the sense of *The clock measured something, over a period of two hours*. And for this sense, *The clock measured two hours* has indeed no passive. So *The clock measured two hours* has two distinct meanings, one (barely acceptable) which appears also in *The clock measured for two hours* (also barely acceptable), and one normal one which appears also in the normal *Two hours were measured by the clock*.

Finally, since the use of words can (with some difficulty) be extended by analogy or by definition, it is in many cases impossible to exclude a word from the domain of a grading preservation (which will be stated in 8.13 to constitute a transformation). Thus if someone were to use *run* in the sense of *run up* (as *to run up a bill*, or *to run up a certain number of hours out of a time allotment*), he might say *They ran (up) two hours* and this could have a passive *Two hours were run (up) by them*. But the passive would be only for this sense of *They ran two hours*, and not for the one which is a grading preservation (transform) of *They ran for two hours*. In general, then, the domain of a grading preservation is a set of *n*-tuples defined not by a list of words in a particular position, but by the fact that the set of *n*-tuples participates in stated other transformations.[6]

The set of word *n*-tuples which preserve grading in two sentence forms is therefore characterized in most cases by syntactic properties, as above. At worst, the justification for distinguishing such a set of *n*-tuples is residually syntactic (e.g. to complete a set of grading preserving form-pairs). Each set of *n*-tuples, or rather the word class (in one of the *n* positions) which characterizes the *n*-tuple set, will almost always be found to have a semantic property (e.g. the *be* verbs, the unit names); but semantic properties can often be found also for other word sets which do not play a role here.

We therefore require that the grading preservation hold not necessarily over all word *n*-tuples of a sentence form, but over a nontrivial syntactically characterizable subset of these. This reduction of our requirement is especially useful in the case where we can partition the *n*-tuples of a sentence form *A* into subsets such that in each subset the grading is preserved as between *A* and another sentence form *B*; but the grading in the set of all *n*-tuples of *A* is not preserved in *B* because the order of the subsets of *A* (in respect to the gradings) may not be fully preserved in *B*.

The existence of restricted domains makes a general definition of transformations difficult; and there may be difficulties in discovering the precise domain of some particular transformations. However, the fact that many

[6] Cf. Henry Hiż, 'Congrammaticality, Batteries of Transformations, and Grammatical Categories', *Proceedings, Symposium in Applied Mathematics* 12, American Mathematical Society, 1961, pp. 43–50.

transformations have no restriction as to word subsets gives an initial stock of well-established transformations for a language which helps in determining various remaining transformations.

Hitherto we have considered sentences the grading of whose word-choices is preserved in some other set of sentences. However, for all sentence forms A, except the relatively short ones, the grading-preserving relation will be found, not only as between A and some other sentence form B, but also as between A and certain sets (pairs, triples, etc.) of sentence forms. Thus the form $N_3 N_2 t$ be Ven by N_1 (e.g. *Wall posters were read by soldiers*) has this relation not only to $N_1 t V N_3 N_2$ (*Soldiers read wall posters*) but also to the pair $N_2 t$ be Ven by N_1, $N_2 t$ be $P N_3$ (*Posters were read by soldiers, Posters were on a wall*), or the pair $N_1 t V N_2$, $N_2 t$ be $P N_3$ (*Soldiers read posters, Posters were on a wall*). Similarly, the sentence form N_1's Vn $V_{sst} N_2$'s An (as in *His arrival caused her lateness*) has this relation not only to N_2's An t be $P_{ss} N_1$'s Vn (*Her lateness was because of his arrival*) but also to the pair $N_1 t V$, $N_2 t$ be A (*He arrived, She was late*).[7]

What makes this manageable is that it will be found that there is a small number of elementary sentence forms, and that a small number of (trans-formational) physical differences suffices to connect every sentence form to one or more of these elementary ones. All sentence portions which appear as substructures in pre-transformational linguistics (e.g. the compound noun *wall posters*) are relatable to sentences (e.g. *Posters are on a wall*) in a regular way and with preservation of grading for the word-choices involved.

8.1.3. *Transformational Analysis*[8]

We now take two sentence forms A and B, of a given n word classes or morpheme classes or subclasses (as variables), where A and B differ in some

[7] *Vn, An* represent nominalized verb and adjective (*arrival, lateness*); V_{ss} is a verb whose subject and object are nominalized sentences.

[8] The system of partially ordered expansional equations, which provide an apparatus for deriving a sentence from an elementary sentence, is presented in 7.8 above, with bibliography. The possibility of deriving or synthesizing sentences from axiomatic (elementary) ones, in a grammar which is seen as an axiomatic theory of the language, is outlined in Z. Harris, *[Methods in] Structural Linguistics*, Univ. of Chicago Press, Chicago, p. 373. Transformations proper were developed in the service of discourse analysis, and a first list of English trans-formations appears in Z. Harris, 'Discourse Analysis', (1952) *Language* 28, 1–30. How transformations are a necessary development out of problems of traditional grammar, and the sentence algebra that results therefrom, is presented in Z. Harris, 'Co-occurrence and Transformation in Linguistic Structure', *Language* (1957) 283–340, reprinted in *Papers in Structural and Transformational Linguistics*, Reidel, Dordrecht, pp. 390–457. Linguistic and logical properties of transformations have been studied by Henry Hiż in many papers funda-mental to the subject, including the works cited in n. 6 above and in ch. 11 n. 12 and ch. 12 n. 15. Transformations as well as constituent analysis ('phrase structure') have been adapted to a generative system of grammar for creating sentential word-sequences, in a major series of publications by Noam Chomsky, beginning with his *Syntactic Structures*, Mouton, The Hague, 1957; 'A Transformational Approach to Syntax', in A. A. Hill, ed., *Proceedings of the Third Conference on Problems of Linguistic Analysis in English*, Univ. of Texas, Austin, 1962; *Aspects of the Theory of Syntax*, MIT Press, Cambridge, Mass., 1965.

fixed morphemes or small sets of morphemes (as constants of the forms), or in the order or omission of certain classes, and where in each form the sentences produced by each set of n values, one for each variable of the form, are graded as to acceptability or discourse environment. Since a particular sentence form may have two or more gradings for the n-tuples of values of its variables (i.e. may be syntactically ambiguous, as in *Frost reads smoothly*), we will use the term propositional form for a sentence equipped with a particular grading of the n-tuples (and having only one of the ambiguous readings, if there are such). We define a transformation $A \leftrightarrow B$ between the two propositional forms A, B, over all n-tuples of their values or a syntactically characterizable subdomain of them, if the grading of the n-tuples is approximately identical for the two. When we define a set of propositional sentences (propositions), where each proposition is a sentence with a position in a grading, and is identified by an n-tuple of values in a particular propositional form, then each transformation takes each proposition A_i of one propositional form A into the corresponding proposition B_i of another form B. Each grading preservation gives thus a partial transformation in the set of propositions.[9]

A modification of this definition can be made for the case where the relation holds between A and a whole set B of sentence forms (end of 8.1.2).

Transformational analysis is thus not primarily an indicator of the structure of each sentence separately, but rather a pairing of sets of sentences, and so of the corresponding sentences in each set, preserving sentencehood (which includes preserving approximate grading of acceptability as sentence).

Transformations generate an equivalence relation in the set of propositional forms (and in the set of propositions), and impose a partition on them. Each proposition obtained by a given n-tuple in one form may be called a transform of the corresponding proposition obtained by the same n-tuple in the other form. If one proposition is a transform of another, it occupies in the grading of its set the same position as the other does in the grading of the other set. Transformations are thus defined not directly on sentences, but on propositions, i.e. on graded sentences (sentences as members of a grading).

Given a transformation $A_1 \leftrightarrow A_2$, we now consider the difference between the morpheme-sequences in A_1 and in A_2. Since the various propositional forms which are transforms of each other contain the same word classes, the forms, as word-class sequences, cannot be arbitrarily different from each other. For two transformationally related forms, the difference in sequence of words (or morphemes) and word classes is, in general, any of the following: a permutation of word classes or constants; the addition or

---

[9] The term 'propositional sentence' or 'proposition' as defined here differs from 'proposition' in logic, where it represents the set of all sentences that are paraphrases of each other. However, it is an approach in natural language to the 'proposition' of logic, for paraphrases in language can be defined only on the basis of propositional sentences. Propositions will later be defined as particular sentence pairs (8.1.6.2).

omission of a constant; and only in limited ways, to be discussed below, the addition or omission of a class. And since the individual sentences which are transforms of each other contain the same n-tuple of word values, they add no further difference to the above differences between the propositional forms.[10]

Given a transformation $A_1 \leftrightarrow A_2$ the permutations, additions, and omissions which differentiate a morpheme-sequence (a proposition) in A_2 from its inverse image in A_1 are the same for all the propositions in A_2. This constant difference in A_2, which is associated with each transformation, will be called the trace of the transformation, and may be looked upon as an operation on the morpheme-sequences of A_1 yielding those of A_2. The trace may be, for example: in the absence of an expected operator (as in *and Mary piano* after *John plays violin*); or in the absence of a required argument (as in *He reads all day* from *He reads things all day*); or in an extra argument (as in *He stood the box on end* against *The box stood on end*, from *He caused the box to stand on end*); or from the fact that a particular subset of words admits the zeroing of a particular operator (e.g. *He spoke two hours* is zeroed from *He spoke for two hours*, above, whereas *He spoke two words* does not result from such a zeroing).

The importance of the trace is that it is a physical deposit in one member of the transformationally related pair of propositions. Each proposition can be covered disjointly by segments, each of which is a transformational trace or a residual elementary sentence (which can be considered the identity transformational trace). For this purpose, however, we have to accept as segments certain zero arguments (traces of zeroing, which erase the phonemic content of certain morpheme occurrences), and certain change-indicators (traces of permutations and morphophonemic change).

We thus find transformations to be not only a relation among propositions but also a segmental decomposition of propositions.

For convenience in investigating the set of sentences under these transformations, we may define the positive direction of the operation in a consistent way for all $A_i \leftrightarrow A_j$ pairs. Here, two criteria are used, which apply equivalently for some pairs of related propositional forms, and complement each other in the remaining pairs:

---

[10] The one difference between sentences which goes beyond the difference between their corresponding propositional forms obtains if a word of an n-tuple has one form in one sentence and another in its transform, i.e. if the shape of a word changes under transformation (e.g. *-ing*, below). This is covered by the morphophonemic definition of the word; if the change is regular over a whole grammatical subclass of words, it is included in the morphophonemic transformations (below). A similar situation is seen when the regularity of morphemic difference applies in certain cases to families of syntactically equivalent morphemes rather than to individual morphemes. Thus is some cases *-ing* and zero are equivalent nominalizing suffixes on verbs (imposed by certain operators): e.g. *He felt distrust* or *trust*, in contrast to *a dislike* but *a liking*. At the same time *-ing* occurs for all these verbs as a different 'verbal noun' suffix (imposed by other operators): *He kept up his distrusting* or *trusting* or *disliking* or *liking*.

1. We take the arrow in the direction $A_1 \rightarrow A_2$ if the number of sentence forms which include the sentence form or trace present in A_2 is smaller than the number of sentence forms which include that present in A_1: for example, we write *I say this* → *This I say* rather than *This I say* → *I say this* because the sentence nominalizations *Sn* do not exist for the *This I say* form (*My saying this*, no *This my saying*). If we took *This I say* as source, then nominalization would not take place on the source, but would on the permutation to *I say this* and again not on the permutation to *This say I*. However, if we take *I say this* as source, then nominalization takes place on the source but not on the permutations. This is clearly simpler than when *This I say* is taken as source.

2. We take the arrow in the direction $A_1 \rightarrow A_2$ if the number of morphemes (including reconstructible morphemes) in the trace in A_2 is not less than in A_1. In many cases the number of morphemes is observably greater: *It is old* → *It is very old*; *It is old* → *I think that it is old*. In other cases the number of morphemes is the same, and the transformation consists in some other regular change to a more restricted form: *He will come only now* → *Only now will he come*. Finally there are transformations in which the number of morphemes seems to decrease; however, the formulation of zeroing and pronouning shows that the apparently lost morphemes are still present in the derived sentence, but in zero phonemic shape (end of 8.1.4, item 1).

There are also other criteria which agree with these. For example, we take the direction as $A_1 \rightarrow A_2$ if the discourse environments of A_1 are much more varied than those of A_2.

8.1.4. *Survey of English elementary transformations*

The specific transformations found in English fall into a few elementary types or are successions of such elementaries. A check-list is given here as an example of the range of the transformations in a language.

1. **Zeroing of words reconstructible from the environment**. This applies to recurrence of words in specifiable situations (G136), as in *John will come but not Mary* from *John will come but Mary will not come*, and in *John advised me to go* from *John advised me for me to go*. The zeroing is in most cases optional. When the repeated word is a noun it can also in many situations be reduced to pronoun rather than to zero, as *John wrote and he also telephoned* or *John wrote and also telephoned* from *John wrote and John also telephoned*. In some positions a repeated word is always or almost always reduced, as in *John washed himself* from the rare *John washed John* (for the same person).

There is also optional zeroing of indefinite nouns or pronouns, in many positions (G153), as in *He likes to read* from *He likes to read things*, and in *She likes to watch dancing* from *She likes to watch people's dancing*. The grammar is made more regular and compact if we assume in these cases that there was an indefinite noun in the given position. If we do not, then the many verbs like *read* have to be both transitive and intransitive; and verbs like *watch* have to have not only a nominalized sentence (*people's dancing*) as object but also a verbal noun by itself (*dancing*).

Finally, in particular environments, there is optional zeroing of words which are unique or highly favored there (G158). For example, the relative pronouns *which*, *whom* can be zeroed in *the book which I needed* (to *the book I needed*), *the man whom I know* (to *the man I know*). It can also be argued that *I ask* has been zeroed in obtaining *Will he go?* from *I ask: Will he go?* from *I ask whether he will go*, and that *I request you* has been zeroed in *Come!* from *I request you that you come* (this being the only way to explain the *yourself* in *Wash yourself!*). Also it can be suggested that *to come* or *to be here* is zeroed in *I expect him* from *I expect him to be here*; otherwise we would have to say that *expect* can have as object not only a sentence (*I expect John to forget*, *I expect John to come*), but also a noun alone (with possible modifiers), as in *I expect John*. Furthermore, as has been noted, the gradation of likelihoods of nouns as objects of *expect* is roughly the same as that of the same nouns plus *to come* or *to be here* as objects of *expect*. And the meaning of *expect* with a noun object is the same as the meaning of *expect* with the same noun plus *to come* or *to be here* as object.

2. **Permuting word classes** (3.4.2–4). This is limited in English to a few positions or to particular words. One type is the 'fronting' of the object of a verb to before the verb or its subject, and the rarer moving of subject and object: *The teachers this deny*, more commonly *This the teachers deny*, and rarely *This deny the teachers*, against the normal English *The teachers deny this*. This is an alternative to the normal linearization (3.4.1), as is the interruption of a sentence by a secondary sentence (*John—this will surprise you—is coming back*, 3.4.4). The length-permutation and the moving of a modifier to before its host are apparently true permutations, i.e. rearranging of previously linearized words. So seems to be also the permuting of subject and tense after certain adverb-like words, as in *Little will he expect this* from *He will little expect this*, or in *Only then will he speak* from *He will speak only then*. This permutation is accompanied by other changes in such sentences as *Will he go?* from *I ask whether he will go*, and *Did he stop?* from *I ask whether he stopped*.

In contrast, some apparent permutations are not due to any process of moving words, but rather to a complicated zeroing of repetitions. Such are, for example, *These books sell easily* as against *They sell these books easily* (G368), and also the passive (1.1.2).

In these transformations the information is preserved, but there are differences in nuance, especially as to what is the topic and as to the aspect (completive, habituative, etc.) of the predication G304, 364).

3. **Single-word adjuncts**. There are a few words in English about which we can say transformationally no more than that they can be inserted next to a particular word class in a sentence to produce an expanded sentence: e.g. *very* in *He is very slow* from *He is slow*. There is no natural way of deriving the occurrences of *very* from a more general kind of adverb on sentences (as in transformation type 4, below), even though such derivation is possible in French, whence *very* was borrowed.

4. **Sentence nominalization** (G40). This refers to a sentence appearing, in modified shape, as the subject or object of a further verb (G77). As object: e.g. *John draws graffiti* in *I know that John draws graffiti, I asked whether John draws graffiti, I prefer for John to draw graffiti, I saw John draw graffiti, I saw John drawing graffiti, I know of John's drawing graffiti*. As subject: e.g. *That John draws graffiti is a secret, John's drawing graffiti has ended*, or *That John draws graffiti is unimportant*, or *That John draws graffiti surprised me*. When the nominalized verb can take more forms than just *-ing*, we find for example *The growing of marijuana is prohibited, The growth of profits doubled, The growth was a dark brown*. When the predicate of the nominalized sentence is an adjective or noun we find also such forms as in *I know of John's sadness* from *John is sad*, and *John's anger surprised me* from *John is angry*.

There are also cases where the nominalized subject or object is not a whole sentence but only the verb with object: *I can read it, I tried to read it, I tried reading it, He is reading it*; similarly *He is receptive to it, He is a recipient of it*. These can be treated as a case of the nominalized full sentence, above, with zeroing of the subject in the nominalized sentence on grounds of being the same as the preceding subject. Such an analysis fits the meaning, and in some cases also the history of the forms (G289).

There are also bi-sentential predicates, under which both the subject and object are nominalized sentences (G85): *That he seemed angry caused my departure, His coming here proves that he is innocent*. The predicates can be (chiefly): verb *V*, preposition *P*, or adjective plus *P*, or noun (*N*) plus *P*, or *PNP*, all with *be* to carry the tense (e.g. above, and *His departure is with my approval, His departure is due to my suggesting it, His departure is the result of my suggestion, That he wrote at all was only with the hope of her answering*). The meaning of the original sentence is preserved, with the meaning of the added material being added thereto.

5. **Conjoined sentences** (G85). Finally, there are sentences *S* composed of two sentences S_1, S_2, with a conjunction *C* between them. S_1 and S_2 are not nominalized, unless the whole of *S* is the subject or object of a further verb (as in *His writing and her not answering caused a stir*). Many zeroings can take place in S_2, rarely in S_1, in situations that depend on the type of *C* (below). The sequence CS_2 can move to many positions within S_1 and even before S_1: *The concert was given inside because it was raining; The concert,*

because it was raining, was given inside; *Because it was raining, the concert was given inside*. If *C* is *and*, *or*, *but*, or certain conjunction-like words such as *so*, then CS_2 cannot move to before S_1. The conjunction can also be *be P* (or *be AP*, e.g. *is due to*): *His departure was with* (or: *due to*) *my approval, He departed with* (or: *due to*) *my approval*. The PS_2, APS_2 can then move to in or before S_1: *Due to my approval, he departed*.

There is also an important conjunction *wh-*, though it is not generally recognized as such (3.4.4). *Wh-* can join S_1, S_2 provided there is some segment occurring in both. This segment (a noun with its modifiers, or *PN*, or a nominalized sentence) in S_2 is then pronounced into *-ich, -o, -en, -ere*, etc., adjoined to the *wh-*, and the CS_2 thus transformed usually moves to immediately after the segment's first occurrence, in S_1 (and to before it when *which is, who is* is zeroed): given *A man telephoned* and *The same man would not leave his name*, we obtain *A man telephoned, who would not leave his name* and more commonly *A man who would not leave his name telephoned*. A necessary condition is that the repeated segment have the same referent, as above. The *wh-*S_2 is called a relative clause.

Types 1–3 above transform free sentences into free sentences, with type 3 and some cases of type 2 adding material to the original sentence. Types 4, 5 transform free sentences mostly into bound sentences (i.e. component sentences), although conjunctions can be looked upon as transforming two free sentences into one free sentence. In any case they add material to the original.

There are transformations which can be analyzed as products of elementary transformations of the above types. For example, adjectives before nouns are the product of a *wh-* conjunction (type 5) acted on by zeroing of *wh-* (type 1) whereupon the residual adjective is moved (type 2, noted under type 5):

> *A book appeared; the same book is new*
> → *A book appeared which is new*
> → *A new book appeared*.

Another example is the question, which is the product of the following successive elementary transformations:

> *I ask whether John will go or John will not go*

repetitional zeroing:	→	*I ask whether John will go or not*
uniqueness zeroing:	→	*I ask whether John will go*
permutation with changes:	→	*I ask: Will John go?*
uniqueness zeroing:	→	*Will John go?*

This derivation of the question makes it possible to obtain both the yes–no question (*Will John go?*) and the *wh-* question (as in *Who will go?*) from the same transformational sequence on two slightly different kinds of source (G331). Many other transformational relations among sentences have partial similarities which suggest that they are composed in part of the same elementary processes. Such is the passive (1.1.2, G362), which contains the same *-en* as the perfect tense (*The book was written, John has written it*) and the same *by* as in nominalization (*The book was written by John, the writing of the book by John*). Such also is the set of nominalizations based on *-ing* listed in 4 above. The composition of these transformations is resolved by the more powerful methods of 8.2 below.

8.1.5. *Algebra of the set of sentences under transformations*

In transformational analysis of sentences we have two sets whose structure can be studied: the set of transformations and the set of sentences. Each transformation has a domain of sentences on which it can act and a range of sentences which result from its action. A transformation can be identified with the range of sentences it produces: e.g. the passive transformation is characterizable by the set of passive sentences. Then if the domain of some transformation, say the question, includes the passive sentences, we can say that that transformation acts on (the results of) the passive transformation (as well as others). Some transformations (roughly types 1–5 in 8.1.4) are elementary, i.e. not composed out of any other transformations; the remaining ones are the products of successive transformations (acting on the same sentences). A transformation can act on the resultant of another only if its domain is included in the range of that other. Hence certain transformations in the language are partially ordered in respect to the monotonically decreasing domains for which they are defined. And when the analysis of a sentence has more than one transformation operating at a single point in it, either they are simultaneous (on the same operand) or the domains of the successive transformations are monotonically decreasing (G364).

Because transformations are defined as going from a particular domain to a particular range, they offer a simplification in the characterizing of sentences: the process of composing sentences out of elementary sentences through transformations is a finite state process. This is due to the fact that the domain of a transformation is defined in terms of elementary sentences and of the transformations which produced the non-elementary sentences in that domain.

The succession of transformations in forming a sentence can be considered a multiplication, and it is possible to define an associative multiplication on the set of transformations of a language. However, the types of elementary transformations and of the non-elementary ones which are their products are so limited that the devices needed for making the set of transformations closed under this multiplication are *ad hoc*. The structure of the set of transformations would be of interest only if languages contained large numbers of different transformations, actualizing many different combinations of the elementary ones.

In any case, the set of transformations provides interesting structures in the set of sentences, S. In the set S as a whole there is perhaps no transformation (other than the identity) which can operate on all members of the set (which includes all sentences that are already the resultants of various transformations). However, for every member of S there are some transformations which operate on it to produce a member of S. Every member of S has a decomposition by members of the set of transformations into a particular (partially ordered) subset of elementary sentences (or rather, of occurrences of these, since some may appear more than once in the decompositions). A transformation's acting on a sentence does not alter the composition of the sentence as it had been hitherto.

In terms of 8.2, we can define O_{oo}, O_{noo}, O_{oon} (3.1) as binary operations; and we define all the other second-level operators as well as all transformations as unary operations: $S \rightarrow S$. Any unary transformation f which is defined on the whole set S as domain, $f: S \rightarrow S$, is a 1–1 mapping of the set of propositions into itself. The other unaries are isomorphic partial transformations in the set of S, each f sending one subset (which satisfies the conditions for the operand of that f) into another subset. The nominalization of a sentence when it is under a second-level operator is thus not an independent transformation but an argument-indicator (G40).

As to the binaries: a subset of them acts as a binary composition in the set S, mapping $S \times S$ onto S. For example, *and*, and to a lesser extent *or*, may be found (with somewhat different effects in some cases) between any two S, even between question and assertion, etc. (*Will you go, or else I will go*; *I will go, and will you go?*). The rest are partial transformations from a subset of $S \times S$ to a subset of S.

Given a suitable list of elementary transformations in a language, each graded sentence (proposition) in the language can be characterized by a unique ordered product of elementary sentences and elementary transformations, aside from commutativity of certain transformational products. If two transformational products which differ in more than this local commutativity produce the same word-sequence, then the word-sequence is an ambiguous sentence, i.e. it represents two graded sentences (and their meanings).

The special provision for commutative sub-sequences of a transformation-sequence that characterizes a graded sentence can be obviated if we say that what characterizes a graded sentence is not a sequence of transformations but a partially ordered set of transformations, which is identical with the sequence except that each commutative sub-sequence is an unordered set at the point in the sentence that was occupied by the commutative sub-sequence. In the semilattice decomposition below, for instance, the transformations at 12 and 13 are unordered among themselves, but the set of them is linearly ordered in respect to those at 11 and 14.

While the trace of a transformation is the same wherever it occurs in the

partially ordered characterization of a sentence, the operand of the transformational operation is different for each occurrence of it, and can be
stated in terms of its position in the partial ordering.

The transformations impose two different and important partitions on the
set of propositions (graded sentences).

One partition is according to the trace which each proposition contains:
two propositions are (transformationally) equivalent if they contain the
same ordered traces. The relation R of having the same transformation-
product trace (without regard to the elementary sentences on which the
transformations operate) is an equivalence relation on the set P of propositions. The factor set P/R is the set of transformation applications. The
mapping of the set P onto its factor set P/R, assigning to every proposition in
P the product in P/R whose trace that proposition contains, is the natural
mapping of P onto P/R. The propositions which contain the traces of no
transformation are sent into the identity of the set of transformation-
sequences. These sentences are the kernel of the natural mapping, and will
be referred to as sentences of the kernel, or kernel-sentences, K. They
are important because, for each graded sentence, they are the residual
(elementary) sentences S_e under transformations. (This requires that sentence expansion be a transformation, as at the end of 8.1.4.1.).

The other partition is according to the residual sentences which are
contained in each proposition under transformations: two propositions are
in the same equivalence class if they contain the same kernel-sentences and
ordered operations on these. All sentences which are transforms (not only
paraphrastic) of each other contain the same residual sentences. The
statement that sentences A_i, B_i are corresponding members (with same
word choice) of sentence sets A, B which are transforms of each other is
expressed in transformational theory by saying that $A_i \leftrightarrow B_i$ if and only if
there exists a succession of elementary transformations (including the
identity transformation) which sends one of these into the other. Two
occurrences of e.g. *They have reduced wages* are not transforms of each
other if one is derived from *They reduced wages* (via *X has the situation of the
same X reducing wages*) and the other from *They have wages which are
reduced* (via *X has the situation of Y reducing X's wages*). As these transformations have been defined, sentences differing only in transformations
must have the same residual sentences. These residual sentences of each
sentence contain no transformational trace, and are therefore in the kernel
of the natural mapping above. In the partition of the set of propositions into
subsets whose members are transforms of each other, each subset contains
one or more kernel-sentences, which are contained in each sentence of the
subset. These kernel-sentences generate the subset by means of the transformations.

This transformational structure of sentences makes it possible to define
sentences as those objects among which the transformational relation

obtains. A word-sequence is a sentence if it has a transformational relation to a sentence.

We have seen that each sentence has a partially ordered decomposition into elements. The decomposition is unique for each proposition, if analogic transformations are taken as single elements. There are certain restrictions on the combinations that occur in a decomposition; certain combinations occur in no decomposition.

With the elements taken to be kernel-sentences and unary and binary elementary transformations, each proposition of the language can be written uniquely as a sequence of element symbols requiring no parentheses, for example in the manner of Polish notation in logic. Certain sub-sequences are commutative (representing elements unordered in respect to each other).

Because both the words and the transformations are partially ordered, and because a transformation is carried out precisely when the conditions for it are satisfied in the sentence (3.3.4), all this structure forms a lattice. For sentences which contain only one kernel-sentence, the decomposition can be represented as a lattice with the kernel-sentence as null element and the given sentence as universal element. The points represent the kernels and transformations: c is the least upper bound of a, b, if c is the first that can be applied after both a and b (i.e. on the resultant of a, b), with co-arguments linearly ordered, on the way from the kernel-sentence to the given sentence, in this decomposition; and correspondingly for the greatest lower bound. The transformation at the universal element can be taken as the final sentence intonation. The decomposition of a sentence containing more than one kernel is a semilattice; in this case, each conjunction C is oriented, with a distinction of right-hand and left-hand, because S_1CS_2 (which would be the resultant of ordered S_1, S_2) is a sentence related, but not generally equivalent, to S_2CS_1 (which would be the resultant of ordered S_2, S_1).

. The set of all lattices and semilattices, one for each proposition, has certain properties, i.e. certain dependencies among the occurrences of particular transformations in particular relative positions within a lattice. The most obvious one is that for each lattice, whose universal element is an arbitrary sentence, the null elements are always a kernel-sentence. For each kernel-sentence beyond 1 there is precisely one O_{oo} point which is its l.u.b. in respect to one of the other kernel-sentences in the semilattice.

As an example of the decomposition of a sentence we take the following:

> *The adrenal appears more and more as a prime endocrine gland: its importance grows in the animal scale: among the mammals, it has become indispensable for life, its removal leads rapidly to death: its functions are multiple.*

This sentence is analyzed in Figure 8.1.

A characteristic of transformational analysis is that languages are rather similar in their transformational structure, and that given a sentence in one

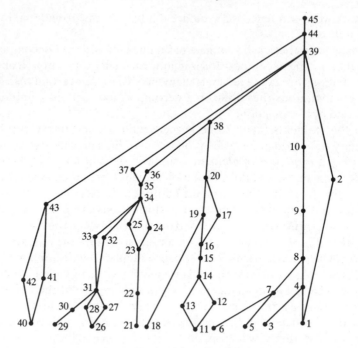

FIG. 8.1. Semilattice of an English sentence.

1. K: *Adrenal is a gland.*
2. mO_{oo}:   *the.*
3. K: *A gland is endocrine.*
4. O_{oo}:   *wh-.*
5. K: *A gland is prime.*
6. K: *A gland is endocrine.*
7. O_{oo}:   *wh-.*
8. O_{oo}:   *wh-.*
9. p: *appears as.*
10. O_o: *more and more.*
11. K: *Adrenal is important.*
12. Pronoun: *Adrenal→it.*
13. p: *Adrenal has importance.*
14. O_o: *Adrenal's importance grows.*
15. p: *Adrenal's importance has growth.*
16. O_o: *Growth of adrenal's importance is along a scale.*
17. m: *Adrenal's importance grows along (or: in) a scale.*
18. K: *Animals are (ranged) in a · scale.*

19. O_{oo}:   *wh-.*
20. mO_{oo}: *the.*
21. K: *A mammal lives.*
22. p: *A mammal has life.*
23. O_o: *Adrenal is indispensable for a mammal's life.*
24. Pronoun: *mammal→it.*
25. p: *has become.*
26. K: *N_i removes adrenal.*
27. p: *N_i effects removal of adrenal.*
28. Pronoun: *adrenal→it.*
29. K: *A mammal dies.*
30. p: *A mammal suffers death.*
31. O_{oo}:   *N_i's removal of it leads to a mammal's death.*
32. z: *N_i's→zero.*
33. mO_o: *N_i's removal of it leads rapidly to a mammal's death.*
34. O_{oo}:   *It has become indispensable for a mammal's life, removal of it leads rapidly to a mammal's death.*

language and its translation in another, the decomposition of each sentence in terms of the transformations of its language will be quite similar. An example is the analysis in Figure 8.2 of the Korean translation of the first part (itself a whole sentential structure) of the sample sentence above.[11]

8.1.6. *Structures in respect to transformations*

8.1.6.0. Introduction

Transformational analysis creates in a language many subsets of sentences, among which various structural and semantic relations hold. A few examples follow.

8.1.6.1. Unambiguous subsets of sentences and decompositions

If the intersection of word subclasses were empty, and if there were no degenerate results from the paraphrastic transformations, there would be no grammatical ambiguity; the only ambiguity in language would be due to the spread of meanings of the words in the elementary entities (kernel-sentences and second-level operators). However, in natural language, it is frequently the case that different classes or subclasses of words, distinguished by their being part of different elementary entities, have some members in common, which reduces the vocabulary needed for a language. This is a source of degeneracies in the set of produced sentences. It can result in the same sequence of words being produced by different transformations on different kernel-sentences.

If we wish to form a language without grammatical ambiguity, we can associate each ambiguous sentence with its transformational decompositions (below), or with a distinguished partial sentence of it (8.1.6.2); in the latter case we have an unambiguous language which consists only of sentences (more precisely, sentence-pairs) and not of analyses. Given a word-sequence which is a grammatically ambiguous sentence, the two or more propositions

[11] The translation and analysis is the work of Maeng-Sung Lee.

Fig. 8.1 *cont.*

35. O_{oo}:	. . . *among mammals.*		40. K:	*A gland functions.*
36. mO_{oo}:	*the* (on second *mammal's*).		41. *Pronoun:* *gland→it.*	
37. z:	*a mammal's*→zero; *the*		42. p:	*A gland has functions.*
	mammal's→zero.		43. O_o:	*Its functions are multiple.*
38. O_{oo}:	semicolon.		44. O_{oo}:	Semicolon.
39. O_{oo}:	semicolon.		45. O_o:	Sentence intonation.

Note: K, kernel-sentence; m, morphophonemic change (7.2); p, preverb (G289); z, zeroing.

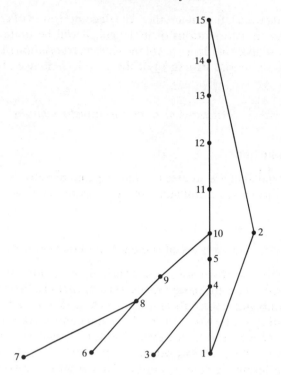

təuktə,   pusin ʉn cuŋyoha n nœpunpi sən ʉlo poi ə ciko iss ta
more and more   adrenal   principal   endocrine gland   seem/appear

More and more the adrenal is coming to appear to be a principal
endocrine gland.

FIG. 8.2. Semilattice of a Korean sentence

1. *K*:   *pusin-i sən-i-ta* 'Adrenal is a gland.'
2. *O_o*: *u n* on *pusin* (marks the topic-word of the sentence).
3. *K*:   *sən-i nœpunpi-lul ha-n-ta* 'A gland does internal secreting.'
4. *O_oo*: E-nominalization (equivalent to *wh-*): → *pusini nœpunpilul hanun sənita*
   'Adrenal is a gland which does internal secreting.'
5. *z*:   on the appropriate verb *hata*: → *pusini nœpunpisənita* 'Adrenal is an
   endocrine gland.'
6. *k*:   *sən-i cuyoha-ta* 'A gland is principal.'
7. *K*:   same as 3.
8. *O_oo*: E-nominalization: → *nœpunpilul hanun səni cuyohata* 'The gland which
   does internal secreting is principal.'
9. *z*:   on *hata*: → *nœpunpisəni cuyohata* 'The endocrine gland is principal.'
10. *O_oo*: E-nominalization: → *pusini cuyohan nœpunpisənita* 'Adrenal is a principal
    endocrine gland.'
11. *p*:   *poi-ta* 'appears': → *pusini cuyohan nœpunpisənulo pointa* 'Adrenal appears
    to be a principal endocrine gland.'

which are expressed by the word sequence differ in their decomposition semilattices. These semilattices differ at two or more linearly ordered points, the later-operating point obliterating the difference brought in at the earlier point. For if two lattices have different operations—second-level operators, sameness statements (G96), reductions—at only one point, and yet produce the same word-sequence, it would have to be the case that the different operations at that point in the two lattices introduce the same words in the same positions of the sentences under construction; but these words would be appearing, in the two sentences, as parts of different operations which act in the same way in the order of operations. Whether this is possible, i.e. whether a given language contains two identically operating transformations such that a part of one consists of the same trace as some part of the other, can be seen from the list of transformations for the language. In English, this does not seem to happen. If, then, we disregard this possibility, there will have to be two or more differences between two decomposition semilattices of an ambiguous word-sequence.

Any two decompositions of an ambiguous sentence must be similar to each other (except for the differences specified below); otherwise they would not yield the same word-sequence. In most cases, they differ in a kernel-sentence and also in one or more transformations (which obliterate the difference in the kernel-sentence). For example, in the ambiguous *Frost reads smoothly*, one decomposition (omitting some details) is shown in Figure 8.3.1. N is used here for a disjunction of all N which might occur in this position of the kernel-sentence; the disjunction could be pronouned as *anything*, *things*, etc. After point 2 the resultant is:

Frost's reading of N_2 is smooth.

After 3, with zeroing of the indefinite N_2,

Frost's reading is smooth.

Given 4 without 3, we have *Frost reads things smoothly*; but the combined effect of 3 and 4 is:

Frost reads smoothly.

The other decomposition is shown in Figure 8.3.2. N is used here as in Figure 8.3.1, but it covers a different subset of nouns, and the corresponding pronoun would be *one*. As in Figure 8.3.1, z is the zeroing of an indefinite, but it applies to an indefinite in a different position.

Fig. 8.2 *cont.*

12. *p*: *ci-ta* 'comes to: → *pusini cuyohan næpunpisənulo poiəcinta* 'Adrenal comes to appear to be a principal endocrine gland.'
13. *p*: *iss-ta* 'is . . . ing': → *pusini cuyohan næpunpisənulo poiəciko issta* 'Adrenal is coming to appear to be a principal endocrine gland.'
14. O_o: adverb *təuktə* 'more and more'.
15. O_o: sentence intonation.

5 • O_o: sentence intonation: → *Frost reads smoothly*.

4 • m: *Sn is A*→*SAly* (denominalization).

3 • z: of N_2 → zero.

2 • O_o: *is smooth*.

1 • K: *Frost reads N_2*.

FIG. 8.3.1. Decomposition lattice 1 of an ambiguous sentence.

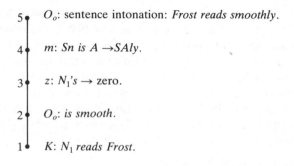

5 • O_o: sentence intonation: *Frost reads smoothly*.

4 • m: *Sn is A* →*SAly*.

3 • z: N_1's → zero.

2 • O_o: *is smooth*.

1 • K: *N_1 reads Frost*.

FIG. 8.3.2. Decomposition lattice 2 of an ambiguous sentence.

After 2, the resultant is:

One's reading of Frost is smooth (or: *goes smoothly*).

After 3, it is:

The reading of Frost is smooth (or: *goes smoothly*).

And after 4, it is:

Frost reads smoothly. (If the denominalization were made without 3, it would yield *One reads Frost smoothly*.)

In some cases, the kernel-sentences of both decompositions are the same, and all differences are in the transformations. This requires that the importation or change of words by some succession of operations in one decomposition be the same as that due to some other succession in the other decomposition. This is possible, since there are several cases in which a particular word or morpheme (e.g. *be*, *-ing*) appears in more than one transformation. Indeed, one of the motivations in defining the elementary transformations was to have each segment of a transformation occur in only one transformation. This result could not be completely achieved in the case of English. Hence it is possible that identical traces may be produced by different successions of transformations, especially if some of them are zeroing, which may zero parts of the trace that is due to preceding transformations. Any such situations which can arise can be determined from an inspection of the list and product table of transformations for the given language.

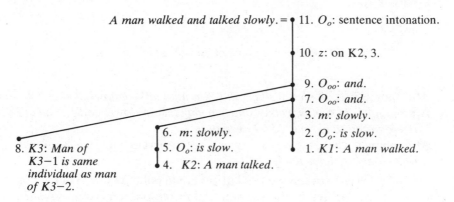

A man walked and talked slowly. = 11. O_o: sentence intonation.

10. z: on K2, 3.

9. O_{oo}: *and.*
7. O_{oo}: *and.*
3. *m*: *slowly.*

6. *m*: *slowly.*   2. O_o: *is slow.*
5. O_o: *is slow.*   1. *K1*: *A man walked.*
8. *K3*: *Man of*   4. *K2*: *A man talked.*
K3−1 *is same*
individual as man
of K3−2.

FIG. 8.4. Decomposition 1 of an ambiguous sentence.

For a simple example, we consider *A man walked and talked slowly.* One decomposition is given in Figure 8.4. The resultant sentences after each point in the semilattice are:

After point 2: *1, A man's walking was slow.*
After point 3: *1, A man walked slowly.*
After point 5: *2, A man's talking was slow.*
After point 6: *2, A man talked slowly.*
After point 7: *1, A man walked slowly and 2, a man talked slowly.*
After point 9: *1, A man walked slowly and 2, a man talked slowly, and 3, man in 2 refers to the same individual as man in 1.*
After point 10, where *z* replaces *K3* by zeroing those words in *K2* which are identical with corresponding words in *K1* (in this case *a man, slowly*) and the permuting of the residue of *K2*: *A man walked and talked slowly.*[12]

The other decomposition is given in Figure 8.5. The resultant sentences after each point are:

After point 3: *2, A man's talking was slow.*
After point 4: *2, A man talked slowly.*

[12] The kernel-sentences are numbered in order of appearance in the lattice.

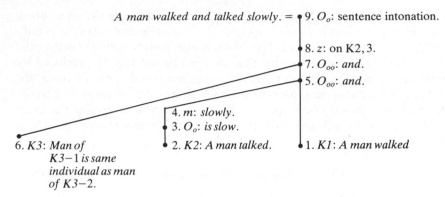

A man walked and talked slowly. = 9. O_o: sentence intonation.

8. z: on K2, 3.

7. O_{oo}: *and.*

5. O_{oo}: *and.*

4. *m: slowly.*
3. O_o: *is slow.*

6. *K3: Man of*   2. *K2: A man talked.*   1. *K1: A man walked*
   K3−1 is same
   individual as man
   of K3−2.

FIG. 8.5. Decomposition 2 of an ambiguous sentence.

After point 5: *1, A man walked and 2, a man talked slowly.*
After point 7: *1, A man walked and 2, a man talked slowly and 3, man in 2*
   refers to the same individual as man in 1.
After point 8, where the zeroable words are in this case only *a man*: *A man*
   walked and talked slowly.

The differences between the two lattices are in point 2 (and 3) of the first
one, which are lacking in the second, and in the differences in repeated
arguments as between point 10 of the first lattice and point 8 of the second.

8.1.6.2. The distinguishing partial sentence

Instead of differentiating the two decompositions of an ambiguous sentence
by the two or more points of difference in each, we can, alternatively,
differentiate them by a distinguished partial sentence in each. In a lattice
presenting a decomposition of a sentence, the resultant at a given point
(i.e. the sentence formed at that point, by the kernel-sentences and the
operations at all lower points) will be called a partial sentence of, or
grammatically included in, the resultant of each higher lattice point.

Two semilattices of an ambiguous sentence have different partial sentences
over a statable section: from the earlier-operating point of difference to the
later-operating point of difference (which cancels the first). Each of the two
semilattices, and hence each of the two propositions which they decompose,
can be distinguished by the least (most included) partial sentence which
does not appear in the other semilattice. If the semilattices differ in kernel-
sentences, the different kernel-sentences are the distinguishing partials.
If not, and one of the semilattices contains a zeroing of a previous operation,
then it is distinguished by the previous zeroable material, while the other is
distinguished by the first partial sentence in which the lack of the zeroable
material is recognizable. Hence for each *n*-fold ambiguous sentence we can

form *n* subsets of sentences, each subset consisting of that sentence and the relevant distinguishing partial sentences. For the examples above, the sentence-pairs are:

1. (*Frost reads smoothly; Frost reads N*),
2. (*Frost reads smoothly; N reads Frost*);
1. (*A man walked and talked slowly; A man walked slowly*),
2. (*A man walked and talked slowly; A man walked*).

We can form a set of subsets of sentences as follows: For each ambiguous sentence, a subset is formed for each different decomposition of the sentence and the differentiating partial sentences of that decomposition, these being in some cases simply the kernel-sentences and in other cases the kernel-sentences carrying a particular transformation. For all non-ambiguous sentences, the subset is simply the given sentence itself. This set of subsets of sentences contains no ambiguities, and maps isomorphically onto the set of propositions of the language.[13] This set of subsets can be obtained from the set of sentences by means of the elementary transformations, and provides us with a set of sentence-pairs (or more generally sentence subsets) containing no ambiguities, which is expressed purely in terms of sentences (subsets of sentences) without explicit reference to the analyses or gradings of the sentences.[14]

8.1.6.3. Graphs of the set of sentences

The existence of a unique decomposition for graded sentences makes it possible to order the whole set of graded sentences in respect to their decompositions, and to investigate various relations among sentences and certain properties of the whole set of sentences.

There are various types of graphs which represent the relations among sentences in such a way that properties of transformations or of subsets of sentences can be obtained from properties of the graph. First we describe a graph which gives the grammatical inclusion relation among all propositions (i.e. graded sentences) of the language.

---

[13] The isomorphism with the set of propositions depends on the fact that no word-sequence can appear twice in an acceptability-graded subset of sentences. If a sentence is *n*-way ambiguous, it appears in *n* different acceptability-graded subsets of sentences, i.e. in *n* partitions of the set of propositions.

[14] Since ambiguous sentences have various transformational decompositions corresponding to the various grammatical meanings, and since the traces of these various transformations affect the word-repetition which is required in all regular *SCS* . . . *CS* and discourse neighborhoods in which the ambiguous sentence appears (9.1), it follows that each *n*-way ambiguous sentence appears in *n* different sets of *SCS* . . . *CS* or discourse environments. The neighborhood therefore differentiates the different meanings of an ambiguous sentence (i.e. in most cases a sentence is not ambiguous in its neighborhood, and can be disambiguated by appeal to its neighborhood), unless the differentiating parts of the neighborhood have been zeroed.

Consider the decomposition lattice of an arbitrary proposition S_1. Here the points are operators (including the null points which select the kernel-sentences, and the universal point which introduces the sentence intonation); the lines connecting the points can be considered to represent a minimal set of independent partial sentences of S_1 sufficient to characterize S_1. Because of the partial ordering of the transformations, some of the partial sentences of S_1 do not appear in the semilattice.

We now construct a graph representing all the partial sentences of S_1, i.e. all the sentences which can be obtained in the course of any order of decomposition of S_1, as follows: each operator point of the lattice is replaced by an element consisting of a directed edge (representing the second-level operator or the transformation) from the end point of the preceding element, to a terminal vertex (which represents the resultant partial sentence); except that the null point is replaced by a vertex alone, and the universal point is replaced by nothing. In addition, each set of n lattice points which are unordered among themselves is replaced by $2^n - 2$ vertices representing all the sentences that can be formed by combinations of the n operations from 1 up to $n-1$ at a time (we draw connecting edges first for each pair, then each triple, and so on, up to the n-tuple of the n points; each set of connecting edges ends in a vertex representing the partial sentence).

In this directed graph of S_1, whose edges are operations, all the vertices of the graph are all the partial sentences of S_1. The graph for each partial sentence S_q of S_1 gives all the partial sentences of S_q, and is a subgraph of the S_1 graph. A particular partial sentence S_p of S_1 may also be a partial sentence of some other sentence S_2. In that case the graph of S_p is a subgraph of the S_2 graph as well.

We now consider the fact that, in the unbounded set of sentences, every sentence is a partial sentence of certain other sentences. It is therefore possible in principle to construct a single such graph for the set of all the propositions of the language. Each proposition is a vertex, and appears only once. It is the terminal vertex for all the transformations (edges) that produce it from its partial sentences, in all the ways in which this can be done, and it is the initial vertex for all the transformations which produce another sentence from it. Because the vocabulary and kernel-forms and transformations are each only finitely many, there is a finite number of edge-sequences $e_1, \ldots, e_n$, such that for arbitrary e_i, $1 < i < n$, we have that e_i reaches a particular vertex A_i after which all that can happen are repetitions of portions of the edge-sequence entering A_i, starting with various preceding vertices; these can obviously be enumerated. We can then connect (by the *and* of logic) all the vertices A_i, $i=1, \ldots, n$, into a vertex U; the *and* of logic occurs between arbitrary S, and is available in natural language. The sentence represented by U is the informationally maximal sentence of the language, because it has the following properties: all further sentences which can be formed transformationally from it contain repetitions, to any

number, of one or more distinguished parts of U; or a referentially unbounded regress of metalinguistic sentences about metalinguistic sentences
about U, its parts, and its extensions. Since the information contained in any
repetition of a sentence part, or a metalinguistic statement about it, is
statable in terms of that sentence part, we can state the informational
content of any further sentence in terms of the parts of U. Since U is finite,
the number of its parts is finite. Except for these foreseeable and enumerable
further sentences built out of U, each proposition of the language appears
once in the graph of U as a distinguished partial sentence of U and its graph is
a subgraph of the U graph. The construction for U does not necessarily hold
for those sentences in mathematical and other symbolisms which do not
have a structurally equivalent translation in natural language.

Put differently: for a particular language at a particular time, it is possible
to construct a graph of the longest (finite) sentence which does not contain
iterations of distinguished parts of itself. It is then possible to describe in a
(finite) adjoined sentence what are these distinguished iterable parts and
what is the informational effect of iterating each of them (to any number of
iterations).[15]

We thus have a directed graph of all the sentences of the language, and
for informational interpretation a finite graph. Any sequence of edges
connecting two sentences is a transformational path which takes us from one
sentence to the other, where a transformation is taken positively when going
in the direction of the edge, and inversely when going against the direction of
the edge.

8.1.6.4. Language-pair grammars

The grammars of all languages are not arbitrarily different from each other.
The similarities in basic structure are sufficiently great as to make it possible
to take the similarly formulated grammars of any two languages and
combine them into a single grammar which exhibits that which is common to
the two languages and relates it to the features in which the two grammars
differ. Explicit and rather obvious methods for doing this can be formulated,[16]
and can be so devised as to exhibit the structural differences between any
two grammars.

In particular, the types of elementary transformations of various languages
are quite similar one to the other. The elementary transformations of a
particular type in one language have approximately the same semantic effect

---

[15] For example, one can state finitely the meaning of any number of iterations of *very* at any
point in which it occurs; one can state in a (finite) sentence the meaning of n iterations of, say,
son's or of *son's daughter's* as a function of n. The sentences which describe the information of
each iterable part can then be conjoined by logical *and* to make one sentence giving the
information that is contained in all iterations of all parts of U.

[16] Z. Harris, 'Transfer Grammar', *International Journal of American Linguistics* 20 (1954)
259–70.

as the corresponding ones of the same type in another language, to the extent that appropriate correspondences can be determined. Furthermore, as was seen at the end of 8.1.5, if a sentence in one language is transformationally decomposed and is also translated into another language, the transformational decomposition of the translation will consist largely of a similar ordering of corresponding kernel-sentences and operations.

This means that in principle language translation can be effected by decomposing sentences in one language, translating the components (elementary sentences, second-level operators, and transformational instructions) into another language, and recomposing the sentence in the other language. Word meaning-ranges correspond (hence translate) better in these components than in the given sentence.

8.1.7. *Transformations as a theory of syntax*

Transformations present a single relation (as method of analysis) which provides a decomposition of each sentence (8.1.4), and also a decomposition of the set of sentences (8.1.5). Given the existence of elementary sentences with their word-choices, it can thus provide a theory of the non-elementary sentences. As a grammatical system, transformations go beyond traditional word-class grammar, providing a grammar that characterizes the actual sentences of a language. They do this not by specifying the word-choices of every sentence but by accepting as given the word-choices in elementary sentences, and then indicating in which sentence structures these choices are preserved. In so doing they reduce word-choice from the unbounded set of possible sentences to the very large but finite set of all possible elementary sentences. This is done within the universe of traditional grammar, i.e. not by investigating the imprecise and unstable word-choices but by showing stable, grammatically regular construction-differences which preserve word-choice. This has proved a large step toward a least grammar (2.2).

The transformational grammar, though more compact than the traditional ones, is more powerful in that it makes many relevant distinctions in grammatical form. Most importantly, transformational grammar shows that a word-sequence is not enough to characterize a sentence, because the same sequence may have resulted from two or more different source sentences, each with a different transformation, thus creating two or more sentences, each with its meaning. In contrast, the word-sequence of an elementary sentence suffices to characterize it, wherefore sentences are best analyzed as transforms of their elementary sentences. Transformational analysis also brings together grammatical structures that are indeed related. For example, *This is proof of what I said* comes out as a transform of *This proves what I said*, instead of the two being unrelated sentences that merely contain etymologically related words. Also, a great amount of multiple classification is avoided. Words which have several different types of object are shown to have a single object-type with transformations on it (e.g. by the side of *I expect John to be here*, the form *I expect John* contains not a different object-type for *expect* but merely a zeroing of *to come*). Nominalized verbs and adjectives (e.g. *absence*, and so most abstract nouns) are verbs and adjectives (*absent*) under a further operator. And much of morphology is treated as residues of transformations (e.g. *His boyhood was harsh* is a transform of some such sentence as *His situation of being a boy was harsh*—as indeed it was historically). The success in doing all this shows that transformations are not an artifice or short-cut of structural description but a real relation among sentences and a real property of language.

The inadequacy of word-sequence as sole characterizer of sentences has led all grammatical models to characterize sentences as (regular) sequences of (regular) sub-sequences of words, these sub-sequences being in general not themselves sentences (e.g. constituents, strings). In transformational grammar these entities which are in general larger than words but in general smaller than sentences are shown to be simply the elementary sentences of the language.

With the preservation of word-choice comes meaning-preservation. Somewhat as two words which have the same selection (co-occurrences) are synonyms, so are two grammatical constructions which support the same selection; and indeed the transformations are found to be paraphrastic, aside from considerations of style and nuance. Transformational grammar also states grounds for a sentence being ambiguous (if it results from different transformations on different sources)—this is in contrast to lexical synonymity and lexical ambiguity.[17] Furthermore, because the domains of many transformations are fuzzy, transformations create various types of marginal sentences, intermediate between the word-sequences which are in the language and those which are not. In these ways transformational analysis brings some meaning-distinctions into grammar, and—what is more important—separates the meaning-bearing arrangements of words (e.g. in the structure of each elementary sentence) from the non-meaning-bearing arrangements.

In respect to the structure of each sentence, any transformations that take place in it are located in the partial order of the words that compose the sentence, since each transformation takes place precisely at the point at which all the conditions for which it is defined have been satisfied.

In respect to the structure of the set of sentences, sentences can be defined as objects which are reached from other sentences by a directed path, which is a transformation defined as being able to operate at the initial point of that path. In some situations, it may be possible to define a grid of transformational differences rather than directed paths; in that case,

---

[17] Differently from lexical ambiguity, this grammatical ambiguity is explicit: an alternative source.

something is a sentence if it differs from other sentences only by some combination of such elementary differences. In this way sentences can be characterized not only by their compositions but also by their relations to other sentences and by their membership in subsets of the set of sentences— so to speak by their behavior in the set of sentences.

Nevertheless, transformations do not constitute a convenient and complete system of inter-sentential relations. They are too limited: they are not the set of all inter-sentence structural differences, because only certain differences between sentence constructions (8.1.0) preserve word-choice. And although each transformation is a mapping from one subset of the set of sentences (the sentences of one construction) onto another (the same-word-choice sentences of another construction), nevertheless the set of transformations is far from being the set of all subset-mappings in the set of sentences.[18] Only for a few statable differences between free or bound-sentence constructions is it the case that (all or a distinguished subset of) the sentences in one construction preserve the word-choice likelihood gradation of (all or a distinguished subset of) the sentences in the other construction. We are thus dealing with a constraint on construction-difference, that is an equivalence (in respect to word-choice likelihoods) among particular constructions, rather than a general property of transformability among sentence constructions.

On the other hand, there are too many transformations for each to be a separate primitive constraint in the language. Once this fact is taken into consideration, it is found that one can factor out from the transformations a small set of elementary transformations: expansions of sentences, such as *I know that John left* from *John left*, and deformations (altered grammatical relations, usually with addition or replacement of certain fixed morphemes), as in *A lynx is what I saw* or *my seeing a lynx*, both of them from *I saw a lynx*. The transformations that are found in a language are either elementary ones or else products (successive applications) of elementary ones. The set of transformations thus has a certain structure in terms of elementary transformations. However, by no means all possible products of elementary transformations are found as transformations in use in a given language. Hence we do not have a composition in the set of elementary transformations, although we have a decomposition of the existing transformations of a language (6.3).

With all the systemic contributions of transformations, they fall short not only in the respects noted above, but also in some not unimportant details. For example, the relation of 'deriving' one sentence from another is too

---

[18] Thus *He painted two hours* and *He painted for two hours* are transforms of each other, but *He painted two boys* and *He painted for two boys* are not. (*Hours* here is in a different subclass of N, namely that of time words, than is *boys*.) *The artist's painting was of two boys* and *Two boys the artist painted* are transforms of *The artist painted two boys*; but *Two boys painted the artist* and *The boy painted two artists* are not.

general. The sentences which are the derivational sources of a transformed sentence are actually present in it as components: derived sentences are simply abbreviations of, additions to, and rearrangements of, their source sentences. In addition, a few issues are left unresolved. One is: what determines the structure and word-choices of the elementary sentences? Another: what determines the particular elementary transformations, and what gives them the power to be paraphrastic to the grammatical structure of their operand (source) sentence (i.e. to sustain the same word-choices)? A third: why can transformations operate in the same way on derived sentences as on elementary ones? That is, what gives the resultant of many transformations (not all, e.g. not the interrogative) the same overt structure as operand for further transformations? These problems are resolved in 8.2.

8.2. *Operator theory*

8.2.0. *Motivation*

Transformations, as we have seen, are word-choice-preserving differences among observed sentences, which can derive one sentence from another, and ultimately from distinguished elementary sentences. But it has been seen that these differences are not sufficiently varied, nor do they combine sufficiently freely, to constitute an overall system of derivations in the set of sentences. We therefore consider an alternative (operator–argument) theory in which transformations are taken to be the effect of something else.[19] In particular, we utilize the limitedness of the set of transformations in such a way as to obtain them as a further step on a more general process that forms elementary sentences directly out of words. To this end, we observe that preserving word-choice is equivalent to saying that any two sentences which are transformationally related to each other are carrying the same base sentence (4.1): not two sentences with the same word-choice but two forms of one sentence with its word-choice. The transformations can then be seen as (a) simply further word-choices adjoined to a sentence already formed, and (b) shape-changes in words that have entered the sentence.

In type (a) we have increments. These increments are found to have approximately the same kind of structure as the sentence on which they act. In English, this structure is roughly a tensed word or phrase (called operator) after a noun phrase or nominalized sentence (called subject or first argument) and possibly before a second and third argument (object). Thus in the underlying sentence *Sheep eat grass* we can have the increment *I know*

---

[19] Cf. M. Gross, *Méthodes en syntaxe*, Hermann, Paris, 1975, p. 9: 'Les transformations pourraient n'être qu'un dispositif expérimental qui permet de découvrir et de localiser des contraintes syntaxiques.'

. . ., producing (as a sentence in its own right) *I know sheep eat grass*, where the status of *know* between its two arguments (*I* and *eat*) is like the status of *eat* between its two arguments (*sheep* and *grass*). (These are 3, 4, 5 in 8.1.4.)

As to the type (b) transformations, which delete words, or insert a fixed word or affix, or permute words, these are found to be the resultants of a few kinds of change in form: primarily reduction of the phonemic shape of words in the underlying sentence, and the moving of certain words toward the beginning of the sentence. The underlying sentence, with its word-choices, remains in the transformed sentence. (These are 1, 2 in 8.1.4.)

The particular development from a transformational to an operator–argument theory is motivated as follows. We note that each transformation from simpler source sentence-forms as domain to more complex resultants as range has more restrictions in the range than in the domain, because each transformation acts on all or a proper (restricted) part of its domain. Thus in the resultants of the passive, *America was left by me last year* is dubious, i.e. somewhat restricted, in contrast to the normal *America was left by the best writers of the generation* and the uncertain *The office was left by me at 5*, whereas in the active source *I left America last year*, *The best writers of the generation left America*, *I left the office at 5* are all equally normal. In the English interrogative, if we compare *Who met her?*, *Whom did he meet?*, *Is he coming?*, *Did he come?* with the sentences of the source domain *He met her*, *He is coming*, *He came*, we see in the interrogative form restrictions in the position of *whom* (compared to that of *her*) and in *is* and the tenses.

This greater restrictedness in the resultants of transformations leads us to seek other pairs of sentence-forms which satisfy the conditions for trans- formation and in which one form is less restricted than the other and is therefore to be considered the source. One such case is the many affixes for which an affixless source can be found. A trivial example is *pro-* before nouns, derivable from *for*: *He is pro-war* from *He is for war* or *He is in favor of war*. Here, '*X* (e.g. *pro-* above) is derivable from a source *Y* (e.g. *for*)' means, in the observational terms of hearer acceptability, that for all sentences of the given form containing *X*, the sentences with *Y* in place of *X* are ordered (as to acceptability) roughly like those containing *X*: *Women are less pro-war*, *Children are pro-air*, *Rocks are pro-universe* have among them much the same acceptability ordering as do *Women are less for war*, *Children are for air*, *Rocks are for the universe*. In most cases, the sentences with *Y* replacing *X* are a subset of the sentences containing *X*. Another example is *un-* before adjectives. Under this condition—before adjectives—*un-*has the same combinations as *less than* rather than as *not*: *untrue* (but not *unfalse*), *unhealthy*, (but not *unsick*), *unlike* (but not *undifferent*; hence also *unscholarly* for *unlike a scholar*), are replaceable by *less than true* (but no *less than false*), etc.; hence we would consider these occurrences of *un-* to be reductions of *less than*. A suffix example is *-less*, e.g. in *hopeless*, derivable from a free word *less*, 'without' (e.g. in a presumed source *less hope*), which is still visible in *a year less a day*. Less obvious is the derivation of *-hood* from *state of (being)* or the like, as in *boyhood* from *state of being a boy*, where it is known that *-hood* is a suffixal form of a Middle English free word *had* which in free position must have been replaced historically by some such word as *state* (G 175–6).

In establishing such derivations, we consider the current affixes, restricted in their positions and in the words to which they are attached, as reductions of currently existing free words lacking these restrictions. As another example, note that when we establish adequate derivations for the interrogative forms, e.g. *I ask whether he came* as source of *Did he come?*, we find that the source is free of the special restrictions of the interrogative (G 331).

As we establish least restricted sources from which the various more restricted forms can be derived, we note that these least restricted forms have a common structure. The structure consists of certain words, chiefly (1) verbs, (2) adjectives, (3) prepositions, and (4) conjunctions, all of which will be called operators, appearing after or between (a) nouns or (b) sentences (including nominalized sentences), which will be called their arguments; (1a) *John coughed, John poured water*, (1b) *John's coughing stopped, John knows she left, John's coughing caused her to leave*, (2a) *John is tall*, (3a) *John is for war*, (3ba) *John left for Europe*, (4b) *Her leaving is because of John's coughing, She left because John coughed*.

This structure is so widespread that it becomes of interest to take every sentence form that does not fit this structure and seek to derive it from one that does, even if no restriction is circumvented by the derivation. An example is the deriving of adjectival adjuncts and relative clauses from secondary sentences: e.g. *I saw a tall man* from *I saw a man who was tall* from *I saw a man; the man was tall*, where the secondary-sentence intonation, written as a semicolon, is the conjunction between the two sentences (type 4b above).

To see an extreme example of a word that does not fit this operator–argument structure, consider *yes*. This word seems to be outside the grammatical classes and their co-occurrence conditions. However, note the following: to the question *Did you close the door?* a person who has done so can truthfully answer *Yes*, or its equivalent here *Yes, I closed the door*. But if, upon hearing the question, he closes the door and then says *Yes*, or *Yes, I closed the door*, he is dissembling and not quite truthful. This means that the *Yes* is not related to the tense of the answer (*I closed the door*, which he had indeed done before answering), but to the tense of the question: *Yes* could be said truthfully only if the door had been closed before the question was asked, since the question was in the past tense. Hence *yes* is a referential word (5.3), containing a repetition of the question (and not a newly made *I closed the door*). *Yes* is thus an O_o (information-giving) operator on the question-sentence, something like *I say yes to your saying 'Did you close the door?'*. It is thus not unrelated to the operator–argument structure.

It turns out that we can find operator–argument sources for almost all words and sentences of the language, with few definite restrictions on word-combination other than those of the operator–argument structure. Furthermore, the transformations that produce the derived sentences out of the sources are found to consist almost entirely of reductions in the phonemic composition of words, specifically of those words which add little or no information to their operator or argument in the sentence. (Nevertheless, various languages have important and common transformations (deformations) which are not readily analyzable as reductions, such as are seen in *The organizing of the event was his*, and *Her cooking is mostly of vegetables*, or in pairs such as *John is the author, The author is John*.) We thus approach a situation in which, first, the only restriction on words in the source sentences is the operator–argument relation—i.e. this relation creates sentences—and, second, virtually all other sentences can be considered to be merely reduced (or otherwise transformed) forms of their source sentences. This is the system described in Chapters 3–5.

8.2.1. *Operator–argument analysis*

Given these considerations about sentences, we can state a partial-order structure which is common to all S. The dependence relation among the words of a language creates a partial ordering on the words in a sentence. In the unreduced sentences, each occurrence of a word must be accompanied by an occurrence of words which satisfy the dependence: in *Mussolini thought Germany would defeat England*, the occurrence of *thought* depends on two words: one which depends on nothing further (here *Mussolini*) and one (*defeat*) which depends on further word-occurrences; and the occurrence of *defeat* depends on the occurrence of two words which depend on nothing further (*Germany, England*).

Furthermore, a single occurrence of a word (or argument) cannot satisfy the dependence requirements of two different word-occurrences (as operators): the argument has to be free for the operator, i.e. not an argument of any other operator-occurrence. This is seen when we consider the components of a sentence. If in a given sentence a depends on m_1, m_2 and b depends on n_1, n_2, and c depends on a, b the question is whether one of the m might be the same word-occurrence as one of the n. The occurrences of a and b had to be accompanied by their arguments before c operated on them. But before c did so, each of a and b with their arguments constituted independent sentences $m_1 a m_2$ and $n_1 b n_2$. No word-occurrence could have been common to these two sentences before c acted on them. If, in the extended sentence formed by c, one of the n is the same word as one of the m, that n may be zeroed but its zeroed occurrence would still be satisfying one of the dependence requirements of b, while the m, unzeroed, satisfied only a and not b.

Compare (1) *I locked the door after washing* with (2) *I locked the door after washing myself*. In (1) it might seem that *I* serves as first argument (subject) for both *lock* and *wash*; but in (2) we see from *-self* that *wash* had its own first argument, which was zeroable as being a repetition of the one for *lock*. Thus auxiliary-like verbs ('preverbs', G289) such as *take* in *I'll take a look*, and even *can* in *I can go*, and *have* in *I have gone* are not taken here as some special expansion of the following verb, but are analyzed as being operators with their own subject, which is however the same as the zeroed subject of the following verb (which is the second argument of that *take, can, have*).

We thus arrive at a theory not merely of sentence transformation, or of derived sentences, but of sentence construction *ab initio*: it deals not with relations among already-extant elementary sentences, but with the formation of word-sequences, first into elementary sentences and then in the same way (i.e. by satisfying the dependence) into all base (unreduced) sentences, and finally with the reduction or deformation of these sentences to yield the other sentences. This theory represents first a dependence process of word juxtaposition (operator on argument), with inequalities of the likelihood (from zero up) for individual words to occur with particular other ones. On these word juxtapositions there acts the shape-changing process (kindred to morphophonemes, 7.4), which consists primarily in reducing words whose likelihood in the given operator–argument situation is very high. These shape-changes on words, when suitably chosen, can produce the same 'derived' sentences as can the transformations on sentences. The 'derived' sentences are thus the 'same' sentence, with the same meaning.

When this is done, it is seen that the cumulative construction of a sentence, and the likelihoods in word-choices, and the word changes, are all defined in terms of the finite (though large) set of operator–argument word-pairs (or triples) rather than on the unbounded set of sentences. It is seen that the high likelihood on which word-reducibility depends is apparently definable, in a complex way, in terms of word-pair frequency in specified sets of discourses, whereas the transformations and the well-definedness of the set of sentences depended on subjective evaluations of sentence acceptability. And since the changes are only in the shape of a word (even if to zero) and its position, but not in its having been chosen, or in its operator–argument relation to other words in the sentence, it follows that the choices and operator relations of the words in the sentence—on which the meaning of the sentence depends—are not altered by the changes, something that does not clearly follow in the case of transformations.

The two characteristic processes, of operator–argument dependence and reduction in the shape of high-likelihood words, leave room for additional processes or entities which may appear in one language or another. For one thing, the operator–argument dependence creates a partial ordering of words. Since speech is in essential respects linear, mapping the operator–argument order onto a linear one leaves room for cases of alternative linearization; some of the apparent permutations of words are due to this. For another thing, the high likelihood is only a necessary condition for reduction. The absence of an a priori sufficient condition leaves room for various languages to favor or disfavor reductions in particular circumstances: for example, English does not normally zero the subject of an independent sentence, even if it is indefinite or repetitional (no *Is quiet* or *Are coming?*). The fact that all vocabulary choices are either operators or arguments, even if in reduced form, does not preclude the existence, in many languages, of additional phoneme-sequences that have a grammatical status, especially such as serve to indicate the operator–argument status in the given sentence, and such as are determined by the course of constructing the sentence (e.g., the *-ing* which shows that a verb is appearing as argument of a further verb). In addition, the phonemic distinctions and phonetic composition of words are arbitrary in respect to the operator–argument relation, leaving room for minor structures and processes (e.g. assimilation) involving these.

There are also processes that are independent of the two characteristic ones, but whose effects can be reinterpreted mostly or entirely as though they were due to the two characteristic processes above. One such process is morphology, found in many languages, whereby words have affixes indicating plural, tense, feminine, diminutive, or the like. Such affixes may go back to pre-syntactic times, but once they are embedded in sentences that are constructed by the operator–argument relation they can be analyzed in the latter terms, i.e., for example, *ten stones*, as though from *stone and stone up to ten*, or *tigress* from *female tiger*. Another at least partly independent process is analogy, whereby for example a derived form (affix or word combination) may be used with a given word A because it is commonly used with some similar word B, even though the underlying operator–argument form existed for B and never did for A. Again, we can analyze the case for A as though it came from the required operator–argument form on A. The justification is that the underlying operator–argument form in B is what gave the derived form in B its grammatical ability to carry its meaning, and if the latter exists also for A it is by reference to a similar operator–argument relation posited for A.

It is seen from this survey that while the two characteristic processes suffice to produce and characterize all sentences of a language, a few other processes may be in effect, but in a way that is limited and that leaves the basic processes as framework.

The methodological novelty in the operator–argument theory is to speak not of the restrictions in word combination but of the constraints that produce these combinations—constraints both in the building process and in the reduction. A single constraint, repeated, builds both the elementary sentences and the ones that these expand into. The applications of the constraint in the building of an utterance are partially ordered, and the reductions in the utterance can be located in this partial ordering, so that the utterance is built and reshaped stepwise. Given the operator–argument word-pairings with normal likelihood, we obtain the actual sentences of the language.

A further step here beyond transformational theory is that the reshapings are not an arbitrary catalog of whatever is needed to derive all observed sentences from their sources, but a particular kind of reduction and position-change. Then for each given sentence that does not have the directly visible operator–argument (predicational) form there can be found a paraphrastic unreduced sentence having that form and differing from the given sentence only by certain ordered applications of the stated reshapings. The unreduced (base) sentences are less constrained than the reduced forms, since reductions are additional constraints in shape and often also in the conditions they require for taking place. Hence the grammar of the unreduced sentences is simpler.

The relation of the paraphrastic sentences to the base ones is that of reshapings of an original shape rather than the general relation of a derived sentence to a source, since the original unreduced sentence remains present in all reshapings, only with stated phonemic reductions or limited position changes: the source of the sentence is inside that sentence itself. In respect to structure, the base sentences and the reduced ones are not quite comparable. For the base sentences, there is a single structure which fully accounts for each of them. In the reduced sentences, however, there are various unrelated local constructions affecting various parts—plural, tense, paradigms, clauses—with certain subsets of sentences having constructions that differ from anything else in the grammar, as in the interrogative form, or in sentences of the type of *It is time that he left*.

Thus traditional grammar does not offer a single structural system adequate for all sentences. As to the theories of strings and of transformations, these center on important features of all sentences—contiguity and derivability respectively—but they do not systematize all that is formal about the utterances of a language. In contrast, the operator–argument theory presents a single small set of constraints which produce all sentences from the words of the language, with provision for distinguishing the normal ones from the rare and the marginal, and for having such other factors as contiguity, change, and analogic formation. The theory deals entirely with the data of sentences, as objects or events. It is based on sentences as said: and it is extended to those that 'can' be said, by potentialities arising from constructing a least grammar for those that are indeed said, normally or rarely. This means that the structure of language which expresses most generally the relations—or constraints—of the sentence-parts is not necessarily quite the same as the composite structural guides which determine how a speaker makes a sentence or how a hearer understands it.

In certain respects, operator theory is an end-point of least-grammar theories of syntax, since it shows that the meaning-bearing arrangements of words have the regularity of a mathematical object, while the reductional-transformational rearrangements are not meaning-bearing and do not have such regularities.

8.2.2. *Adequacy of the analysis*

We consider now the adequacy of the operator–argument theory. There are certain data situations which would falsify the theory if they were found in language, but they seem to be always avoidable. One such is if word combinations are entirely random, or could not be described by any regularities. A less crucial, and encapsulatable, problem would arise if there were operators on N—the lowest arguments—which did not create a sentence. This would be the case if it were impossible to derive the plural from an operator on a sentence instead of from some operator on a noun, or if it were impossible to derive *and* between nouns from *and* between sentences. Yet another (perhaps circumventable) situation that would falsify the theory would be if an operator on an operator could be different from an operator on a sentence: that is, if an operator could act differently on a verb without the presence of that verb's arguments than with them (in the former case the verb would be appearing without satisfying its argument-requirement). Even more problematic would be the ability of an operator A under a further operator B to take an argument X which A could not otherwise take, since A had to have its argument before B operated on A. In contrast, no problem arises from A refusing an argument X when A is under B; for then we would simply say that when A is carrying X, then B does not select A (i.e. B does not operate on any occurrence of A that carries X).

A serious but circumventable difficulty would arise if there were words which could take, after being reduced, additional co-occurrents (selection within its operator or argument classes) that they could not take before being reduced. This would not accord with some statements in the theory. It would also mean that the base does not carry all the information in the language. If such an apparent situation is found, one would try to show that the reduction contains some additional word, in zeroed form, which is responsible for the added co-occurrents.[20] In contrast, no problem arises if a word X, after being reduced, loses the ability to take certain co-occurrents A that it could take before being reduced. For this effect can be obtained simply if the domain for the reduction does not admit A, i.e. if the reduction takes place in X only when X does not occur with A.

The present theory arrives at a mathematical formulation for the partial-ordering relation, though not for the reductions. There are certain situations

---

[20] This analysis is needed because some words when under particular operators change their further co-occurrents (3.2.2). Hence one would seek a zeroed intervening operator in the reduction.

which would obstruct the mathematical formulation—unless they can be shifted into the reductional description. One of these is the case of operators rejecting certain words in their argument class. The only serious example of this in English is that of the auxiliaries, *can*, *may*, etc., which reject any second argument whose subject (first argument) is not the same as their own subject. In *John can drive* we say that the second argument of *can* is *John drives*, with *John* requiredly zeroed as a repetition (the argument cannot be *drive* alone, since *drive* must have its subject with it). But the second argument after *John can* must not be *Tom drives*. Such an exclusion violates the rule that an operator can take any member of its argument class, even if very rarely. One can only seek some other operator (e.g. *able*) which is not limited in this way (even though *I am able for Tom to drive* may never be said, on semantic rather than grammatical grounds), with *can* being an optional variant when the two subjects are the same. (This is not the history. Some auxiliaries had been verbs which fairly recently accepted a different subject in their second argument. However, we can say that when they ceased to accept this they ceased to be operators in their own right and became variants of the semantically closest verb which was not thus restricted. Such analyses can be made with good reason in some other cases of restricted words.)

In many languages we also meet certain situations which detract from the advantages of the theory, but which present a problem only if they are widespread. Even so, the problems can be circumvented. The chief example is of sentences which cannot be obtained by known reductions from an operator–argument structure: e.g. certain borrowed foreign phrases, and partially frozen expressions (such as *the more the merrier*), and linguistic jokes (such as *let's you and him fight*). One might say that these constitute a small set of locally asyntactic word-sequences which function like an addition to the vocabulary. Or one can create *ad hoc* transformations to derive them from related regular sentences (perhaps by the actual historical route). Somewhat similar is the case of any morphology which we cannot derive from an operator–argument structure, or which one might consider to involve too costly a derivation, e.g. the plural suffix. Here one might simply say that the language contains pre-syntactic N pairs with added singular–plural meaning (and with the plural-agreement restriction), e.g. *stone–stones*, with one or the other occurring in a given N position according to the meaning that is intended.

Languages whose vocabulary elements are not in general free words do not interfere seriously with the operator theory. In Semitic, most words can be constructed in a regular way from consonantal roots possessing dictionary meanings, intercalated with vowel 'patterns' possessing grammatical meanings. This construction does not preclude the operator–argument analysis if we can reasonably choose one root plus vowel combination as the base word, and consider the various shifts to other vowel patterns as various operators on the base; it is indeed reasonable to suppose that such may have been the distant history. Such a solution is clearly appropriate in Semitic, where the simplest verb form can be the source from which other verb forms and related nouns are obtained; in addition, some simple nouns would be taken as sources for verbs.

More generally, the operator theory does not intrinsically require that the source form of operators and arguments be free words. The fact that to some extent affixes can be derived from free words makes it possible to see the similarity in operator–argument structure between languages which have morphology and languages which do not. But it is not essential for the present theory.

The problem is different if we wish to construct words like the English *conceive, perceive, contract,* etc., let alone *dazzle, frazzle,* etc., out of two free words, and more generally out of an operator–argument relation. There is not enough regularity to permit this in English. Furthermore, the small number and encapsulated nature of these composite words allow us to consider them as indivisible entities in the syntax; their composition is then a semantically relevant pre-syntactic structure in the vocabulary.

Similarly, if nouns consist of a stem plus a required case-ending, we would take the ending, at least in the nominative, as a required indicator of argument position. Then for the operator–argument analysis the stem would function as a free word, though in being said it would carry the indicator. Likewise, if nouns have gender affixes with no consistent male–female pairing, the affixes could be considered argument indicators; but if they have paired male–female affixes, one of those affixes (depending on morphophonemic convenience) would be considered as (male or female) operator on the base noun (which, in the absence of this operator, would mean the other gender and would contain the other affix as noun indicator rather than as gender indicator).

Another aberrant situation is that of reductions which are required rather than optional. When the required change *a* follows on an optional one *b*, we can say that the ordered pair *ba* together is optional. In (1) *the glass which is broken,* the zeroing of *which is* is optional, since (1) exists; but once *which is* is zeroed certain residues must move to before the host, yielding (2) *the broken glass;* the two changes together change (1) optionally into (2). When the required change acts directly on the source sentence, we cannot make it part of an optional event: *John thinks that he is right* must be derived from (3) *John thinks that (the same) John is right,* but (3) is not normally said. The ultimate sources, the elementary base sentences—what were called the kernel-sentences in transformational theory—indeed do not have required reductions, but the extended sentences of the base, such as (3), may. This means that not all base sentences are actual sentences of the language, even though complicated *ad hoc* restrictions would be needed if we were to formulate a grammar that would routinely exclude (3). As long as required reductions hold for only a few grammatical situations, such as in (3), we can still claim that the base is a subset of the sentences of the language (i.e. that all source sentences exist in the base)—aside from the few required reductions —and that all information statable in language is statable in its base subset. If, however, required reductions in base sentences were many and diverse, this claim would have to be given up; the base would then become a grammatically regular set of forms that underlie the sentences of the language rather than being a subset of it.

Another problem can arise in the case of analogy. In this process, given a form A_i, together with A_j differing from it by certain reductions (or additions or replacements of words), and given B_i (but no B_j), there is formed a new B_j having the same overt difference from B_i. In most cases, B_j is then understood as though it had the same reductions or other grammatical differences from B_i as A_j has from A_i. In some cases, however, the words of B do not admit of the changes made in A, or B_j may have been made on the model of A_j without B_i existing in the language. If these latter difficulties are very numerous, the explicatory value of the present theory for the individual sentences of the language is decreased.

There are also certain properties which are common in languages, but which are not essential to the adequacy of the theory, although some of them increase its utility or simplify its recognition of sentence analysis. One is the fact that reductions are ordered in respect to the operator–argument ordering. Another is the availability in many languages of operator indicators and argument indicators; e.g. case-endings are argument indicators for the zero-level arguments N. A third is having different words in the different dependence classes N, O_n, O_o, etc., with only few words multiply classified. If a language had a great many words which are members of several classes, then the classes, which in any case are defined as relations among word-occurrences ('word tokens'), lose the support of being distinguished by their word members ('word types').

Some features of traditional grammar are readily describable in operator–argument terms. Such are pronouns (as repetitions); the infinitive (as an argument indicator making operators into arguments); the status of adjectives, prepositions, and conjunctions (as operators); the interrogative and imperative (as metasentential operators); the absence of *to* under auxiliaries such as *can* (as zero variant of the infinitive). Other features are more difficult. Such are, in English: the various nominalizations; the borderline beween 'indirect' objects and adverbial modifiers as in *rely on something* (object), *sit on something* (object or adverbial phrase?), *stand on something* (adverbial phrase); the requirement for *a* before 'count' nouns in the singular. There are also problems with many individual words whose use is restricted or specialized in various irregular ways: e.g. *afford* (*He can afford to buy it*; no *He affords to buy it*), *supposed to* (which differs in selection and meaning from *suppose*: *I'm supposed to be there tonight* is not paraphrastic to *People suppose me to be there tonight*). In some cases the difficulty of a regular description is due to a construction being in flux, historically.

The regularities which can be stated, secondarily, in terms of the operator –argument structure may fail to hold in some cases. For example, each operator when combined with its arguments forms a sentence. In almost all cases the resultant sentence can stand by itself, in suitable linguistic or situational context. However, there are in English certain operators which produce sentences that are not comfortable unless some further operator acts upon them. This happens in English for many intransitive verbs. Thus *enter* operating on *John* yields *John enters*; but this sentence would not normally be said by itself (except in stage directions). Instead we have *John is entering*, *John enters before everyone else*, and the like.

For some features of grammar, either the operator–argument source or the derivation from it is felt to be artificial, though technically grammatical; but precisely these sources have grammatically explanatory advantages. In English this is the case for the tenses (G265), for the plural and quantifiers (G244), for the passive (G362), for *not* (G221), and to a lesser degree for affixes, referentials, possessive *'s*, *the*, and the demonstrative pronouns.

All of these difficulties are treated either by admitting special forms or special domains of known types of reduction, or by creating certain of the base (source) sentences or intermediate sentences of the derivation as reconstructed sentences which no one would ever actually use in speaking (and never or hardly ever in writing); in extreme cases the reconstruction may be marked by a dagger (†). For example, for *John drives slowly*, which is not in operator–argument form, we reconstruct *John drives*; *said John's driving is in slow form* (or even . . . *is in a form; said form is slow*); to take an extreme example, for *John entered* we reconstruct *I say that John enters*; *John's entering is before my saying that John enters*. These reconstructed sentences are not simply operator-form paraphrases selected *ad hoc*: they are operator–argument constructions which differ from the given (attested) sentence only by established reductions and linearizations, applied in the conditions which these reductions and linearizations require. Thus they differ from attested sentences in the same way that attested sentences differ among each other. No line can be drawn between the reconstructed sentences and the normal ones, and only with great difficulty can grammatical rules be stated that would exclude the reconstructed forms. Indeed, any grammar based only on sentences which would be generally accepted as normal will have odd lacunae and oddly limited rules, the more so in that the set of sentences is not well- defined. But if we state a system of constructions and relations—let alone one with reasonable mathematical characterization— which is just sufficient for all 'normal' sentences, we will unavoidably admit many sentences which no one would ever say, either because they are far too labored, or because the availability of reduced forms has overshadowed the unreduced, or because the vocabulary needed for the unreduced form is not (now) available.

In some cases historical replacement has led to this vocabulary lack, as in the reconstructed base *state of (being a) child for childhood*, where *-hood* is the modern form of a once-free word meaning 'state'. But in many cases it is not a question of historical loss. Rather, it is a matter of the partial independence of syntax and vocabulary. True, the syntactic structure is arrived at as a relation among words in utterances. But when that structure is stated, it is not in respect to specific words—which in any case can drop out of the language or enter into it—but in respect to the ability of variously related words to fill that structure. In particular, the N, O_n, etc., classes of the present theory are defined by certain dependence properties of words, not by a particular list of available words. Given a particular sentence, and a reconstructed one from which the given can be reached by known reductions, the reconstructed form may be peculiar because a particular word in the given sentence does not normally occur (or no longer occurs) in the position it is supposed to occupy in the reconstructed sentence, or does not have normal likelihood in respect to the other words in the reconstruction. This is easiest to see when the word had indeed occurred there and had then been replaced. For example, at the time when the percursor of *-hood*; (above) had still existed as a free word, we would have had by the side of the compound noun *child-hood* (the vowel in it was then *a*) also *the had of (being a) child*, just as by the side of *birth-rate* we have *the rate of birth*; the stressed *had* was the free variant of the unstressed *-had*. When, however, *had* ceased to be used in free-word position, the word that replaced it there—*-state* or the like—became the free-word suppletive variant of the unreplaced suffix *-hood*, yielding the reconstruction *the state of (being a) child*.

Because the operator–argument theory refers everything in sentence structure to just a few principles—and ones which can be considered reasonable, given the use made of language—the theory has an explanatory value, showing how the properties of the grammar and those of particular forms—even of some peculiar ones—arise from these first principles. For example, we see why derived sentences have the same general structure as do base sentences (8.2.1, G77); how cross-reference from one word-position to another is managed without a system of word-addressing (5.3.2); why grammatical constructions are nested one in another (2.5.4) and why permutations are limited as they are (3.4.3); the distinction between the zero reduction of a word and non-occurrence of that word (G136, 150); why certain transformed (reduced) sentences, including metaphors, do not support further transformations which their apparent form would admit (G405); why some of these require modifiers (e.g. not *He peppered his speech* but only *He peppered his speech with asides*); and why many word subclasses are fuzzy as to membership (e.g. the auxiliaries, the reciprocal verbs). In respect to particular constructions, we see, for example, how tense (G277) and plural (G244) can be fitted into the rest of the grammar, and why certain peculiar uses occur in them (ch.5 n.5); how compound tenses fit in grammar (G290); how 'topic' permutation takes place (3.4.2, G109); how the relative clause is formed, and modifiers in general (G227); what in a sentence can be asked about and what can carry a relative clause (G121); how yes–no questions and *wh-* questions have a common derivation (G331); the grammatical basis for the distinction between use and mention of words (5.4); how we obtain the peculiar grammatical properties of the comparative (G380); how the special limitations of the passive arise (G364); how peculiar sentences arise such as *It is true that John left*, *It is John who came* (G359); how certain words such as *any*, *but*, *only* come to have uses that are almost opposites (G327, 401); why one can say *He truly came on time* or *He will undoubtedly come* but not *He falsely came on time* or *He will doubtfully come* (G306).

As an analysis of language, the operator–argument theory has certain overall advantages:

First, it covers a language more fully than has been expected of grammatical theories. It can in principle yield not just the sentence forms but also the actual sentences, and this with the aid of a list not of elementary sentences but just of normally likely operator–argument pairs; to prepare such a pair list, with the aid of many fuzzy subclasses of words, would be a tremendous task which no one would undertake, but it can in principle be done. A word-sequence which is *n*-ways ambiguous on grammatical grounds (e.g. G143) is obtained not once, as a single sentence, but *n* times, as *n* sentences originating from different sources via different reductions which degenerately yield the same word-sequence. The theory can also account for various kinds of nonce-forms, daggered reconstructions (as above), joke forms, and fragments, as particular kinds of departures (chiefly by extending the domain of a reduction) from the main, but not well-defined, body of attested sentences. At the same time, the theory does not overproduce, does not yield word-sequences which are not sentences. The product of the theory includes every contribution to the substantive information of a sentence, as a fixed interpretation of each step in producing the sentence. And it does this entirely in accord with the apparatus of the grammatical theory (especially selection), without appeal to any mental structures.

Second, the operator–argument theory is intended to be maximally unredundant in that everything that is constructed is then preserved under further construction. In particular, the existing vocabulary is utilized more fully, as when affixes are obtained as positionally restricted reductions of free words (which are already attested otherwise), or when pronouns are obtained as reduced repetitions of a neighboring word. The existing constructions are utilized maximally, for example in keeping multiple classifications to a minimum, or when the intransitive *read* is obtained by second-argument zeroing from the otherwise existing transitive *read* (*He reads all the time* from *He reads things all the time*). The meaning-interpretation is simplified by showing the reductions and alternative linearizations to be substantively paraphrastic. All this means that fewest constraints have to be stated in order to obtain the sentences of the language and to relate them relevantly to each other.

Third, the theory provides an effective procedure for characterizing utterances. It does so by locating all its events in the finite though large set of operator–argument constructions rather than in the unbounded set of sentences (let alone whole utterances). And it retains its effective property by having only stated types of deformations—primarily reductions and alternative linearizations—and by having one necessary though not sufficient condition for determining the domain of almost all reductions, namely low information (high expectancy) in respect to the immediate operator or argument. When grammatical events are formulated in such a way that the relevant environment for them is the operator–argument pair rather than the sentence, then the reduction of a word can be based on its low informational contribution, ultimately referable to its relative frequency in the given position, rather than on the acceptability of the sentence as reduced.

Lastly, because the events of construction and of deformation are uniform in their basis—and because nothing extra-linguistic is presupposed—it becomes possible to compare structurally, and to measure in respect to complexity (and the cost in departures from equiprobability), each analysis of a sentence (especially if more than one analysis is possible), and also the sentence structure itself, and even the whole language.

Most generally, one reaches a fundamental pervasive structure consisting of a few conditions (constraints) on word combination, with the successive or simultaneous satisfaction of these conditions producing the specific constructions of language occurrences. What is gained here is not only the simplicity of the structure, but also the finding that each of these elementary constraints makes a fixed contribution to the meaning of every sentence in which it appears.

The major cost of reaching a simple and regular syntactic system has been found to be moving many form-differences from syntax into morphophonemics, i.e. into the phonemic shapes of morphemes (words). However, the phonemic shapes of words are 'arbitrary', i.e. they can only be listed since there is no regularity which would enable them to be derived. And various morphemes in various languages have more than one phonemic shape (7.4). Therefore when we replace morpheme-combination differences by phonemic-shape differences (e.g. by saying that *-hood* is a suffixal phonemic shape of *state*, or that different declensions are just matched sets of different phonemic shapes of the case-endings), we are only adding to existing lists of the phonemic shapes of words; we are not bringing a new degree of freedom into the grammar. Furthermore, the total phonemic content of words is neutral to syntax: it does not correlate with the meanings or the combinability of the words. Hence moving a form-difference into the stating of phonemic shapes of words is moving it out of the stating of syntax—so long as this moving preserves the selection (combinability) of the words affected, e.g. so long as the selection of -*hood* as an independent suffix is included in the selection of *state* of which -*hood* is to be considered a variant.

In evaluating any theory of syntax, certain limitations must be kept in sight. There exist additional processes of considerable importance in the formation or availability of sentences in a language, or in the way speakers understand the sentences (i.e. relate them to others). Such processes are: the phonemic similarity among syntactically related words, morphology, borrowing, historical change in phonemes and in use-competition among words, and above all analogy. These processes are presently external to any systemic syntactic theory, but they affect considerably the form of many sentences which the syntactic theory has to account for. Since the present book offers not a survey of syntactic description but rather a theoretical framework, it suffices here to note that given the forms which result from these other processes, we can reconstruct for them operator–argument 'sources' in the same general way we use for attested sentences in the language. The forms are as though derived by the known types of deformation from the known types of sources. This is the case for all established sentences except some listable, usually frozen, expressions.

Some features of language remain effectively outside the syntactic framework: for example, the phonemic composition of words, and the quasi-morphemic composition of certain words (e.g. the *sl*-, *gl*- words, the -*le* words, and the *con*-, *per*-, -*sist*, -*ceive*, etc., set). Other features are only tangentially disturbing to the operator–argument structure: for example, popular etymological perception of particular words, morphological paradigms over classes of words, and the intentional departures in vocabulary and syntax seen in argot, linguistic jokes, and in some literature.

There remains one analysis which is outside current syntax but which would simplify the application of the operator–argument construction. This is factoring words into syntactically more elementary word-components. In general, syntax starts from the existing vocabulary. Any detailed syntactic theory will have to admit some regularization of the vocabulary, such as combining restricted suppletive words into a single word (*is, am, are* into *be*; -*s*, -*z*, -*en*, etc. into a plural morpheme), recognizing some suffixes as variants of free words, etc. The present theory goes beyond this in order to give operator–argument sources for words that do not seem to accord with operator–argument structure (e.g. *the* from the phrase *that which is*, or *which*, *who*, etc. from *wh*- plus a pronoun; also *able* or the like as source for *can*). Such analyses are not based on any semantic criterion, which is not controllable, but on their usefulness in reconstructing appropriate operator–argument sources by means of established types of reduction. There are also considerations of systemic simplicity. For example, English has very few O_{nnn} operators, whose argument-requirement is a triple N, N, N. One such is *put*: we have *He put the book on the table*, *He put the book down*, *He put the book (in a place) where they could see it*, but not an independent *He put the book*. One could decompose *put* into the operator–argument construction *place . . . so that*, with *He put the book on the table* reconstructed into *He placed the book so that the book was on the table*. If such decomposition were reasonable for all O_{nnn}, we would eliminate this little-used operator class.

Replacing a word by a phrase or operator–argument construction, as above, can be carried out just on the exceptionally irregular or restricted words. One might also think of carrying it out in a systematic way, in order to modify both the vocabulary and the details of syntax into a maximally simple joint system. This last may be virtually undoable and perhaps of little value for the whole language, but it becomes a matter of great interest in the sublanguage of science (10.3).

8.3. *Sentence semilattices*

The considerations of 8.2 show that the words of a sentence have the relation of a branching process in which the dependence of a on $b, \ldots, d$ is represented by each of $b, \ldots, d$ branching out of a, with $b, \ldots, d$ linearly ordered (because likelihood is not preserved among the different orderings of co-arguments, as in *John ate cheese* and *Cheese ate John*). More specifically, for all word-occurrences a, b in a sentence S, we set $a \geq b$ for the case where either b is the same as a, or where b is reachable from a by a sequence of dependings. When the length of this sequence is not zero, $a > b$; and when it is zero, b is identical with a. Here a is an upper bound of a subset Z of word-occurrences in S if $a \geq c$ for all c in Z, and it is the least upper bound (l.u.b.) of Z if $a' \geq a$ for every upper bound a' of Z. The word-occurrences of a base sentence have the join-semilattice property, namely that every two occurrences have a l.u.b. in the sentence. If $a > b$ and there is no c such that $a > c > b$, the a is said to cover b, or a is the (immediate) operator on b, and b is an argument of a. Here there is the added property that the word-occurrences covered by a are linearly ordered; they are called co-arguments of each other: e.g. *John* and *cheese* in *John ate cheese*. The cover relation is the previously described dependence relation among word-occurrences in a sentence, and is the one elementary relation of sentence construction. Also, every subset of S which contains its l.u.b. is a component sentence of S. A join-semilattice property of word-occurrences is that no two occurrences a, b have a greatest lower bound other than a or b themselves. This semilattice differs in various respects from the transformational lattice of 8.1.5.

The actual sentences are a fixed linearization of the semilattice; in certain situations, there are also alternative linearizations which differ from the normal one in stated simple ways.

A major problem arises when one sentence (the 'primary') is interrupted by another (3.4.4). We see this first of all when the interruption is an independent sentence, as in *John—it was late—left hurriedly*. Here we can say that the interruption is represented by a separate lattice, and is located at a particular point in the linearization of the sentence. If the interruption starts with a conjunction (an O_{oo}), as in *John—only because I suspect him— began accusing me*, we can only say that it is a conjoined sentence that had been permuted to this position. (This can also be said for the preceding case, if we take the dash as a conjunction.)

The problem becomes more complicated when the interruption is a relative clause. It was noted in 3.4.4 that the relative clause is a separate secondary sentence which begins (in its linearization) with an argument that is the same as a word ('the antecedent') in the primary sentence. If the relative clause were taken as a conjoined sentence permuted to inside the primary, we would need a complex addressing system to indicate the location of the antecedent, to after which the relative clause is permuted. This addressing capacity is needed, even without the permuting, to tell us which two words (one in the secondary and one—the antecedent—in the primary) are the same (usually with the same referent). Languages do not use such complex addressing systems, and have obviated the need for them by having the two 'same' words appear in easily stated fixed positions (linearly or in the partial order) in respect to each other. Clearly then we have to take the relative clause and the sameness statement as interruptions of the primary, at the point in the linearization (or the partial order) at which they appear. Once we recognize such a structure, we can consider the reduction-instructions of 5.2 to be also secondary metalinguistic sentences which interrupt the primary sentence at the point in which the conditions for the reduction have been satisfied by the sentence.

The various secondary sentences are all separate sentences with their own lattice structure, but they all take effect on the primary sentence at a particular point of it. The point in question can be identified in the semilattice of the primary sentence, but it is transparent to the operator–argument relation expressed by that lattice.[21]

8.4. *Individual language structures*

The operator–argument theory of language fails to provide a complete description of the grammar of each language. This failure results from three

[21] The solution to this problem, namely the transparency of the metalinguistic material in respect to the primary-sentence lattice, comes from Danuta Hiż, and in a somewhat different way from Henry Hiż,

limitations. One is that the theory contains only a necessary but not a sufficient condition for reduction, so that the specific reductions and permutations in each language have to be listed. The second is that individual languages may have many and prominent regularities which are not germane to the operator–argument structure, such as phonemic similarities within or among particular subsets of words, or morphological paradigms. The third limitation is that some languages have created local restrictions on the general system, by letting certain sentence-making steps become required rather than optional: in a given situation, a particular word may be required (e.g. repetition of subject under *can*, *may*), or a reduction may be required (e.g. zeroing of this repeated subject).

Differences in these respects give languages different grammatical features (aside from the differences in vocabulary—in phonemic content and combinatorial ranges). Some ('inflectional') languages have interrelated systems of affixes, others have non-systematized affixes strung onto host words or onto stems, and yet others have no explicit affixes. For the operator–argument theory the affixes are of little moment, since they are equivalents of free-word operators on the host words, but for the grammar of the individual language these differences are important. Some languages also have for all or many words special forms which are foreign to the operator–argument relation, but can be shown not to violate it. Thus in languages which lack enough free words, because most words have stems or roots which do not occur without one or another of a set of attached morphemes, one can (as noted above) consider the combination of stem with one attached morpheme to constitute the base free word, with the other forms obtainable from it by known operations of the grammar. All such features, including the fixed phonemic (and syllabic) properties which words have in some languages, are important regularities in the description of the language, but secondary here to its operator–argument and reduction structures.

In addition to these word-structure properties, many languages have word subclasses with important regularities which again do not violate the operator–argument and reduction systems but rather are in many cases by-products of them. Many of these subclasses are fuzzy in their properties (e.g. the 'human' class in English) or in their membership (e.g. the English auxiliaries). In certain cases, members of a subclass constitute a required or virtually required participant in sentence composition, above all in the case of grammatical paradigms: tense and person, with aspect in some languages; also noun-case, gender, plural, and agreement. Many of these overtly required constructions create grammatical categories for the language. Such categories may apply semantically even to simple words which have no explicit construction, if the absence of the construction means a particular value of the category (e.g. if a word without plural affix is singular, in meaning and in agreement, rather than being unspecified as to number).

Furthermore, languages differ in transformations, i.e. in the way sentences are combined and changed, and in overtness of the forms used therein. Some languages have regular affixal forms for making an operator into the argument of a further operator (infinitive, participle, gerund). Some have very visible and unchanging forms for certain transformations (e.g. the Semitic conjugations), or have specialized transformations requiring a complicated reductional path (e.g. the English construction seen in *It was this that I said*, G359). Many languages have regular ways to make almost any operator or argument in a sentence into the subject of a paraphrastic sentence, but they differ in that in some languages some of these rearrangements are particularly frequent, made in particularly conventionalized and overt ways.

All of these additional and overt regularities, plus the avoidance of certain constructions (e.g. of zeroing the main subject in English), give particular languages and families of languages unique structural appearances (1.5, end). To the extent that they are intertwined in form or in grammatical function, they may give the language the appearance of an overall grammatical character or strategy ('genius'). These regularities and intertwinings may increase through analogy and conventionalization, yielding the diachronic effect which Edward Sapir called 'configurational pressure' or drift in the language.

What is ultimately, if not always overtly, simple in language is the set of basic and universal processes, especially the operator–argument process that makes sentences out of vocabulary. The final description of an individual language—resulting from all the processes, similarities, and events in it—is not simple and not universal. What makes the simple structure basic is that the other, individual, features can be shown to operate as additions to it, or can be transformed as though they are products of it.

The total regularities and similarities in the language come into play in the problem of how to analyze each particular sentence, especially by the a priori and objective procedures of computer programs. To formulate a regular procedure for discovering (recognizing) the composition of a sentence, and its grammatical relation to other sentences, is a complex endeavor, over and above the general sentence-synthesizing grammar. It is not simply a matter of taking the inverse of the steps that produced the sentence, both for essential problems of process-inversion, and because not a few of the synthesizing steps are degenerate, a consideration which goes unnoticed in the sentence-producing system. That is, two different reductions on two different operator–argument combinations may yield the same word-sequence, which is then ambiguous. Faced with the ambiguous sequence, one has to reconstruct both reductional paths from the two sources, and then choose between them (disambiguate). In many situations, the degeneracy is only local: one of the reconstructions does not fit the other (later) material in the sentence and has to be abandoned. But many globally degenerate ambiguous sentences remain. In analyzing written rather than spoken material, additional difficulties appear, because various grammatically-relevant intonations are not adequately distinguished in writing, and some punctuation such as commas is used erratically.

For these and other reasons, a procedure for analyzing sentences has to combine its own logic with the inverse of sentence-production, and with everything that can distinguish which productional path has been taken in the given sentence. Such distinction is possible because of various factors: for one thing, to a large extent operators and arguments are different words (as against being the same word used in different relative positions), which facilitates their identification. For another, many languages have affixal or other indicators that mark the operator or argument status of a word-occurrence, in some cases even indicating the extent of the sequence (if it is a sequence) that is thus marked. Thirdly, most or all reductions leave a trace (which may be degenerate) in the resultant form: either a specific entity (e.g. the interrogative intonation), or an operator's having an argument of a different class than it should have (G19, 108), or an otherwise un-grammatical word-sequence (G30, 150). Also, most relations are *in situ*— the word simply receives a different phonemic form, often zero; position-changes (so called permutations) are in many languages few, and limited as to how far the word can move (3.4.3).

Given these considerations, it is clear that sentence analysis (recognition) is a major task in itself, and requires its own strategy. For each language or family of languages, one may need particular detailed strategies and algorithmic presentations of how the particular reductional traces fit into the operator–argument reconstruction.[22]

Partly related to the difference between the theory of sentence production and the strategy of sentence analysis is a difference which arises in the actual use of language, namely between the speaker and the hearer. In the first place, the user's perception of the language, informally or as a result of learning traditional grammar, does not have to be the same as the system of constraints described above. People can remember and hold in view a much larger stock of relations and structures than the few required in the operator–argument theory. They also tend to avoid the great number of repetitive small steps such as are used in this theory's processing of a sentence, both in building up a sentence and in changing its form. For these reasons, it is not surprising that people seem to operate grammatically with a rather large stock of words and constructions, even though many of these can be derived theoretically from a much smaller stock. This is especially common in the case of derived words and derived sentence structures whose derivation is very complex, or involves delicate distinctions or obsolete forms (e.g. source-words or intermediate structures, necessary for

---

[22] Cf. ch. 4 n. 1.

a derivation, that have gone out of use). Certainly, speakers of English, following frequent combinations and overt similarities and regularities, use such a word as *childhood*, or the tenses, or the auxiliary verbs (*can*, *may*), or the passive form, all as individual words or structures, without considering that the complexities that these bring into English grammar can be avoided by deriving them in particular ways from simpler forms.

In addition to this, the speaker's relation to what he is saying differs in important respects from the hearer's relation to what he hears. For one thing, the speaker has little reason to note the traces left by the reductions which take place in his sentences, whereas these traces are important for the hearer. Indeed, the speaker need not be aware if his words have homonyms (or, in writing, homographs), and if the reductions in his sentences are degenerate either locally or globally over the sentence; but the hearer has to note them when there is danger of ambiguities, and he has to know what in the environing material can disambiguate them. Failing disambiguation, the hearer may not know what the speaker meant. There are also differences in respect to reference. One issue is whether the reference is to the same individual or not. The speaker can say *I saw men enter and, later, leave* either if he knows them to be the same men or if he does not know, or if he knows them to be different. He says *I saw men enter and, later, men leave* only if he does not know them to be the same or does not want to impart that information; if he says *I saw men enter and later I saw them leave* he is saying they were the same men. In the first case the hearer does not know which information the speaker meant. More generally, when the speaker uses a free pronoun instead of a noun, he knows which previous occurrence he is referring to, whereas the hearer has only the pronoun to go by and in many cases cannot be sure which other word is being referred to.

Most generally, the speaker's sentence is both partially and linearly ordered. The grammatical relations and the likelihoods (hence also meaning-specifications) and the reductions are all in respect to the partial ordering, but at the end there is a linearization. This is not to say that the speaker first formulates the sentence in a partial ordering: he may think in elementary sentences which are already linearized; unintended Spoonerisms may show if the grammatical partial ordering affects the order in which words have been chosen by the speaker. In any case, the hearer meets only the linear form; if the partial ordering has reality for him (as grammatical relations), it is only at the point where he has completed his understanding of the sentence. It has been shown (by Lashley[23]) that hearers try to process the grammatical relations in the sentence as they hear it, in the order in which the words appear; they may have to reanalyze the beginning of the sentence if later material in it does not fit their analysis up to that point.

---

[23] Cf. ch. 3 n. 4.

IV

Subsets of Sentences

9

Sentence Sequences

9.0. *Motivation: constraints between sentences*

Chapter 7 established the sentence as an utterance-segment within which are contained the major constraints, those of operator–argument relations. Chapter 8 showed that sentence structures form a system of relations among sentences. Chapter 9 introduces the first globally new constraint in language, beyond the constraints that created sentence boundary. This is a relation that holds only over domains larger than a sentence. In Chapter 9 the domains are continuous speech segments larger than one sentence. (In respect to this relation, single sentences are not merely short discourses, for the discourse introduces a new relation.) In Chapter 10 the domains are sets of sentences that have in common a speaking-situation: these domains can be defined (a) externally (e.g. articles in a particular research line, or conversations when people are involved in a particular joint interest), or (b) structurally (if the sentences in a corpus of discourses show the same kind of structural relations that are found in discourses that satisfy (a)). As to meaning: much as the operator–argument structure of the one-sentence domain means predication (saying one word about another), so the pluri-sentence structure means subject matter (saying more than one thing about the same referent or the same class of referents).

Thus further, pluri-sentence, relation is reached by considering two new problems: what about constraints that are not contained within the limits of a sentence? That is, are there constraints that reach among successive sentences? And what relations hold among sentences not throughout the language but specifically in certain subsets of sentences: when they are in the sentence succession that constitutes a spoken or written discourse; and when they are all in one subject matter (sublanguage)? We will see that in all these cases there is indeed a further constraint, but not on the major word-class combinations. Rather, it is on word-choices within these.[1] Specifically there is a demand to repeat word-choice in conjoined sentences (9.1); and a demand to repeat the operator–argument relation among words of particular subsets in discourse (9.3) and in sublanguages (ch. 10). In the case of

---

[1] There is no further constraint on word classes because sentence boundaries were reached and defined precisely as the point beyond which no constraint was carried over stochastically from the preceding word classes (7.7)

discourse, this yields a double-array structure, as against the linearized partial-order structure of single sentences.

9.1. *Conjoined sentences*

A sequence of even just two sentences, if they are conjoined by connectives (a subset of O_{oo}), reveals a constraint. This constraint can be considered as the likelihood constraint imposed by each of these O_{oo} on its pair of sentential arguments. To set the stage, we note that certain conjunctions demand certain minimal differences between their two argument sentences: *and* may be found between two identical sentences (e.g. *I welcome you and I welcome you*, only slightly unusual); *or* expects at least one difference (*John will write it or Mary will write it*, *We will write it or we will type it*, with *John will write it or John will write it* being quite peculiar outside of logic examples); *but* expects at least two differences (*John will write it but Mary will sign it* or . . . *but Mary will write it too*, whereas there is no *John will write it but Mary will write it* except in special cases where differences of expected kinds have been hidden by reductions, e.g. in *John wrote it, but John wrote it sloppily*, or in semantically opposite operators (which conceal a zeroed contrastive sentence) such as *I studied it but I forgot it.*

We now consider the general question of constraints on sentences under various conjunctions and other binary second-level operators (O_{oo}). No clear constraints have been noted in respect to specific vocabulary; but under most binaries it is expected that there be at least one difference between the arguments (except e.g. *We will arrive when we will arrive*). The major constraint here is that there be at least one word that occurs in both sentences, except in so far as one occurrence was removed by reductions. To see this, consider $S_1 O_{oo} S_2$, where S_1, S_2 are the arguments of O_{oo}. For a given $S_1 O_{oo} S_2$ sentence, its acceptance as something unexceptional rather than surprising or peculiar is either roughly equal to, or else lower than, the lesser of the acceptances of S_1 and S_2 separately. Thus in (1) *We stayed at home because we were tired*, we find that S_1, S_2, and $S_1 O_{oo} S_2$ are all equally unexceptionable. In (2) *We stayed at home because today is Tuesday*, S_1 and S_2 are unexceptionable, but $S_1 O_{oo} S_2$ is odd, or leaves the hearer waiting for something more unless that explanatory something is already known to the hearer. And in (3) *We stayed at home when vacuum consumed the chair*, both S_2 and $S_1 O_{oo} S_2$ are peculiar while S_1 is unexceptionable. When $S_1 O_{oo} S_2$ has lower acceptance than all its components, as in (2), we note that its acceptance can be raised by conjoining an unexceptionable S_3 which contains a word of S_1 and a word of S_2--for example, *and Tuesday is the college's at-home day* in the case of (2).

This suggests that to retain in a resultant sentence the level of acceptability of its two components (i.e. the degree to which they are unexceptionable) we need word-repetition among the components. There are cases where the unexceptionability is retained without word-repetition; but then we can always find zeroable sentences which would provide the word-repetition and which we can reconstruct as having been present, but well known and therefore zeroed. Thus given (4) *We stayed at home because John was tired*, where $S_1O_{oo}S_2$ is as unexceptionable as S_1, S_2 but without repetition, we can assume some explanatory zeroed sentence connecting *John* with *we* (e.g. *John is one of us*, or *John is our son*).

The zeroed adjoined sentences which provide the missing repetition are in many cases matters of common knowledge, or definitions, which are zeroable as adding nothing to the information in $S_1O_{oo}S_2$. For, example, in (5) *I wrote it on Tuesday because I had to hand it in Wednesday morning*, the conjoined repetition-supplying sentence would have been *and Wednesday is the day after Tuesday*, or the like (2.2.3, 2.3.2). When our supplying the presumed zeroed sentence $O_{oo}S_3$ raises the acceptance of $S_1O_{oo}S_2$, as in (2), it is because the $O_{oo}S_3$ of (2) was not known to the hearer, whereas in (4) and (5) it is known. Thus a sequence $S_1O_{oo}S_2O_{oo} \ldots O_{oo}S_n$ is generally as acceptable as the least acceptable S_i in it if each component S_i contains at least one argument or operator which is also present in some other component S_j (before reduction). The above sequence can be shortened without loss of acceptability by zeroing any S in it which is common knowledge, at least to the speaker and hearer. Omission of any other S may make the sequence less acceptable than any of its components. The fact that sentence-connecting has this property leads to important results about the structure of discourses and the nature of argumentation, as will be seen below.

9.2. *Metasentences*

In the reconstructed sentences that provide word-repetition under O_{oo} we see that to a sentence S_i we can adjoin (by an additional O_{oo}) sentences which talk about this S_i, and which do not appear because they are known and have been zeroed. This is possible because metalinguistic sentences, which talk about sentences and their parts, exist (5.2), and secondly because the ability of a metalinguistic sentence to refer to one preceding it in the $S_1O_{oo}S_2$ sequence is supplied by the addressing (referral) power which is contained in the relative-clause procedure that adjoins one sentence to another (3.4.4, 5.3). Metalinguistic sentences which refer to the 'host' sentence to which they are adjoined will be called metasentences. When they speak about a word, they are referring to an occurrence of that word in the host sentence, with the sense and selection which it has in that occurrence. For example, *The kite flew high; in this sentence the word kite refers to a bird, not a toy*; also *Everybody wanted the funny hats, but the dolls* (stressed) *wanted only the girls, where the stressed word dolls is the object and the word girls the subject of the word wanted*; and *He asked for cardizem; the word cardizem here is the name of a medication* (usually reduced by 'abus de langage' to *He asked for cardizem, which is the name of a medication*).

The metasentence adjunctions can fill many functions for the host sentence. They can state that a word-occurrence X in one position is the same word (with the same selection, i.e. in the same sense), or that it refers to the same individual, as a word-occurrence Y elsewhere; in this case the metasentence can be replaced by reducing X to a pronoun (possibly zero) of Y. For example, *I saw some high-flying kites, which are hard to keep up* from *I saw some high-flying kites; kites—the preceding word refers to the same as the prior word—are hard to keep up*; note that neither the *which* nor the sameness statement that is its source would be said if the prior occurrence of *kites* referred to the bird 'kite', since the predicate in the second sentence is in the selection only of the 'toy' occurrence of *kite* (5.3.3).

The metasentences can state that the speaker is talking about something, in which case the word for that thing in the host sentence can be pronouned as a repetition from the metasentence, which is then zeroed; this gives a demonstrative (deictic) meaning to what is constructionally just a repetitional pronoun, as in *This coffee is cold* from, say, *I am speaking about some coffee; this coffee is cold* (5.3.4). The metasentences can also supply grammatical and dictionary information about all parts of the host sentence. When they do, the sentence—host plus its metasentences—becomes a self-sufficient universe requiring understanding only of the given metalinguistic material (grammar and dictionary) but not of the host sentence itself, since the latter is.explained by the metasentences.

That this is indeed the case has been remarked above (5.2), namely: we can take an arbitrary sequence of English phonemes that satisfies English syllabic structure, and can make it an English sentence if we adjoin to it appropriate metasentences, for example ones stating that an initial subsequence in it is the name of a scientific institute, and that the middle subsequence ending in a (spelled) *s* is a term for some laboratory technique, and that the final segment is some science object (e.g. a chemical compound). This means that when we hear a sentence which we accept and understand as English, it is because the metasentences which performed the above kind of task for it had been zeroed as being common knowledge. Thus a sentence of a language is simply a phonemic sequence of it with known or guessable, usually zeroed, metasentences identifying its successive segments as entities permitted there by the metalanguage and usually known by the hearer.

Since different grammatical and dictionary metasentences may apply to the same sentential phoneme-sequence, the zeroing of them may leave the phoneme-sequence ambiguous. This takes place when the metasentences state different reductions on different operators and arguments, which happen to be degenerate as to their resultant phoneme-sequence. Another kind of degeneracy results when metasentences are erroneously zeroed; for example, if we zero a metasentence which identified a word in the host as the name of an object or relation rather than as referring to the object or relation, we would obtain the effect of 'abus de langage' as above.

9.3. *Discourse analysis*

If we now move from $SO_{oo}S$ sequences to the general case of connected discourse, i.e. of long utterances, we find a property that goes importantly beyond the word-repetition seen in 9.1. In each discourse we can find two or more particular subclasses of words appearing in many sentences, in operator–argument relation to each other, or in certain (not all) other grammatical relations. The subclasses are not necessarily known in advance; they are established for each discourse from the word-pair repetitions that are found in it. This is a property that seems to be satisfied by all discourses, even rather aimless ones, but is not in general satisfied by arbitrary collections of sentences (for example, one composed of the first sentence on every fiftieth page of an encyclopedia). The various grammatical relations in which the word-class-pair repetition is found are approximately paraphrastic to each other, and are found to be transforms of each other (8.1). This property is different from the word-repetition of 9.1 in that it specifies repetition of two or more word classes (not specifically of the same word) in certain grammatical relations to each other.

In practice: given a discourse, we look for sentences or sentence-segements which have a sequence of two or more words in a grammatical relation (chiefly operator–argument, or host–modifier) that recurs in various sentences, in the same relation or in transforms of that relation. We may thereupon reconstruct the sentences or segments containing the transformed relation, so that a canonical sentential form is selected for all segments containing that word-sequence. Thus, given *Lymphocytes produce antibody* and *Antibody is produced by lymphocytes*, we might transform the passive back into the active, to obtain two occurrences of *Lymphocytes produce antibody*. We then seek segments in which one of the words in the original sequence appears in the given relation to other words (not present in the original sequence): e.g. *Lymphocytes contain antibody*. This enables us to suggest that the new words (e.g. *contain*) are in the same equivalence class as the words they 'replace' (in this case, *produce*).

To show how this analysis proceeds, we take a complete article, quoted below. In Figure 9.1 transforms of its sentences are aligned into a table; the sentences were transformed in a way that would maximize their alignment.[2]

---

[2] Z. Harris, (a) *Discourse Analysis Reprints*, Papers on Formal Linguistics 2, Mouton, The Hague, 1963, pp. 20 ff.; (b) 'Discourse Analysis', *Language* 28 (1952) 1–30 for the general method; (c) 'Discourse Analysis: A Sample Text', *Language* 28 (1952) 474–94.

The Structure of Insulin as Compared to that of Sanger's A-Chain[3]

by

K. Linderstrøm-Lang *and* John A. Schellman

1 The optical rotatory power of proteins is very sensitive to the experimental conditions under which it is measured, particularly the wavelength of the light
2 which is used. Consequently single measurements of optical rotation do not give an adequate description of rotatory properties, though they have often proved very useful in the characterization of proteins and the detection of changes in
3 structure in solution. The diversity of the factors which affect optical rotation is in many ways an advantage, for the variation of each parameter yields a separate source of experimental information.
4 One of the authors (J.A.S.) has for some time been engaged in a study of the rotatory properties of several proteins including the effect of temperature,
5 wavelength, pH and the denaturation reaction. One phase of this research, the dependence of the rotatory properties of proteins on wavelength, is recorded here because it is of special importance to the problem at hand.
6 In agreement with the observations of HEWITT it was found that all the polypeptide systems which were investigated obeyed a one term Drude equation within experimental error (usually less than 1%);

$$[\alpha] = \frac{A}{\lambda^2 - \lambda_c^2}$$

7 where $[\alpha]$ is the specific rotation and λ the wavelength of the measurement. A and λ_c have a certain amount of theoretical significance but will be regarded here as empirically determined quantities.
8 The dependence of the optical rotation on experimental conditions results from the fact that A is in general a function of temperature, pH, ionic strength,
9 denaturation, etc. λ_c, on the other hand, varies only as a result of drastic changes in the protein system, in particular denaturation by urea or guanidine or titration
10 to a pH between 10.5 and 12.0. The values of λ_c and $[\alpha]_D^{20}$ are given in Table I
11 for a number of proteins and related substances. (A_{20} may be obtained from
12 $[\alpha]_D^{20}$ by means of the relation $A_{20} = [\alpha]_D^{20}(\lambda_D^2 - \lambda^2)$. λ_c is not dependent on temperature within experimental error (approximately $\pm$ 50A).
13 The results for gelatin are taken from CARPENTER AND LOVELACE.
14 [Table I was inserted here.]
15 [Note to Table I] Except for insulin the specific rotations were independent of moderate changes in protein concentration.
16 The substances in Table I are listed in the order of descending values of λ_c.
17 The table has been divided into two groups to emphasize an obvious pattern, namely, all the substances in Group A are native globular proteins, all those in Group B are not.

---

[3] Reprinted from *Biochimica et Biophysica Acta*, 15 (1954) 156–7. Table 1 in this paper has not been reprinted here, because it is not in language form, and only details the information given in this paper.

18 The entries in Table I are too few to permit any certain generalizations but the
   indication is that λ_c provides a measure of the presence of secondary structure in
   polypeptides, having values above 2400 A for the ordered configurations of
   native proteins, and values less than 2300 A for polypeptides in disordered states.
19 We are here subscribing to the view that heat and chemical denaturation result in
20 the unfolding of the hydrogen bonded structure of a protein. Clupein naturally
   falls into the latter group because the internal repulsion due to its high positive
   charge makes folded molecular forms unstable.
21   The presence of the oxidized A-chain of insulin in Group B was not expected.
22 In fact it was hoped that the A-chain would exist as an α-helix in solution and
   would therefore serve as a model substance of known structure in the study of
23 denaturation. Instead it was found that the A-chain possesses rotatory properties
   which resemble those of clupein very closely, but do not resemble those of insulin
24 itself. Most striking is the fact that the specific rotations of clupein and the A-
   chain are virtually unaffected by strong solutions of urea and guanidine chloride.
25 Ordinary proteins, including insulin, undergo changes in specific rotations of
26 100% to 300% under these conditions. These results suggest that the oxidized A-
   chain is largely unfolded in aqueous solution and are in agreement with the recent
   finding that the peptide hydrogen atoms of the A-chain exchange readily with
   D_2O, whereas those of insulin do not.
27 A detailed report of this work will appear later.

   In the table in Figure 9.1, column 1 gives the sentence number in the original,
column 2 gives the sentence number transform. Each following column contains
an equivalence class; that is, the entries in each column have been shown to be
equivalent to each other by the procedure described above. It turns out that
throughout the whole article, every sentence of the transform consists of the
HRLK equivalence classes (or of two or three out of these four), with intervening
verb equivalence classes.

   The interpretation is, of course, not that all members of an equivalence class
are synonyms of each other, but that the difference for the given discourse
between any two members X_1, X_2 of one equivalence class X corresponds to the
difference between the corresponding members Y_1, Y_2 of the class Y with which
they occur. That is, the discourse compares X_1, Y_1 with X_2, Y_2, or equivalently,
X_1, X_2 with Y_1, Y_2.

The word subclasses in question can be determined in the following
equivalence relation (transitive except for subscript, and symmetric and
reflexive) on word-sequences:

---

[4] After the equivalence classes have been set up and their relative occurrence studied, we
may find reason to say that in a particular case $a \neq_o a$: i.e. that a particular occurrence of the
morpheme-sequence a is not discourse-equivalent to the other occurrences of the morpheme-
sequence a. (This differs from the case of two sentential word-sequences being identical in form
but not in derivation, i.e. source: see 8.1.5.) The $a \neq a$ case will happen if we find that accepting
the equivalence in this particular occurrence forces us to equate two equivalence classes whose
difference of distribution in the double array described below has reasonable interpretation.
Such situations are rare; and in any case the equivalence chain has to start with the hypotheses
that occurrences of the same morpheme-sequences are equivalent to each other in degree zero.
The equivalence relation then states that a is equivalent to b if for some n, $n > 0$, we have $a =_n b$
by the formula given here.

S	P	C	H	V	R	U
T	1		Insulin	has	structure	
	2		Sanger's A-chain	has	structure	
1	3		Proteins	have	optical rotatory power	
	4	wh	Proteins	have	optical rotatory power	
	5	particularly	Proteins	have	optical rotatory power	
2	6	Consequently			rotatory properties	
	7	though				
	8	and				
	9				structure	
3	10		Proteins	have	optical rotation	
	11	In many ways				
	12	for				
4	13		several proteins	have	rotatory properties	
	14		several proteins	have	rotatory properties	
	15		several proteins	have	rotatory properties	
	16		several proteins	have	rotatory properties	
	17		several proteins	have	rotatory properties	
5	18		proteins	have	rotatory properties	
	19					
	20	because				
6	21					
	22		the polypeptide syst.	obeyed within experimental error	a—	one—
	23		the polypeptide syst.	obeyed within usually less than 1%	a—	one—
	24		the polypeptide syst.	obeyed within usually less than 1%	$[\alpha]$ the specific rotation	=
	25	all wh	the polypeptide syst.			
7	26					
	27	but				

FIG. 9.1. Analysis of a text, (Part I).

L	W	K	*Metadiscourse*	S
			compared with 2 compared with 1	T
	very sensitive to measured under very sensitive to	the experimental conditions the experimental conditions the wavelength of the light which is used		1
	changing in	solution	are not given adequate description by single measurements of optical rotation the latter has often proved very useful in the characterization of proteins . . . in the detection of 9	2
	affected by	a diversity of factors a diversity of factors	is an advantage the variation of each parameter yields a separate source of experimental information	3
	including the effect of including the effect of including the effect of including the effect of	temperature wavelength pH the denaturation reaction	One of the authors (J.A.S.) has for some time been engaged in a study of . . .	4
	depending on	wavelength	One phase, 18, of this research is recorded here it is of special importance to the problem at hand	5
term− term− A	Drude− Drude− $\div(-\lambda_c^a)$ $+$	equation equation $\lambda^2)$ the wavelength of the measurement	22–4 was found, in agreement with the observations of Hewitt were investigated	6
A, A,	λ_c λ_c		have a certain amount of theoretical significance will be regarded here as empirically determined properties	7

S	P	C	H	V	R	U
8	28 29 30 31 32	results from fact in general	(proteins	have)	optical rotation	
9	33	on the other hand				
	34	in particular				
	35	or				
	36	or				
10	37	and	a number of proteins	have values of	$[\alpha]_D^{20}$	and values of
	38		a number of related substances	have values of	$[\alpha]_D^{20}$	and values of
11	39	by means of relation			$[\alpha]_D^{20}$	from^{-1}
	40				$[\alpha]_D^{20}$	=
12	41					
	42					
13	43		gelatin			
14	44a ff.		(The substance . . .	has value	of $-[\alpha]_D^{20}$	and value
			Insulin	has value	—	and value
			—	has value	—	and value
			—	has value	—	and value
			—	has value	—	and value
			—	has value	—	and value
			β-lactoglobulin	has value	—	and value
			—	has value	—	and value
			—	has value	—	and value
			—	has value	—	and value
			—	has value	—	and value
15	45		(proteins	for^{-1})	the specific rotations	
	46	but	insulin	for^{-1}	the specific rotations	
16	47		the substances			have value
17	48					
	49		substances proteins		native globular, all	are
	50		substances proteins		not native globular, all	are

FIG. 9.1. Analysis of a text, (Part II).

L	W	K	Metadiscourse	S
A A A A	dependent on is a function of is a function of is a function of	experimental conditions temperature pH ionic strength denaturation, etc.		8
λ_c λ_c λ_c λ_c	varies only as a result of varies as a result of varies as a result of varies as a result of	drastic changes in the protein system denaturation by urea denaturation by guanidine titration to a pH between 10.5 and 12.0		9
λ_c λ_c			given in Table I given in Table I	10
A_{10} A_{20}	$(-\lambda_c^2)$ +	$\lambda_D^2)^{-1}$	may be obtained	11
λ_c λ_c	is not dependent within experimental error on is not dependent within approximately ± 50A on	temperature temperature		12
			has results taken from Carpenter and Lovelace	13
of λ_c — ⎫ — ⎬ Group A — ⎨ — ⎩ — ⎫ — ⎬ Group B — ⎨ — ⎩	under under under under under under under under under under under	conditions) pH 3,2% — — — — denaturation by 7M urea — — — —	(is) Table I	14
	were independent of were not independent of	moderate changes in protein concentration moderate changes in protein concentration		15
of λ_c			listed in descending order in Table I	16
  in Group A in Group B			The table has been divided into two groups to emphasize an obvious pattern, 49–50	17

S	P	C	H	V	R	U
18	51					
	52	but				
	53		polypeptides	have	secondary structure	whose presence is provided a measure by
	54		proteins		native with ordered configurations	have values
	55		polypeptides	in	disordered states	have values
19	56					
	57		a protein	has	unfolding of the hydrogen bonded structure	
	58		a protein	has	unfolding of the hydrogen bonded structure	
20	59		clupein			naturally falls into
	60	because	(clupein	has)	unstable folded molecular forms	
21	61		the oxidized A-chain of insulin			is present in
	62					
22	63	In fact				
	64		the A-chain	as	an α-helix	
	65	and therefore	the A-chain as model substance	of	known structure	
23	66	Instead				
	67		the A-chain	possesses	rotatory properties	
	68		clupein	possesses	rotatory properties	
	69	but	insulin itself	possesses	rotatory properties	
24	70					
	71		clupein	has	specific rotations	
	72	and	clupein	has	specific rotations	
	73	and	the A-chain	has	specific rotations	
	74	and	the A-chain	has	specific rotations	
25	75		ordinary proteins	have	specific rotations	
	76	including	insulin	has	specific rotations	
26	77					
	78		the oxidized A-chain	is	largely unfolded	
	79	and				
	80		the A-chain	has	peptide hydr. atoms	
	81	whereas	insulin	has	peptide hydr. atoms	
27	82					

FIG. 9.1. Analysis of a text, (Part III).

L	W	K	*Metadiscourse*	S
			The entries in Table I are too few to permit any certain generalizations the indication is 53–5	18
λ_c of λ_c >2400 A <2300 A of λ_c				
	led-to by led-to by	heat chemical denaturation	We are here subscribing to the view 57–8	19
Group B	made by	internal repulsion due to its high positive charge		20
Group B			61 was not expected	21
	would exist in would serve in	solution the study of denaturation	64–5 was hoped	22
			67–9 was found resembling X very closely resembling X very closely not resembling X	23
	virtually unaffected by virtually unaffected by virtually unaffected by virtually unaffected by	strong solution of urea strong solution of guanidine chloride strong solution of urea strong solution of guanidine chloride	The fact 71–4 is most striking	24
	undergoing changes of 100% to 300% under undergoing changes of 100% to 300% under	these conditions these conditions		25
	in exchanging readily with not exchanging readily with	aqueous solution D_2O D_2O	Results 71–6 suggest 78 results 71–6 are in agreement with recent finding 80–1	26
			A detailed report of this work will appear later	27

$a =_o b .=. a$ is the same word-sequence as b[4]

$a =_n b .=. env\ a =_{n-1} env\ b$ (read: $env\ a$ is $_{n-1}$-equivalent to $env\ b$, or $env\ a = env\ b$ in degree $_{n-1}$),

where $a, b, \ldots$ are word-sequences, and $env\ a$ is the remainder of the sentence (or of a transform of the sentence) which contains a; that is, $env\ a$ is the sentential environment or neighborhood of a, and is itself a (possibly broken) sequence of morphemes. Here $env\ a =_{n-1} env\ b$ is taken to mean that at least some part of $env\ a =_{n-1}$ the corresponding part of $env\ b$, and that all other parts of $env\ a =_{m<(n-1)}$ the corresponding parts of $env\ b$. That is, $n-1$ is the highest subscript of equivalence between any part of $env\ a$ and the corresponding part of $env\ b$.

The subscripts n are intended only to show a small ascending chain of environment equivalences. And the formulas above are not included to suggest a decision procedure for discovering a structure (let alone the true structure) of a discourse. Rather, they constitute an orderly record of how and at what cost a particular structure of that discourse can be obtained.

The equivalence $a =_o b$ is used only when we can find a chain of equivalence with ascending subscripts. To find such an ascending chain, it is usually necessary that a and b occur corresponding in grammatical positions within their respective sentences, or within the transforms of their respective sentences. Ubiquitous words like *the, in, is* will usually not satisfy this condition, and are therefore useless as a base for a chain of equivalences.

The equivalence $a =_n b$ between particular word-sequences may be reached by more than one chain. The degree n of the equivalence between them will be understood to be the lowest subscript of $a=b$ in any chain in which $a=b$ appears.

The purpose of this chain of equivalences is to create out of the word combinations—or more simply the word-occurrences—in the discourse certain word subclasses which will be found to occur repeatedly in a constant (under transformation) grammatical relation to other equivalence classes in the sentences of the discourse.

The possibility of achieving this result is enhanced by further grammatical analysis of the sentences of the discourse. For example, if the discourse contains *Antibody appeared in the lymphocytes* and *The lymphocytes secrete antibody*, where the procedure above cannot be applied directly, one can choose the passive transform of the second sentence so as to make it *Antibody was secreted by the lymphocytes*. Then *antibody* and *the lymphocytes* are subject and object, respectively, of the two operator-entities *appeared in* and *was secreted by*. Here *antibody* $=_o$ *antibody*, and *the lymphocytes* $=_o$ *the lymphocytes*, and *appeared in* $=_1$ *was secreted by*. The effect here has been to align *antibody* in both sentences as being the subject of the sentence or of a transform of it, and correspondingly for *the lymphocytes* as object.

Furthermore, decomposing the sentences by breaking them at O_{oo} points into their component sentences reduces the complexity of the environment which the procedure has to take into account, thus enhancing the ability to build up equivalence classes. Since the transformations, and more generally the reductions and position-changes, do not change the operator–argument selection or the substantive information in the sentences, any chain of environment-equivalences of word subclasses established for transformed sentences will be valid also for the original (untransformed) text. And since the grammatical relations, and the reductions and other transformations, have been established for the language as a whole, and are not being proposed *ad hoc* for the discourse which is being examined, there is nothing circular in applying them. One might however go beyond the whole-language transformations in one case: one could use any synonym or classifier of particular words as an aid in furthering the building up of their equivalence classes, but only if it is clear to everyone involved that these are indeed synonyms and classifiers of those words in the subject matter and style of language use of which the given discourse is a case.

This method segments the discourse into recurring sequences of certain equivalence classes. These classes and the sentence types made of them constitute a special grammar of the particular discourse. Thus, although the discourse has to satisfy the grammar of its language, it also satisfies a grammar of its own based on its particular word-choice co-occurrences. If we take a translation of the given discourse into another language, we find that the sentence grammar is then that of the other language, but the word-choice subclasses and their sentence types are much the same as in the original discourse.

It is possible, in a discourse, to collect all the sentences which have the same classes, so as to see how the members of a class change correspondingly to the change in members of the other classes, and so as to compare these sentences with sentences having in part other classes. In many cases it is possible to obtain a rough summary of the argument; and it is possible to attempt critiques of content or argument based upon this tabulation of the structure of the discourse.

If we consider many discourses of the same kind, we find that their analyses are partially similar. For example, in scientific articles it is characteristic that we obtain a set of sentences in the object-language of the science, and on some of them a set of metascience operators which contain words about the actions of the scientist, or operators of prior sciences (e.g. logical relations). In the table the last column (and some of column 3) is of this nature.

The result of discourse analysis is essentially a double array, each row being a sentence of the transformed (i.e. aligned) discourse, and each column a class of sentence segments which are equivalent by the discourse procedure. All further analysis and critique of the discourse is based on the relations contained in the double array. The discourse dependence of later sentences in a paragraph upon the first sentence is different from the dependence of S_2 upon S_1 in $S_1O_{oo}S_2$ (9.1). In a discourse, the later sentences are not independent; it may become possible to extract from them certain factors of modification which might be considered to operate on segments of the first sentence to produce segments of the later sentences. These factors are in general independent of the first sentence, and it is these factors then that constitute the second dimension. It is in this respect that discourse, as a two-dimensional structure, differs from language, which is a one-dimensional structure even in its $SO_{oo}S$ word-repetition requirement.

In the case of standardized types of text, and in the case of certain literary pieces, such as carefully constructed short stories, the double array may exhibit the compositional structure of the text, and may even locate and reveal some of the points on which the composition turns.[5]

In the double array which is created by a repetition of a sentence type made of particular word subclasses, one can see various differences in the successive members of a word subclass, i.e. in the successive rows for a particular column. One may be able to see some regularity, or some explainable feature, in the way different columns change in respect to each other, for successive rows. One may also be able to see other grammatical features such as preferences in the use of pronouns, synonyms, reductions, and other transformations in respect to the double array; reference, as noted in 5.3, is always within a discourse, and may be affected by discourse structure. In cases of extreme regularity it may be possible to describe a double array as in part the resultant of two axes: first, the top row introducing the sentence type, and then a generalization of how successive rows differ from their predecessors. In longer discourses we generally find not one but several different sentence types, using in part the same word subclasses but in different grammatical combinations, each repeated to create its own double array. (In the example above, sentence 61 belongs to a different type—covering cols. 4, 7, 8—than 64–81, which covered cols. 4–6, 9, 10.)

The existence of the double array, or the position of a particular row in it, does not confer any value of truth or consistency upon the individual sentences making up its rows, nor does it empower us to derive for the discourse further sentences that would fit or extend the inter-row differences seen in the array. Nevertheless, in some cases the array can suggest educated guesses as to which unsaid sentences might fit into the discourse, or even why particular sentences that could fit into the array were not included in the discourse.[6] Going beyond this, there is a possibility of additional conditions on the double array which give it the power to identify some of the rows as conclusions from preceding rows, or as having other semantic relations such as consistency with the preceding rows. This arises from the fact that certain

---

[5] Cf. the analysis of James Thurber's 'The Very Proper Gander' in Harris, *Discourse Analysis Reprints*, Papers on Formal Linguistics 2, pp. 20 ff.

[6] An example of the latter is in Harris, 'Discourse Analysis: A Sample Text', *Language* 28 (1952) 474–94.

changes within a column (e.g. substituting classifiers for ordinary members of the word subclass), or correlated changes within different columns in successive rows, or the action of particular O_{oo} successions between successive rows, can separately or together give direction to the double array. Then, from being just a set of occurrences of a sentence type, it becomes a progression in the sentence type, consisting often of a first sentence with various modifications of it in the successive rows, which semantically support certain later rows as conclusions or suggestions from the earlier rows. This additional power will be considered more fully in Chapter 10 in conjunction with the power gained from the sublanguage condition.

We have seen, in considering discourse, that despite the grammatical importance of the sentence there are structural relations in language which are not encompassed in the structure of the sentence. Not even a further step beyond the sentence, namely a relation between sentence structures, suffices to yield the discourse property. Rather, the latter is a repetition which is indeed between sentences, or more precisely between operator–argument entities (as are all grammatical events), but is not a relation of sentence structures. Indeed, the repetition is seen fully only when we undo transformations in some sentences and add transformations to others until we maximize the alignment of the repeated word subclasses. The double array extends the sentential relations among the words of the sentences into the more complex relations among the repetitions, thereby extending the 'aboutness' of the sentence—the operator being predicated about its arguments—to the more complex aboutness of a discourse. The $SO_{oo}S$ repetition and the double array show that in order to discuss anything more than a single unadorned one-sentence observation, one cannot merely string various independent sentences but must have grammatically interrelated repetition of words in small subclasses. It follows that, as noted above, a discourse is not simply a longer sentence, or an arbitrary sequence of sentences, nor is a sentence simply a short discourse.

10

Sublanguages

10.0. *Introduction*

Certain proper subsets of the sentences of a language may be closed under some or all of the operations defined for the language. That is, carrying out a grammatical operation on a sentence of that subset yields another sentence of the same subset. Such subsets constitute a sublanguage of the language, because they satisfy the grammar without involving any sentence that is outside the subset. It will be seen that the structures of these sublanguages have important new properties of their own, and also throw further light on the structure of language as a whole.

The sublanguages are found not by using new methods or by asking new questions with the existing methods, but simply by using the existing methods on bodies of data whose provenance is restricted in the real world. The dependence relation of 3.1.1, and the likelihood-based reductions of 3.3, can have different effects in different data conditions. When carried out on arbitrary samples of utterances of a language, they yield the zero-, first-, and second-level word classes. When carried out on each discourse (any extended sequence of sentences arising in the real world) separately, they yield the discourse word classes and sentence types of 9.3. And when carried out on the utterances of distinct social or occupational types of situation (without regard to the sentences' occurring in a sentence sequence), they yield sublanguages with specific word classes and sentence type (10.3–5). The distinct social or occupational situations are in most cases recognizable as dealing with distinct subject matters, most characteristically in the various sub-sciences, but also in such situations as law cases, opinion surveys, or even heraldry.

There is no particular limit to the number of possible sublanguages in the open-ended whole language, or to the overlappings among them. But this does not mean that the whole language is merely an envelope of the sublanguages, or that its grammar is just a washing-out of the differences among the sublanguages. The language as a whole has properties which do not hold for any of its sublanguages. In particular, every sublanguage except the base (10.1) has an external metalanguage which is a subset of the whole language but not of that sublanguage, whereas the metalanguage of the whole language is always a subset of a whole language (10.2). To this is related the fact that sublanguages have the dependence of 3.1.1, but not in general the dependence-on-dependence of 3.1.2.

A sublanguage can differ from the whole language by omitting some grammatical properties of the language. This is possible because many of the constraints that create grammar are satisfiable independently of each other. A sublanguage may happen not to contain the conditions for a given constraint. The subset of sentences which forms a sublanguage must satisfy the whole-language grammar, but it can do so vacuously in the case of some of the constraints, which are thereupon omitted in the sublanguage grammar. And it can satisfy in addition certain other constraints which are washed out in the rest of the language.

In Chapter 10 we will see sublanguages based on grammatical conditions (10.1–2), and those based on subject matter (10.3–4), also formula-sublanguages of science which can be considered distinct linguistic systems (10.5). There will be also comparisons of natural language to other systems of a linguistic character, specifically mathematics (10.6), and to entirely different systems of symbols, chiefly music (10.7).

10.1. *Grammar-based sublanguages*

Some sublanguages are composed of sentences which satisfy certain grammatical conditions that are not satisfied by all other sentences of the language. One such sublanguage consists of the elementary sentence, i.e. those formed by N and $O_{n \ldots n}$, all under just the O_{oo} (conjunctional) operators of the language (or under *and* alone): for example, *John sleeps*, *John sells art*, *John puts the book on the shelf*, all under *and*, *because*, etc. Such a sublanguage may be of interest because it leaves out the O_o operators and those which mix N and O in their arguments: the O_{no} (e.g. *think* in *I think he left*), the O_{on} (*surprise* in *My entering surprised the dog*), and the O_{noo} (*attribute* in *I attribute his appointing you to my phoning him*). Since these operators have particular semantic characters, it is possible that in certain applications language may never need to use them.

Another grammatically conditioned sublanguage is the base language (4.1), that is, those sentences which do not result from reductions, position-changes, and other (transformational) shape-changes. This sublanguage is important because it suffices for all the substantive information that the whole language carries, yet is (differently from the whole set of sentences) transparent in its form: the hierarchy of operators and arguments comprising each base sentence can be read off mechanically from its word-sequence. If it is desired to make this sublanguage directly readable as English, or whatever language the base is in, while retaining reasonable transparency for processing by humans or by computers, we can add to the base sentences the effects of the required and the most common transformations: e.g. for

English, tense, plural, *the/a*, referential indication, and possibly the relative clause in the base sentences. It may be noted that, since the base structure is more universal than the reductions, the base sentences of one language translate more easily into those of another than do the transformed sentences, so that practical considerations are involved here.

There are also various distinguished subsets of sentences which have grammatical and informational properties but fall short of satisfying the conditions of being sublanguages. Such are classifier sentences, definitions, and 'general sentences' such as those that express laws of a science. For example, in classifier sentences it is not always possible to distinguish formally or semantically between predicates indicating classification and predicates indicating other properties (e.g. in *Cats are mammals*, *Cats are hunters*), or to specify the ultimate objects of classification (e.g. *A horse is a mammal*, *A roan is a horse*, *A stallion is a horse*, *The stallion is a roan*).

In all cases of grammatically based sublanguages, sentences can be usefully characterized not only by their decomposition but also by the sublanguages (or more generally the characterized subset of sentences) into which they fit. To a lesser extent, this holds also for the subject-matter sublanguages, and for the subsets of sentences which do not quite make the sublanguage grade, as above.

10.2. The metalanguage

A third sublanguage satisfies both the above property of having special grammatical additions and omissions, and also the property of dealing with a closed subject matter. This is the metalanguage. As noted in 5.1–2, the metalanguage of a natural language is a natural language. More precisely, one can form in any natural language various sentences that speak about that language (or about any other), about the combinations of parts in occurrences of the language, and the like. There is no way to define or describe the language and its occurrences except in such statements said in that same language or in another natural language. Even if the grammar of a language is stated largely in symbols, those symbols will have to be defined ultimately in a natural language.

The importance of the fact that the metalanguage is in the language has been noted in 2.1. Here we consider the fact that not merely is it in the language but it constitutes a sublanguage of that language. It is distinguished grammatically from the rest of the language by the fact that its zero-level-argument (N) positions can be occupied not only by words but also by sentences, and indeed also by otherwise non-occurring phonemic sequences (*The sequence of phonemes, or of words, he went is a sentence*; *The phoneme-sequence pxk is not an English word*). However, it also has the subject-matter property (10.3) of being restricted to a small, relatively closed, set of words as its own vocabulary of operators. With these it forms its own set of sentence types, as can be seen in 5.1. Like other subject-matter sublanguages it need not have grammatically relevant differences, or grading, of likelihood among members of a subclass (10.3), though it can describe the likelihood gradations that are found in the colloquial language. The fact that the metalanguage of a natural language constitutes a sublanguage does not conflict with the fact that its sentences are at the same time sentences of a natural language, satisfying its grammar. The metalanguage succeeds in describing the operator–argument structure that creates sentences only by stating it in sentences that themselves have that structure.

The metalanguage itself uses only the operator–argument relation and a subset of words with individual selections (co-occurrence likelihoods). It is thus a part of the base of the language (4.1). Within the metalanguage one can not only characterize the whole base of the language, but also define the reductions for the language. Thereafter, the metalanguage can itself use those reductions to make the sentences of the metalanguage less cumbersome.

The relation of the metalanguage to language may appear circular: the metalanguage is part of language, yet can generate the whole language without being reachable except via language. This circularity is avoided by the natural history of language acquisition (whether for an individual's first language, or second language, or for the species' development of language initially). That is, the gross structure of language, in particular the operator–argument relation and the selections of various words, can be recognized from observing the gross regularities in occurrences of language (speech or writing); or it arises by conventialization in the course of language development (12.3). This suffices to establish a basic grammar and vocabulary for the metalanguage. In turn, the metalanguage then suffices to characterize (or generate, if we will) the sentences of the language.

The internal metalanguage of a natural language L_0 can be obtained if to each sentence of L_0 we attach by some grammatical connective a sequence of conjoined metalinguistic (5.2) sentences (of L_0) which state the grammatical properties (and, if desired, the definitions) of each word in the given sentence and of all grammatical relations among these words in that sentence. For example: *In the sentence John likes to read, John is a word in* N, *like is a word in* O_{no} *with ordered arguments John and read, and read is a word in* O_{nn} *with ordered arguments John and things, and -s is the operator-indicator on like, and for . . . to is the argument-indicator on read, and the words for John are zeroed* qua *repetition while the word things is zeroed* qua *indefinite.* As has been noted, these metalinguistic appendages are implicitly present on every sentence of L_0: if we did not know them we would not understand the sentence or know that it is a sentence of L_0; and because we know them they are zeroable, leaving the given sentence *John likes to read.* If we consider just the metalinguistic sentences on all sentences of L_0, we obtain a separate subset of sentences which specify the grammatical status of the words of L_0 and of their grammatical relations.

Taken together these metalinguistic sentences, or rather summaries of them, constitute the metalanguage L_1 of L_0. Their primitive arguments (N) are the elements of L_0 or classes of these. Their first-level operators (on N) are classifiers and relations of these. The structure of the sentences of L_1 is the partial order of dependence among their words; but the dependence-on-dependence relation is not required since the membership of N is specified directly, the other classes being defined in respect to it. Also the relative likelihood of word combinations plays no role; and reduction is not necessary —we can choose (inconveniently) to use only unreduced metalinguistic sentences. One grammatical property that is essential to the metalinguistic sentences is the capacity to name elements and their relative location in sentences, i.e. the existence of classifier and locational operators at the first level. This is what makes it possible for the metalinguistic sentences to refer to words and their locations in stated sentences or in all sentences of L_0; in L_0 this capacity is used only for cross-reference and for reductions.

In addition, those grammatical properties of L_0—whether essential to L_0 or not—which are not part of the structure of the metalanguage can be defined by means of sentences of the metalanguage. Thus, dependence-on-dependence, and likelihoods of word combinations within the dependence relation, and also reductions, and alternative linearizations, can all be defined for L_0 in metalinguistic sentences which do not themselves exhibit these structural features. In respect to these secondary features, the meta-linguistic sublanguage has a simpler structure than the language which it describes. Rather than speak of the metalanguage as being in general richer than the system it describes, we see that in this case it is simpler, because it is an extract from the whole language, containing just enough structure to describe the language itself. And rather than the metalanguage being prior to natural language, the existence of the metalanguage is made possible by the prior existence of the language itself, because only the dependence-on-dependence relation, which is unique to natural language, assures the arising of a dependence system out of word use that lacked dependence.

It is also possible to look upon the sentences of L_0 (including those of its metalanguage) as being generated by discourses in that metalanguage, rather than to obtain the metalanguage as appendages to the sentences of L_0. Thus *John likes to read* can be formed from the sequence of conjoined metalinguistic sentences that follows it in the example above.

The relation of a natural language L_0 to its metalanguage, whether starting from L_0 or starting from the metalanguage, recurs in the relation of the metalanguage to its own metalanguage in turn. There is necessarily an infinite regress of metalanguages. But it will be seen that this is the case only for their referents; beyond a certain stage, their formal composition remains constant. To see this, we take the grammar of L_i as a discourse written in the metalanguage, L_{i+1}, of L_i. As noted above L_1 has a structure of its own: its zero-level arguments include all entities in utterances of L_0; its first-level predicates include, for example, *differ phonemically*, *is a phoneme*, *is a word*, *operates on*, *occurs with*; its second-level operators include *highly likely*, *is more frequent than*. The grammar of L_1 is written in L_2, which is the metalanguage of L_1; like everything which is not a whole natural language, L_1 has an external metalanguage, and L_2 is a sublanguage of L_0, not of L_1. L_2 in turn has a structure of its own. Its zero-level arguments are not the L_0 entities themselves but the classes and relations of them as established in L_1, hence in effect the predicates and operations of L_1, such as *phonemic distinction*, *grammatical events*, *occurrence with*, *high likelihood*; its operators include *is partially ordered*, *is less redundant*, but not, for example, *is more frequent than*, because, differently from whole languages, the relative frequency of words in a sublanguage (here, L_1) has no grammatical relevance to the sublanguage (10.3–4). Proceeding further, the grammar of L_2 is written in L_3, the metalanguage of L_2. Here again we find certain differences. The arguments in L_3 include the predicates of L_2 such as *classification*, *redundancy*; in addition, certain operators such as *is a word*, *occurs with*, will be found in all grammars including those of the metalanguages. After stating these elements as a grammar of L_3, written in L_4, we find that the grammar of L_4 is formally identical with that of L_3. Its referents are different, in that the grammar of L_3, in L_4, refers to sentences of L_3, while the grammar of L_4 (written in L_5) refers to those of L_4. But the kinds of argument and operator classes needed to describe L_4 are not different from those needed to describe L_3. Thus, although there is an infinite regress of referents in the meta-languages, the form of their grammars remains essentially unchanged after a certain point. The reason for this syntactic result is that L_1 has the form of a factual description—the grammar of a (natural) language, while L_2 has the form of a theory, describing the syntactic structure of such factual descriptions, and L_3 has the form of a metatheory, describing the syntactic structure of theories, which then repeats in L_4 and thereafter.[1]

This means that the language of metatheory, that is, the sublanguage which presents the structure of a theory, itself has a metalanguage which is structurally identical with itself, rather than being a proper subset of it as is the case for a whole language, or external to it as is the case for subject-matter sublanguages. To generate a whole language it may therefore be

---

[1] The chain of metalanguages has a certain similarity to the chain of translated sentences in the story about the mathematician who knew no English and asked a friend to translate an article of his into English. In the published form three lines appeared under the translated title:

I thank X for translating this paper and its title.
I thank X for translating the above line.
I thank X for translating the above line.

The change of referent in successive occurrences of *above* enabled the author to copy line 3 himself from the translated line 2.

possible to start with L_3, defining it in an L_4 structurally identical to it, and then use it to define L_2, thence L_1, and finally the language L_0.

10.3. *Subject-matter sublanguages*

Many sublanguages are composed of sentences which deal with a more or less closed subject matter—one in which a limited vocabulary is used and in which the occurrence of other words is rare. Whereas the closure of the grammatical sublanguages above arose from specific grammatical constraints, the closure here arises from the limitations of word use due to the limitations of the subject matter, which are sharp enough to constitute constraints on word selection when one is speaking within the subject matter. The meta-language has both of these properties.

The limitations of use are found by much the same method as the equivalence classes of discourse analysis (9.3), except that here the environment needed for the equivalence is specifically the operator and co-argument, if any, on a word, or the arguments under it, since that is the main scope within which selection is defined (3.2). The reason for this difference from 9.3 is that here we are discovering the grammar of a language—a sublanguage in this case—rather than seeking the maximum alignment of words in a text. In the sublanguage material, what we obtain from the equivalence classes in respect to the operator–argument environment is various word subclasses which co-occur in operator–argument relations, i.e. in base sentences, but also less transparently in reduced sentences. That is, we obtain various sentence types, each a particular combination of word subclasses. This is the characteristic property of subject-matter sublanguages; it holds whether the sentences are in discourses or occur singly. It is sharpest in science, especially in the 'hard' sciences, where words and sentences drawn from outside the science are both irrelevant and absent.

A detailed description of the word subclasses and sentence types in a sublanguage constitute its grammar. If we now ask how a sentence can satisfy both the grammar of a sublanguage and the grammar of the whole language, the reason is that the subclasses of the sublanguage are each inside a main word class (or a class with modifiers) of the whole language, as they could not but be. In the sublanguage of biochemistry, for example, the sentences *The polypeptides were washed in hydrochloric acid* or *The proteins were treated with acid* would be cases of a particular sentence type, say $N_{mol}O_{sol}N_{sol}$ (where O_{sol} includes *is washed in*, etc.; *mol* is for molecule, *sol* for solution); in English they are cases of N_2ON_1 (more precisely, the passive of $N_3O_{nn}N_2PN_1$: *We washed the polypeptides in hydrochloric acid*; the subscript numbers distinguish corresponding N). The difference between the two representations involves the fact that English O_{nn} allows also the sentence (1) *Hydrochloric acid was washed in polypeptides*, for whatever meaning that would have; whereas in terms of the biochemistry grammar, (1) would be $N_{sol}O_{sol}N_{mol}$, which is not a sentence type there.

In the grammar of the sublanguage we find particular subclasses of arguments and operators, which combine in various ways to form particular sentence types. These latter necessarily satisfy the partial ordering of words according to their argument-requirements, and so are subclasses of the whole-language sentence structures (e.g. $N_{mol}O_{sol}N_{sol}$ a subclass of $NO_{nn}N$). In the sublanguage all the operator–argument sequences, with possibly a few exceptions, satisfy one or another of its sentence types. In the grammar of the whole language the word subclasses and sentence types of the sublanguage do not appear, because when the sentences of the sublanguage are considered not separately, as a separate universe of data, but together with arbitrary other sentences of the language, the O_{sol} words are not found to be restricted to ordered $N_{mol}N_{sol}$ pairs in the whole-language data, and do not constitute separate classes. Thus the sublanguage has constraints which do not apply in the whole language.

In addition, the whole language has constraints which do not apply in the sublanguage. Since the sentences of the sublanguages are in the language, and must therefore satisfy its constraints, this additional property is possible, as has been noted, only if the whole-language constraints are satisfied vacuously by the sublanguage sentences. This arises if certain operator classes (e.g. certain kinds of O_{no} such as *know*, *hope*, with their largely human subjects, or certain repetition-requiring O_{oo}) or certain transformations (including stylistic ones) never occur in the sublanguage sentences.

It follows that the sublanguage grammar contains constraints that the language sentences violate, and the language grammar contains constraints that the sublanguage sentences do not satisfy. Hence while the sublanguage is a proper part of the language, the grammar of the sublanguage intersects the grammar of the language (although the intersection of the two is essential to the sublanguage's being a language) rather than being a sub-grammar of it.

The differences between language and sublanguage structure are important not only as showing what new forms can be attained in sublanguages, but also as showing the capacities of linguistic systems beyond that which has developed in natural language.

The greatest difference arises in the different word subclasses and sentence types to which the sublanguage is restricted, within the operator–argument sentence structure of the whole language, and in the possibility of new argument-requirement types among the operators.

In addition, there is the overriding fact that the metalanguage of the sublanguage is outside the sublanguage, since the word subclasses of the metalanguage will not in general be among those of the sublanguage. The externality of its metalanguage releases the grammar of the sublanguage from being tied to the least description of (sublanguage) constraints, for now we can always define a class or its environment or other relation not by its constraints but by an external metalinguistic statement. Thus it is possible to have multiply classified words—elements which in certain environments belong to one class and in others to another, the distinction being made in the metalanguage: in the immunology example below (10.4) it will be seen that in some environments *wh-* is a form of the special time-connective in the bi-sentential sentence type of that science, while elsewhere in the science it is the general subordinating connective that it is in the whole language.

As in discourse grammar, but not in the whole language grammar, a sublanguage can have phrases as members of a word class. Even in the whole language it is possible for certain phrases to be considered as members of a word class. For example, *providing, provided, provided that, with the provision that*, all appear as conjunctions, classifiable under O_{oo}: *I will go, providing he will go too*, and correspondingly for the others. But all of these also have an analysis consistent with their parts, without which they could not exist in the sentence. We can see this for *with the provision that* if we begin with:

> *I will go, with a provision; the provision is that he will go too.*
> → *I will go, with a provision which is that he will go too.*
> → *I will go with the provision that he will go too.*

Similarly we can see this for the conjunctional *provided that*, if we begin with:

> *I will go if it is provided that he will go too.*

And we may see it for the conjunctional *providing that*, if we begin with something like:

> *I will go, if my going provides that he will go too.*

In all these forms, the starting-point conforms to the meaning of *provide* and the operator structure of the sentences. With various zeroings, which are not all common but are here supported by the broad selection of these phrases (i.e. the range of sentences connected by them), we obtain the shorter forms and the more general meaning of 'only if' rather than specifically of 'provide'. We end up with phrases which do not have quite the meaning of their words, and where the operator–argument relations are not obvious, but where the whole phrase can be regarded as a new member of O_{oo}. Similarly in the case of sublanguages (but not for science languages, 10.5) there has to be a 'correct' operator–argument structure to account for each part of the phrase, but the semantic connection of the phrase to its words can be tenuous, and in effect we simply have various phrases as members of word subclasses, defined as such in the metalanguage of the sublanguage. This includes the possibility of considering, as in the methods of 9.3 and 10.3, that both *appear in* and *is secreted by* are members of the operator class on the ordered pair *antibody*, *lymphocyte*, disregarding the whole-language fact that one is passive.[2]

The external metalanguage is a factor in most of the other differences: first, a by-product of the external metalanguage is that the sublanguage need not constitute a mathematical object by itself. Its word subclasses need not be defined entirely by their interrelations, because it is possible to define at least one of them, usually the N subclasses, by listing the members of these subclasses (in the metalanguage statements).

Secondly, related to the above is the fact that the set of subclasses of the sublanguage is as though closed. Although new subclasses and sentence types can enter the sublanguage at any time, one can at any time analyze its sentences in respect to its currently specified word subclasses. The analysis is considered to fail for any sentence that cannot be analyzed in these terms, whereupon a new subclass may have to be defined. What is observable before analysis is that the sentences of the science seem to be in a nearly closed vocabulary: in fields where this is the case, sublanguage analysis is easier.

Thirdly, likelihood differences and gradations among members of one subclass in respect to each member of its operator or argument subclass are not relevant to the sublanguage grammar, though they may correspond to the meaning-differences among those members.[3] This arises from the fact that such relations as synonymity and classifier status among the words, and possibilities of factoring the scientific terms into component terms, and also any useful difference in meanings, can all be stated in the metalanguage, so that the sublanguage does not need to have a combinatorial basis for them.

Lastly, since synonymities and classifier relations are to a large extent specifiable in the metalanguage, there are fewer ambiguities in the base

---

[2] This is also justified by the fact that passive constructions such as *is secreted by* may not have in the sublanguage the stative environments and meaning that they have in principle in the language as a whole. Whereas the restriction of selection and meaning that such constructions as the passive, or *provided*, *providing*, should have may be unnoticed or disregarded in various sentences of the whole language, in sublanguages they may be systematically disregarded so that there they have to be accepted not as grammatical constructions but as phrasal members of a word subclass.

[3] The sublanguage subclasses are not merely an extreme case of whole-language combinatorial likelihood differences. Rather, they provide a classification of the sublanguage vocabulary into subclasses in respect to combinability, in a way that fits the informational framework of the subject matter. In contrast, the differences in likelihoods of combination that remain among the words within a subclass (after synonymy has been eliminated) reflect differences in meaning of the various terms within the current classification knowledge. Thus the sublanguage classes are specializations of the dependence relation (3.1.1), not of the likelihood relation (3.2). An opposite analysis was made for the auxiliaries (3.2.4), whose combinatorial peculiarity (lack of a different second subject) was taken as zero grade of likelihood rather than as dependence. The justification there was that auxiliaries were a small set of operators whose combinatorial peculiarity could be readily 'cured' by supplying a synonymous source that was regular; they did not constitute, as in the sublanguage, a part of a global combinatorial subclassification of words that proved semantically useful.

(unreduced) form of the sublanguage than there are in the base of the whole language.[4]

Aside from the differences between sublanguage and whole language, we can also consider the relations among subject-matter sublanguages, because the fact that they are all subclass specializations within the operator–argument structure makes inter-system relations specifiable, much as one can specify the cost in respect to the operator–argument system of one another of a given sentence. The transformational analysis of 8.1 has already introduced the notion that sentences can be identified not only by their components but also by their relations to other sentences, as transforms of them. The sublanguage analysis shows that sentences can also be identified by another inter-sentence relation: by having the same sublanguage properties.

Sublanguages can have various relations to each other. First, some sublanguages may largely share the same word subclasses, others may have partly similar structures, somewhat as certain fields of mathematics are abstractly equivalent. Secondly, sentences of one sublanguage may have special relations to another sublanguage. For example, certain sentences of one sublanguage may be arguments in certain sentence types of another: in *Digitalis affects the heart conractility* (an example of a common sentence type in pharmacology), the second argument (*The heart contracts*) is an example of a sentence type in physiology. This particular syntactic relation in sentences of different fields creates a relation of 'prior science' between sublanguages, where one science acts upon events of the other. One may even think of large parts of language as being covered by various sublanguages, possibly overlapping in their sentences. But in any case, the whole language—colloquial, general, literary, and so on—is not merely an envelope of sublanguages. There is always a complete, independently standing language over and above all sublanguages. Informationally, this whole language may have a relation to the perceived world (11.6) similar to that of a science language to its science. But there may be little to be gained from considering the structure of the whole language to be itself a sublanguage structure.

It is also of interest to consider the similarities between the structures of subject-matter sublanguages and discourse structures. The main similarity is that both have special word subclasses and sentence types, but for partly different semantic reasons. Discourse has these because of the constraints due to maintaining a topic, to giving information and discussion beyond separate observations. The sublanguages have these structures because all

---

[4] Pure synonyms, which rarely exist in whole natural languages, are distinguishable in sublanguages. Within a particular research line for a given two words, if the likelihoods of their combining with other words become more similar as the corpus of discussion increases, then the two given words are synonymous for that research line. This does not indicate that they are synonymous elsewhere. For example, in the immunology material of 10.4, *agglutinin* and *gamma-globulin* are synonyms of *antibody*, although their specific meanings, and their uses elsewhere, are not identical.

the sentences deal with an interrelated set of information. Both tend to maximize the structural similarity within their restricted universes, and in both cases the external structural status of sentence-segments outweighs the importance of their internal structure, so that words can freely belong to different subclasses, and the subclasses can have whole phrases, or transparent segments such as passive verbs, as members—all this defined in the meta-language, which is external not only for sublanguages but also of course for each discourse.[5]

The main difference is that a sublanguage does not have the double array structure of discourse; it is, like the whole language, naturally represented linearly, whereas discourse is not. Sublanguages have less of the characteristic repetition than discourse has, and usually more sentence types. Discourse analysis maximizes the similarity among its sentences via the equivalence chain of 9.3, more assiduously than does sublanguage analysis. The sublanguage grammar allows for new sentences that can be made within its sentence types, whereas the discourse structure allows only for further sentences that fit the content of the given discourse.

Most sublanguage material comes in discourses, not in isolated sentences. Combining sublanguage and discourse analyses yields the greatest results, especially in applications. If scientific texts are analyzed by both methods, it becomes possible to organize their information, to summarize it, and to retrieve specific items. Also, in the discussion or argumentation sections of articles it is possible to seek what conditions on the successive sentences enable earlier sentences to serve as support for later ones. Since all of these applications are made possible by the regularities of repetition and inter-word relations within a subject matter, it is not excluded that in some subject matters the regularities may be so complex as to defy any useful utilization.

10.4. *Science sublanguages*

Among possible subject-matter sublanguages certain ones stand out as by far the most important. These are the science sublanguages, each consisting of the sentences and discourses of the practitioners of a sub-science, issued in the course of research and discussion. They are important because their subject is important, and because they throw light on the structure of the science itself, or in any case on the structure of its information. They

[5] In both sublanguage and discourse analysis, maximizing similarity is largely a matter of aligning sentence-segments. In discourse, we seek the maximal alignment of the sentences. In sublanguage analysis, we try to get segments that have similar combinations into similar positions. Thus in the immunology material of 10.4, when we find verbs V_1 that occur between A and C (*Antibody appears in lymphocytes*, AV_1C) and verbs V_2 that occur between C and A(*Lymphocytes contain antibody*, CV_2A), we form a class V in formulaic order AVC which represents the V_1 verbs with argument order $A \ldots C$, and represents V_2 verbs with argument order $C \ldots A$.

constitute large and complex systems, and not merely a collection of sentences containing scientific terms; and they have room within their grammars, in distinguished ways, for disagreements and change—necessarily, since the grammars are made from the sentences of science discussion, which indeed contain disagreements and changes.

Attempts to investigate the language of science, implicitly as a means of organizing its information and its conclusion-drawing, were made chiefly by the Logical Positivists of the Vienna Circle,[6] using the tools of mathematical logic and, later, probabilistic considerations. The difficulties encountered there can be avoided if we use, instead, the methods of word-combination analysis. As has been seen for discourses (9.3) and for sublanguages (10.3), the repetition and classification of word-combination in such subsets of language create special grammars which yield much of what was sought as the language properties of science.

In order to see what a science sublanguage is and how it is determined, an example is given here. The field analyzed is from the early years of immunology, and covers particularly the research into what cell in the lymphatic system is the producer of antibody. In this line of research, lasting from the late 1930s to the late 1960s, there had been a controversy as to whether the lymphocytes or the plasma cells of the lymphatic system produced the antibodies. It was finally resolved by definite evidence for both cells that they produced antibody, and by recognizing that these names were being used for different stages of the same cell-line. This situation was of interest to our present purposes, because one can check whether the analysis of the sentences in the various articles shows their stand in the controversy, and what each article contributed to it, and whether the analysis shows the relation of the resolution of the controversy to the earlier controversial statements. This served as a test of the ability of the analysis to represent the information, and the differences in information, in the sentences of the sublanguage.[7]

We find here several subclasses of zero-level arguments, most of whose members form a rather small vocabulary of terms of the science, and a few subclasses of first-level operators (mostly verbs), some of whose members are also science terms while others are drawn from general English. There are also various subclasses of first-level operators, in subsidiary sentences as modifiers (adjectives), some of which are science-specific and others general English. The latter include classes of quantity and time words. The second-level operators, such as the conjunctions and the metascience segments, are drawn from general English, and only in certain cases show science-specific

---

[6] See in particular R. Carnap, *The Logical Syntax of Language*, Harcourt, Brace, New York, 1934.

[7] The full report of this work is in Z. Harris, M. Gottfried, T. Ryckman, P. Mattick, A. Daladier, T. N. Harris, S. Harris, *The Form of Information in Science: Analysis of an Immunology Sublanguage*, Boston Studies in Philosophy of Science 104, Kluwer Academic Press, Dordrecht, 1989.

restrictions. All these appear together in a few sentence types specific to the science.

In broadest tems, the grammar for the immunology material turned out as follows (meanings are given here for convenience, but were not used):

The words for 'antigen', marked G, were either synonyms of it or names of various antigens.

Words for 'antibody' (A) included local synonyms such as *gamma globulin*, indicators of antibody presence such as *agglutinin*, or names of specific antibodies. It should be noted that most of these synonyms would not be synonymous with *antibody* in other research—*gamma globulin* is a broader term than *antibody*—but locally, in these texts and word environments, there was no difference, and indeed in most cases what is meant, for example by *gamma globulin* here, is just *antibody*.

There is a class of bodies, body parts, and animal names (B): *rabbit*, *footpad*, etc.

Inject (J) is used between G and B; there are also local synonyms for *inject*. In this material *immunize* was used synonymously with *inject*, and *normal* was used to mean *not injected*.

There is a large class of 'tissue' names (T) with many different tissues and organs which have to be distinguished by subscripts: T_l *lymph*, T_n *lymph nodes*, T_s *spleen*, and so on. They all have certain common environments, with differences among them in respect to detailed environments.

There is a smaller class of cell names (C): C_y for *lymphocyte*, C_z for *plasma cell*, C_b for *blast cells*, and quite a few other names, some proposed in the given article just for particular properties that were being reported.

There is a class (V) of the verbs that occur between A and C (or A and T): V_i for *in* (synonymously: *found in*, *present in*, *appear in*; also *contain* in the inverse order C-A) as in *Antibody appeared in the lymphocytes* (AV_iC_y); V_p for *produced in*; V_m for *pass through*; V_s for *secreted by*, as in *Antibody is secreted by the cell* (AV_pC).

There is a large class (W) of words and adjectives after T or C, naming properties or changes of tissue or cell, with different subscripts: *tissue is inflamed*, *cell is mature*, *cells proliferate*, etc.

There is a class (U) of verbs which occur between *antigen* and *tissue* or *cell*: *antigen moves to tissue*, *perishes in tissue*.

Finally there is a class (Y) of verbs between *cell* and *cell*: *is similar to*, *is called*; there is also Y_c *changes into*, *differentiates into*; and some more special subsets such as Y_u *contaminated with*.

While a subscript on X indicates a subclass of X, a superscript on X (seen below) indicates a modifier attached grammatically to X.

The major sentence types, with some indications of their examples were identified as follows:

GJB, for sentences of the type *Antigen is injected into a body part* or *an animal*.

GU^{ft}TT, similarly for *Antigen moves from some tissue to some tissue* (the superscript indicates *from . . . to*).

TW and *CW*, for *A tissue* (or *cell*) *has some property* or *undergoes a change*.

AVC for *Antibody appears in*, *is produced in*, or *is secreted from a cell*.

CYC for *Some cell is similar to*, or *is called*, *some cell*.

CY$_c$C for *A cell develops into another cell*.

In the cell-transfer studies, in which antigen is injected into one animal, whereafter lymphocytes are injected (transferred) from that animal to another, with antibodies then being sought in the second animal, an additional sentence type is found:

CI^{ft}BB, for *Cells are injected from an animal into another animal*. The *I^{ft}* with two *B* differs from the *J* above, even though *injected* appears in both subclasses.

There is a special conjunction, internal to a particular sequence of sentence types, which is seen or is implicit in almost all occurrences of the pair *GJB* and *AVC*. This is *thereafter* and its synonyms, marked here by a colon. It often carries a time modifier, e.g. *three days*, *five hours* as in *GJB:^{t}AVC*, for *Antigen is injected into a body part; three days later* (or the like) *antibody appears in cells*. Also, *Antibody appeared three days after injection of antigen*, and the like. This conjunction takes different grammatical forms (e.g. *to* in *The cell contained antibody to the antigen* and even the relative clause *wh-* (10.3)); all of these forms synonymously connect *AVC* (or *CW* or *TW*) to *GJB*.

Metascience material, giving the scientist's relation to the information of the science, can be separated off. Mostly, this is isolated immediately, as the highest operator in a sentence: O_o and O_{no} (e.g. *as was expected* or *X and Y have shown that*). But there are also cases where metascience operators combine with operators of the science language proper, as in *Antibody is found in lymphocytes*, which could be analyzed as either from (roughly) *It was found that antibody is in lymphocytes* or else from *We found antibody in lymphocytes*.

The relation of this sublanguage structure to the information of the science can be seen in various ways. We note first how the known change of information through time is seen in the change of word subclasses and sentence types in the successive articles of this period. The class *C* (*cell* in contrast to *T*, *tissue*) first appears at the very end of the earliest article, and thereafter becomes very common. The class *S* for intra-cellular ultrastructures appears only midway in the series and becomes common as more cellular structures become discernible. In *C*, lymphocytes and plasma cells appear early, and later we find a welter of different names for lymphatic cells as more and more cell distinctions are made; many of these names appear only in particular articles and not throughout. In the subclass V (operators on the pair A, C), the earlier articles have primarily the member V_i (*is found in*, etc.) but the later articles increasingly have V_p (*is produced in*), as evidence of actual production becomes obtainable. Finally, the whole subclass Y (operator on the pair C, C) puts in a relatively late appearance, both in its member *is the same as*, *is called*, which is used to match various new cell names that appear in different articles, and in its member Y_c (*changes into*, *develops into*), when it becomes recognized that some cell names represent merely later stages of what other cell names represent.

In sentence types, the earlier AVT (*Antibody is found in tissue*) is later replaced by AVC (*Antibody is found in a cell*), where the neighboring sentences show that this is an abbreviation or conclusion from AVT_1; CWT_1 (i.e. *Antibody is found in a tissue*; *certain cells abound in that tissue*). Later yet the AVC occurs without a distantly connected AVT: the antibody is found directly in the cell. Finally, the later articles introduce a new sentence type C_1YC_2 (*what has been called by cell name C_1 is the same as*, or *develops into*, *what has been called by cell name C_2*); this becomes necessary as the cell names proliferate, and especially when it is realized that the plasma cell is not an alternative to the lymphocyte but a transformation-product of it.

Next, we note how the controversy is reflected in the sentence structure. The disagreement appears first in the fact that while some papers have exclusively (1) AV_pC_y (*Antibodies are produced in lymphocytes*), others have AV_pC_z (*Antibodies are produced in plasma cells*) to the exclusion of (1) but frequently accompanied by $AV_p{}^rC_y$ (*Lymphocytes have a role in production of antibodies*—as against actually producing them). This superscript r (*have a role in*) is in itself an indication of the controversy: it has the same position (on V_p) in the AV_pC sentence formulas as the negative superscript $\sim$ which denies production, and as the absence of $\sim$ which asserts production; and its meaning—judging from the English words used and from the different relations of AV_pC_y, $AV_p{}^{\sim}C_y$, and $AV_p{}^rC_y$ to their neighboring sentences— is somewhere between denying and marginally admitting production. Finally, the disagreement appears explicitly when some of the papers that contain AV_pC_z also have $AV_p{}^{\sim}C_y$ (or else AV_pC_y under negative metascience operators): e.g. *The experiments clearly refute the idea of antibody production by mature lymphocytes* and *Antibody is not produced by small lymphocytes*.

It is relevant to the 'which cell' controversy that several differences in experimental conditions were common in sentences involving plasma cells as contrasted with lymphocytes. In the formulas these differences appear as: J^3 rather than J^1 or J^2 (J^3 indicating more massive and repeated injection of antigen), $^{t+}$ rather than $^{t-}$ ($^{t+}$ indicating longer time before antibody finding), and $V_i{}^+$ rather than $V_i{}^-$ ($V_i{}^+$ indicating larger amounts of antibody found). Thus characteristically $GJ^3B{:}^{t+}AV_i{}^+C_z$ as against $GJ^1B{:}^{t-}AV_i{}^-C_y$. In addition, we find AV_sC_y but not AV_sC_z: *Lymphocytes (but not plasma cells*) *secrete antibody*, which can account for their containing less antibody. All this is relevant to the issue of whether lymphocytes produce antibody, and it is visible in the formulas. There is also a relevant consideration that is not visible in the formulas, but rather requires a combining of these C_z formulas with other C_z formulas: this is the evidence that the C_z are large cells with organelles in which antibody could be stored, which could account for the plasma cells containing more antibody than the lymphocytes. In contrast, the C_y are small cells with little internal structure which were widely considered (without adequate evidence) to be the end of their line. The resolution was that lymphocytes produced and secreted antibody, then grew into plasma cells which continued to produce, and store, antibody.

One might ask why the sublanguage analysis is needed to show the disagreement when it is expressed so clearly. The answer is that in the sublanguage sentence types, such phenomena as disagreement can be found on grammatical grounds. The immediate purpose here was to show that the disagreement which we see in the *Lymphocytes produce* (or: *do not produce*) *antiboody* sentences, explicitly in the use of *refute* and *not* and implicitly in *have a role*, appears upon inspection in the superscript position on the V_p in the formulas. Furthermore, in that position, after the sentences of the articles have been mapped onto *AVC* and the other sentence types, the disagreement can be located a priori and even mechanically, without much knowledge of the meanings of words: if the texts contain a sentence formula both with the tilde ($\sim$) and without it, then there is either a disgreement or a difference in experimental conditions. Thus when the sentences of scientific articles are mapped into sublanguage formulas, as can in principle be done by computer programs, denials or doubts or hedgings about a given statement if present can be found in a priori statable locations of the formula for that statement, and so for other items and aspects of information. This makes possible various human and machine inspections and processings of information, and also a formulation of the structure of information in the science.[8]

Lastly, we come to how the resolution of the 'which cell' controversy appears in the sentence structures. The information structure that turned out to be involved in the resolution appears early in the series of articles, when a new sentence type *CYC* is introduced. It is introduced for two new kinds of information: *Y* without subscript is introduced for *is called* and *is the same as*, for the proliferating cell names; and Y_c is introduced for *changes into*, *develops into*, noted above. Implicit in the need for these sentence formulas is the fact that cell names are artefacts of the scientist's work, and

---

[8] More generally, considerations of formula structure may suggest questions about the interpretation of data. For example, the mere fact that we find both *GJB:AVC* and *GJB:CW*, i.e. that the colon can connect *GJB* to either *AVC* or *CW*, could raise the question as to what precise relation—of time or of cellular activity—holds between *AVC* and *CW*, and what mechanism represented by the colon has the effect of leading, at least initially, to one outcome rather than the other, or one outcome before the other.

that one named cell changing or developing into another (CY_cC) is the same fact as a cell undergoing certain development (CW). When evidence accumulated that both plasma cells and lymphocytes produced antibody, the controversy was resolved in terms of Y_c: namely by $C_yY_cC_2$ (*Lymphocytes develop into plasma cells*). As before, this is not to say that one would have known in advance that plasma cells are a later stage of lymphocytes: the need to have introduced the CYC sentence type does not imply the $C_yY_cC_z$ case of that type. But it shows that the resolution can be recognized structurally in the sublanguage, and is related to earlier developments in the sublanguage; perhaps awareness of the sentence type might have made it easier to consider the possibility of the $C_yY_cC_z$ case.

In the standardized ways of presenting scientific results, the articles have different sentence types in different sections: e.g. in the Materials and Methods section as against the Results section, with certain word subclasses and also sentence types being common to both sections and certain ones specific to each section. The Discussion section, in contrast, is not formed out of new sentence types but mostly out of the Result sentences with special constraints. These constraints involve primarily quantifiers, classifiers, certain secondary operators (superscripts, in the formulaic representation), conjunctions (more generally, O_{oo} operators), and finally metascience segments (O_{no} and O_o operators). Much of the vocabulary of all these further constraints is common to discussion in many sciences, although certain classifiers and secondary operators (such as the *have a role in* above) are specific to the given sub-science. The science sublanguage structure has specifiable positions in which we find expressions of the disagreements, the uncertainties, and the change in knowledge. In the present case, the disagreements are found primarily in the membership of the C subclass as second argument of V_p. The sharpest uncertainty is in the r superscript (*have a role in*), in superscripts such as $+$ and $-$ for imprecise quantifiers (*much*, *little*), and in secondary operators (as superscripts) and metascience operators that explicitly express uncertainty (*likely, it may be that . . .*). Change in knowledge is expressed by new members of existing word subclasses, by new subclasses, and by whole new sentence types such as CYC; also by changes in the uncertainty markers above.

A detailed examination of the sublanguage of the immunology articles shows that here, as also in other sciences, the boundaries of the sublanguage have yet to be defined. There are several ways in which we can judge the boundaries of the early-immunology sublanguage, in each case satisfying the requirement of closure under the set-theoretic conjunctions *and*, *or* of the whole language, and under the reductions and other transformations of the whole language.

In the first place, the difference in sentence types between the Methods and the Results sections could be considered grounds for constructing two sublanguages, but this is undesirable because of their close and complex interconnections. Differently, there are distinguished portions of many sentences, occupying distinguished sections of their semilattices, which could be considered separate sublanguages. These are above all the meta-science segments which operate on the science sentences proper (e.g. *Rich demonstrated that* in *Rich demonstrated that the acute splenic tumor cell was identical with the lymphoblast*), and the quantifier words in all their occurrences. In these cases, each class of segments has its own interrelated vocabulary, common to its occurrence in all sciences and not specific to the immunology material; its relation to the specific science material in the sentences is much the same as it is in other sciences. One could therefore say that each science sublanguage, in somewhat different ways, draws upon these more general sublanguage systems in constructing its sentences.

A different situation is seen if we consider closely related research lines. For example, by the side of the search for which cell produced antibodies, there was another body of research to see what happens when one animal (the 'donor') is immunized and certain of its cells then injected into another animal. This second line of research, which threw much light on the first one, turned out to use largely the same sublanguage as the first but with certain additions, mainly the $CI^{ft}B_1B_2$ sentence type (*Cells were injected from the donor animal to another animal*). This additional sentence appeared inside the $GJB{:}AVC$ sequence of the 'which-cell' articles, to form $GJB_1{:}\ CI^{ft}B_1B_2{:}AVCB_2$. One can see how the differences and relations between lines of research are reflected in the differences between their formula structure. A more general case of this situation will be seen in the relations between neighboring sub-fields (10.5.3).

10.5. *Formulaic languages of sciences*

10.5.0. *Introduction*

The way science sublanguages differ from whole natural languages suggests that it may be of interest to look at them as distinct linguistic systems rather than as sublanguages. This view is supported by the fact that when in a given science we analyze articles written in different natural languages, we obtain essentially the same sublanguage structure.[9] Thus the consistent structure obtained from all the articles of a field, in whatever languages, is not necessarily a sublanguage of any one natural language. Furthermore, if we are freed of the need to state the grammar of the sublanguage in a way that is

---

[9] Thus in the immunology study of n. 7 above, analysis of articles written in French yielded substantially the same formulaic representation as in the case of English articles. It is also relevant that many articles written in one language are read and discussed by scientists who speak other languages, and that the peer pressure which is exerted on article writers to adhere to precise and standard formulations can come from foreign colleagues as well as from same-language colleagues.

maximally similar to the grammar of its whole language, we can maximize the regularizations in describing the sentence types of the science information even if this decreases the similarity to the natural languages in which the articles were written.

We then obtain a representation of the sentences of the science consisting of word classes and subclasses (the latter indicatable by subscripts), which combine into particular formulaic sentence types and formulaic sentences,[10] with no synonymy in the vocabulary and no stylistic synonymy in the transformations—that is, only such transformations as are needed for cross-reference and for aligning the information into tabular structures. The formulaic sentences then constitute a canonical form for the information in the science; and the grammar of least constraints which they satisfy constitutes a new linguistic system, differing somewhat from the set of natural languages, and similar in some respects to mathematics. We thus obtain a representation of science sentences by sentential formulas free from the specific words of language: an example was seen in 10.4.

10.5.1. *A distinct linguistic system*

When we consider science languages—i.e. the formulaic representation of the sentences of science (such as the $GJB{:}AV_p{\sim}C_l$ of 10.4)—as distinct linguistic systems, we see certain major differences from natural language. It is true that the formulaic sentences, like the word-sentences of natural language, are projections of partially ordered argument-dependence relations, with the interpretation that the operator is predicated about its argument; without this they would not constitute a linguistic system at all. But whereas the word classes of natural language are formed by dependence-on-dependence, yielding nothing smaller than the N, O_o, etc. classes, the word classes of a science language are formed by a best fit of how certain individual words (or phrases) appear as operators on other individual words. In this the relation is similar to likelihoods in the operator–argument construction, but it is not individually graded as in natural language. Rather, it is a yes–no distinction: a particular word occurs either sometimes or never as operator on a particular other word. In this it is a dependence, like the relation that creates the large N, O_n, etc. classes of the whole language, although here it creates the subclasses and sentences of a science language.

Since these special classes are (as subclasses) within the big classes of the whole language—G (*antigen*), A (*antibody*) within N; J (*is injected into*), Y (*is present in*) within O_{nn}—the sentence types formed by them are subtypes within the basic operator–argument sentence structure of the whole language. Thus we have in the science language something new: a number of different sentence types distinguished by their word classes, but all having the same

---

[10] It should be clear that the formulas are simply canonical representations of sentences.

operator–argument, i.e. subject–verb–object, structure. Just as the special word classes of the science have characteristic meanings ('cell' for C, 'cell or tissue events' for W), whereas the large classes of the whole language have only a general grammatical meaning (of noun, verb, etc.), so the sentence types have characteristic meanings (cell–antibody relation for AVC) whereas the basic structure of language has just the general meaning of predication. The sentence types may combine into larger sentences, called macro-sentences; this itself is a novelty for language. Thus in the immunology language, the GJB (*Antigen was injected into the footpads of rabbits*) and AVC (*Antibody was found in lymphocytes*), which appear separately, also appear frequently, especially in the Results section, conjoined by a colon (*thereafter*, etc.) in $GJB{:}AVC$. This statement form, which is the main one in the research line analyzed in 10.4, can accept an inserted GUT or $GUTT$ (*Antigen arrives by the lymph stream*), to make a full macro-sentence-type $GJB{:}GUT{:}AVC$. In the donor research there is a necessary infixed sentence: $GJB_1{:}CIB_1B_2{:}AVCB_2$. Some sentence types may thus have to be defined as being expandable in particular ways.

There are additional differences between science languages and natural language. The vocabulary of natural language is open, and is identified by each word-occurrence having one or another dependence-on-dependence relation to other word-occurrences in its sentence. But the vocabulary of a science language is closed or at least characterizable in any one moment or corpus; and any new words are constrained to be characterizable in respect to the existing classes, there being no general dependence-on-dependence criterion which could determine their status. New members of a class or subclass, or a new but related word class, have to fit into an existing or new but related sentence type—each innovation being explicitly either definite or unsettled, as will be noted below. It is possible that the complex relations, and hierarchies of relations, that a science has to take into account may lead to words having different argument-dependences than those that are current in natural languages, and this may already be the case with such words as *respectively*, *ratio*, *rate*, *proportion*, *covary*, *derivative*, and *integral*, in the way their use is constrained in science languages.[11]

[11] A possible example of new grammatical constructions in sublangauges is the following. In whole-language grammars it is not clear whether there is a distinction between (1) 'restrictive' and (2) 'unrestrictive' adjectives (and relative clauses), as in (1) *a white and black cat* contrasted with (2) *a white and black panda*. In English, the most overt distinction is that the unrestrictive relative clause, (2), can be separated off by a comma (here *I will draw a panda, which is white and black*, meaning any panda, since all are such), while a restrictive one, (1), cannot (*I will draw a cat which is white and black*, meaning a particular cat, others having other colorations). In a sublanguage, it may be found that certain adjectives are always restrictive in respect to certain nouns, while others are not. In that case, the distinction can be made by adjective and noun subclasses. The difference between restrictive and unrestrictive combinations affects the analysis of a sentence, in that for (1) further operators are selected by the noun as modified by its restrictive modifiers, but for (2) they are selected by the noun alone without any effect from its unrestrictive modifiers. The difference lies in the order of entry of the modifier and the further operators.

In addition to the special word classes and sentence types, a particular science language may contain special constraints on certain subclasses of words, yielding the effect of special grammatical constructions. One case of this is seen in the metascience segments that operate on the whole science sentence, as in NO_{no} (*We expect that*. . .), O_o (. . . *is demonstrable*), O_{oo} (. . . *indicates that* . . .); here there is a subset of second-level operators which differ in their specific argument-requirements, but have the property of occurring freely on all science-language sentences, and of carrying meta-sentence meaning. Another such subset of operators carries the meaning of evidentiality, as in *may* (*Antibodies may appear*). Another case is that of special O_{oo}, both the science-specific ones that create macro-sentences (in the immunology language: the colon, meaning 'thereafter', in *GJB:AVC*), and any specially relevant connectives or sequences on the observations or conclusions of the science (e.g. *is carried by*, *is identical with*, *is a condition for*).

Finally, an important kind of special construction is the case of those secondary operators which become modifiers of particular word classes, but which fit specially important or even necessary informational categories rather than being just added secondary information. Such is the set of time modifiers on the colon above (the superscript t in 10.4); this is the only position in the immunology language where elapsed time has a recognized standing (but time order is important as modifiers on V_i: in what tissue the antibody appears soonest). Such also is the set of quantifiers in certain sentence types (in immunology: how much antibody appears in a tissue, how many times antigen is injected). And there is the special status of r on V_o to carry the argument that lymphocytes had less than complete ability to produce antibody (10.4). Such special grammatical constructions differ from science to science.

Some comment must be made about the reliability of the methods for determining a sublanguage grammar. There is clearly a need for more experience in discovering these structures, and for methods of testing their adequacy. Evidence exists that within a science field there is clustering of words in respect to the grammatical relations among them. Experience suggests that such clustering will hold in very many specialized subject matters, and perhaps in all sufficiently small subfields of science. Experience also suggests that this clustering creates for each field a few word subclasses and a few 'sentence types' as sequences of these word subclasses. However, short of much additional work, we cannot know whether there may not be fields in which complexities and degeneracies of word-classification and of word-class sequencing would preclude the applicability or the utility of the sublanguage analysis described above.

To the extent that the analysis is applicable, it can be carried out in an objective and reproducible manner. It is true that some knowledge of the field analyzed (as well as of language structure), and some heuristic considerations, may be helpful in carrying out a sublanguage analysis; and these may well differ for different analysts. However, the criteria by which word subclasses and sentence types are to be judged are objective; and their adequacy is tested by seeing if discovery of new classes and types falls off as the corpus of data (texts in the field at the time) increases, and also by seeing if the structure yields useful categorization and organization of the information of the science. Any difference among analyses reached by different analysts would be mostly in details of word subclassification and of sentence type distinction. These should for the most part be encapsulated within parts of the grammar. In any case, such differences should be visible upon inspection of the analyses and should be testable as to their equivalence or difference in what they represent as the informational structure of the field.

10.5.2. *Formulas of information*

The science languages are a more precise representation of science information than are the sublanguages. In the sublanguages, which draw upon whole-language vocabulary and grammar but use it under science constraints, homonymy is inescapable. Thus, in the English immunology sublanguage, we have *Antibody is found in lymphocytes* and *Many lymphocytes are found in this tissue. Found* is used here for two different relations among objects of the science; and in the formulaic representation, these are AV_iC_y and C_yW_iT respectively. In the sublanguage, *found in*, and also *contained in*, are homonymous members of each of the two subclasses V_i and W_i. In the formulaic language, the vocabulary simply includes two symbolic 'words', V_i and W_i.

For the same reason, synonymy (in respect to the subscience) is common in the sublanguage: e.g. *antibody, agglutinin, gamma globulin*, are all undifferentiated members of A in respect to the issue in these articles. However, in the science language, only A itself is a vocabulary element. Thus the sublanguage isolates the synonyms, but the formulaic language eliminates them. It should be understood that when we lose the sublanguage words and retain only the symbol, we lose little of the total information, and may even lose nothing if the retained symbols express enough variations. For the various sublanguage synonyms may have different meanings in the whole language and even in the rest of the given science, but if the methods of 10.1–4 permit them to be represented by the same symbol, then any differences among them are not relevant to the information in the given corpus or subfield at the given time. Indeed, in the example of A above, the members of A have different meanings in immunology, and their use here as undifferentiated words for antibody is a stylistic imprecision that is immaterial to the given research problem. When the various words used are replaced by a single symbol A, the information is clearer, and the correspondence between form and content is improved. The tighter structure of science languages in comparison with science sublanguages thus enables us to tighten the structure even more, for example by removing homonymity and synonymity from the system.

Another tightening of form–content correspondence arises in pluri-symbol words. The sentence formulas require that the words of each sentence be mapped onto one or another particular sequence of symbols. In some cases we find single words which have to be mapped onto a combination of symbols: either a symbol and its superscript (i.e. a word class and a modifier) or else two symbols (e.g. an operator and its argument). For example, *injection* is found to be *GJ* (i.e. equivalent to *antigen injected*), or just *G*. An extreme case of this is seen when one symbol is in the metascience segment and the other is in the science-language sentence proper, for example in *They found antibody in the lymphocytes*, where the formulaic representation would be $M_s M_{no} A V_i C_y$ (M_s: metascience subject; M_{no}: metascience O_{no} verb; V_i: *in*; C_y *lymphocyte*), as against *Antibody was found in the lymphocytes*, which would be $A V_i C_y$. A solution that is applicable in some cases is to segment the two-symbol word into two components of the sentence, e.g. *find* into *establish (that) . . . is present* (M_{no} . . . V_i), as though the sentence were *They established that antibody is present in the lymphocytes*. Such factoring of a word into a combination of other words is done not on semantic grounds (though it should preserve meaning in the environment), but only to regularize the mapping of the sublanguage sentences onto the formulas. Such a solution may not be desirable in another kind of pluri-symbol word, namely the classifiers of sentences. For example, in the immunology sublanguage, *response* is used for antibody presence, but only following antigen entry, not for the background presence of antibody. Thus *response* is a classifier covering the $GJ:AV_i$ portion of a formula. Such classifiers are not directly reflected in the formulas, unless a subscript is inserted to indicate that a classifier—i.e. a general word—is being used. It goes without saying that the status of a classifier is not decided semantically or etymologically but again by the combinations into which it enters. Thus *immunize* is found to be a classifier for *GJ*, not for *GJ:AV*; use of that word reports antigen injection (i.e. an attempt to cause immunity) but does not contain information as to whether antibodies were thereupon produced (i.e. whether immunity resulted).

Mapping the sentences of the science sublanguage onto an explicit and more or less closed formula structure has other advantages and disadvantages. It requires that decisions be made—if possible—in many details of the information in the original sentence. For example, for the formulaic representation of small lymphocyte, we have to decide: is this the name of a particular cell type, hence to be marked by a subscript on the symbol for cell (e.g. C_s); or is it a relevant property of the lymphocyte (one of a recognized set of properties), to be marked by a superscript indicating that the lymphocyte in question has this property $(C_y{}^s)$ carried by restrictive *wh-* (n. 8 above); or again is it simply a secondary fact that the lymphocyte in question was unrestrictively small—permanently or at the time in question—in which cases the modifier *small* would be reconstructed into a secondary sentence (a distinct formula attached here by unrestrictive *wh-*) *The lymphocyte was small* $(C_y W_s)$? The difference between these formulaic assignments can be decisive for how well the formulaic sentence fits into the sequence of sentences that constitute scientific conclusion-drawing (10.5.4(d)). Of course, we may not know which representation is intended or is best. But even to raise the question, and to be aware of the alternative representations and their implications for the sentence-sequences, is a step toward a more aware and efficacious treatment of scientific information in process of accumulation (10.5.4(d)).

For formulas to be obtained in a controlled procedure directly from the texts—or for the texts to be mapped in a regular way onto formulas—assures the relevance of the formulaic representation, and its coverage of the texts, and also its sensitivity to change in knowledge. An example of the sensitivity of sentential formulas to informational change: during the early 1970s, when what was indicated (or understood) by the words *site* in biochemistry was changing from a location in a cell membrane to a receptor molecule there, sublanguage analysis had shown, without appeal to biochemical knowledge, that *site*, which was at first clearly a member of the 'cell-portion' word subclass, was increasingly getting, as predicates on it, words that were otherwise predicates on molecules rather than being predicates on cell-portions; the observations made about 'sites' were becoming increasingly similar to observations about molecules (later established as receptors). The gradual shift in experimental observation was visible early in the details of the formulas. Somewhat as grammatical transformations can in some cases show the direction of an ongoing grammatical change, so the changes in operator–argument environments of a given word can in some cases show the direction of an ongoing change in understanding.

The text-derived formulas also come closer than the text itself to a pure representation of the information. It is known that both the selection of scientific problems and also the views and presentations of the information obtained are affected by socially fostered fashions and expectations, in addition to the direct pressure of scientific developments. The procedure for obtaining the formulas cannot correct for the choice of problem, but it can correct for any language use beyond what is essential for the information, whether it is teleological or analogical terminology used for its suggestive or communicative value, or else dramatic presentation or over-reaching conclusions and the like. The formulas, and the constraints on their conclusion-sequences, also show something about the particular scientific field, and to some extent about the character of current scientific work in general. However, there is here no claim that the formulaic representation gives everything that can be understood from a reading of the natural-language text. It may be that the additional vocabulary and transformations of the text suggest things not captured in the formulas, let alone the fact that the reader's thinking is done in natural language and not in formulas. The formulaic representation here is therefore only for information processing, whether by humans or by machines, whereas in mathematics it can carry virtually the whole discourse.

It does not follow from the advantage of the formulaic language that nothing is gained from the availablity of natural language, even though it is less precise—especially less precise in respect to science information. In certain situations there is need for imprecision, when one is dealing with unsettled questions and with areas where concepts are not fixed because the operations or relations of the science are not adequately understood. The recourse to imprecision is always available, since in any case the information of the science will be written in natural languages—more specifically, in the science sublanguage—with the formulaic representation serving possibly for control, and for applications such as are noted below (10.5.4(b)). The parts of the science sublanguage that are well established in a given period of the science can be mapped onto the formulaic language to which the sublanguage has a close and regular structural relation. The rest remains in the vocabulary of the sublanguage; an example of this is the r superscript in the immunology sublanguage above. The ordinary-vocabulary sublanguage segments are not a disorderly retreat from precision, because in this situation they are encapsulated and have a known location in respect to the formulaic structure, so that one can inspect what is unclear and what is the relation of that to the precise knowledge. It is in this way, and with the availability of its corresponding sublanguage, that the formulaic language of a science is both a closed system and yet open to uncertainty, disagreement, expansion, and change.

The change in a science language is contained within its remaining structure, so that it is not easy to say at what point, if any, the amount of change has become massive. This is not to say that there may not be situations in which massive changes of knowledge and understanding would impose a massive reformulation of a science language, but in such cases, the old science or perception of the world usually disappears from use, often only gradually, possibly to be followed by the beginnings of a new science which fills a rather different niche in human thought and in practical life. Such a situation has the character of cultural change rather than of science-language development. Even within an ongoing science, many views that are erroneous or prove unproductive just fade away, rather than being buried and explicitly replaced by something else.

The formulaic representation also clarifies discourse structure. In a large part of science, what counts is not only the observations but also how the observations were made and in what conditions and boundaries, and how they are juxtaposed and confronted with other observations and considerations. In science languages, this is reflected in the fact that science sentences do not appear alone or in small independent sets, but in rather extended sequences, that is, in discourse. Characteristically, there are regularities and constraints in the discourses, beginning with the general discourse property of repeating word-class relations (9.3). For example, in the standardized styles of scientific articles, the articles in certain experimental fields have three essential divisions: Materials and Methods, Results, and Discussion. It is found, as has been noted, that the word classes and sentence types of the Methods section limit what can appear in the Results section, due of course to the plan of experiment, but visible in the reporting sentences. Also the sentences of the Discussion section are not independent of those in Results; indeed the Discussion consists largely of Results (and other) sentences, modified in certain ways (e.g. by use of classifiers) and arranged under a restricted hierarchy of O_{oo} operators which build an argumentation. Thus the Methods sentences impose constraints on the Results sentences, while the Discussion sentences are constrained to stay close to the Results sentences. In addition, the sequence of sentences within the Discussion follows a special type of discourse constraint (10.5.4).

The informational relations within science texts are thus indicated in the structure of the sentence types (the rows of the double array), the correlated changes in the columns, the local constraints on successive rows (10.5.4), and the more general relations on the sentence types within a discourse (e.g. immediately above). Within each discourse there may also exist more complex paths of informational interconnection among sentences; but these are not reached at present by any a priori method of analysis.

10.5.3. *Boundaries of fields*

In discussing science languages, we have to ask what are the boundaries of a science language and how it is distinguished from others. First, the grammar of the formulaic sentences, and the correspondences for mapping the original sentences onto them, can identify what material in the original sentences does not belong to the science language, and in what way that material relates to the formulaic sentence. This may be useful in fields such as law where there is formalizable legalistic material intermingled with not-readily-formalizable general-language segments (e.g. examples from life) that have at least a content relation to the formalizable material.

Second, the semilattice analysis of the formulaic sentences is a mine of information about the science and its boundaries, because the different science statuses of material in a single sentence appear in different sections of its semilattice. In the semilattices, the metascience operators are upper bounds of the operators in the science language proper. If there are any categories such as time or quantity which draw on a general language of science to modify particular items of the given science, the appropriate symbols will appear as secondary operators (also upper bounds) on the lattice items in question, for example in the t (time) on the colon of $GJB:^tAVC$. If there is a prior science to the science in question, it will appear at particular lower points in the semilattice. For example, in the sentences of pharmacology, there are as noted above O_{no} pharmacological operators whose first argument is a drug noun and second is a sentence (or noun) of physiology: in the pharmacological sentence *Digitalis affects the heart contractility* we have as argument a physiology sentence, *The heart has the ability to contract*. Physiology is here a prior science to pharmacology.

Third, relations among sub-areas of a science which do not take place inside of single sentences are marked by statable similarities between the semilattices involved, as in the similarities between Methods sentences and some of those in the Results section.

Fourth, neighboring scientific fields have appropriate similarities and differences in their formulas. This holds for neighboring research lines within a field (e.g. the 'donor' research, 10.4, end), for neighboring sciences, and also for many successive periods within a single research line. Indeed, earlier and later articles within a single research endeavor show appropriate differences in their formulas, differences which mount up when major advances are made in the field. All this may make it possible to quantify the differences between fields, or periods of a single field, by weighting the differences in word classes and in their memberships, in sentence types and in the special constructions that are reflected by the superscripts in the formulas, and finally in any differences in respect to boundaries (e.g. to prior sciences).

10.5.4. *Information and argumentation*

We can now summarize what kind of information we may obtain from the structure of science languages about the science and its structure, keeping in mind that such terms as 'information' and 'structure of science' are undefined and can only be used loosely. The operator–argument analysis, and the mapping of science sentences onto the formulas of a science language, suggest that the latter can contain almost all the language-borne information in the reports and discussions of the science. This means a better fit than we have in natural language between differences in form and differences in information—loosely, a better fit between form and information. What there is to learn from the structure of science languages can be considered in four structural contributions: (a) the internal structure of all sentential formulas, (b) the different types of sentential formulas, (c) the kind of information that characterizes the given science, and (d) the way the formulas are combined in discourses of the science.

(a) The semilattice structure of each sentential formula isolates and identifies the various sources that contribute to the information of the science: the metascience, the prior science, the procedures and reliability of observation, the quantities. It also instigates clarification of the relation between particular subsets of terms: e.g. in a given research line, as to whether *small*, or *mature*, is (as noted above) a subtype of *cell* (hence a subscript on *C*), or a necessary condition of it for the given experiment (hence a superscript on *C*), or an incidental (non-restrictive) property (to be stated in a separate formula, attached by the *wh-* relative clause).

(b) Since only a few particular types of formula are found in each line of research, they constitute a framework for its information: anything not expressible in the stated types of formula is excluded, except as being a specifiable addition to the field described thus far. Having a framework creates new conditions for the representation of information. Since the information is fitted into the available formulaic structures, we can obtain a tabulation of the information in a text, and know where to look for each kind of information if it is present. This makes it possible, given a tabular representation, to inspect it for any particular kind of information, and to summarize or otherwise process the contents of the text. All this holds only if the formulaic framework has been obtained not by fiat—by someone's understanding of the field—but by the regularization of the text itself, or of a whole corpus of similar texts, in such a way that it constitutes an objective best fit for the information in the texts (as in 10.4). Furthermore, the fact that this framework is not made by one expert or another, but is reached by objective procedures carried out on the discourses of the science, assures in principle that mapping the discourses onto the framework can be carried out by computer programs. Finally, it becomes possible to recognize differences in respect to the framework—imprecisions, disagreements, change through time.

(c) As to the kind of information: the word-subclass relations that appear in the various sentential formulas express the various types of fact and conclusions with which the given science deals at the given time. This specification makes it possible also to identify the kind and amount of difference in subject matter between neighboring sub-sciences, and the amount of change in a science over time. It also excludes from the science any irrelevant material; such material cannot be mapped into any of the sentential formulas. Thus a science language is protected against nonsense, as is mathematics; but, like natural language, it is not protected against error and falsification.

(d) The way the sentential formulas are sequenced in discourse is important for inspectable presentation of methods and data, and above all for controlled and inspectable drawing of conclusions. The latter raises the question of proof and truth in science.

To the extent that this last can be discussed within present knowledge, it is a property not of individual sentences (formulas) of a science language, but of constrained sets and sequences of such sentences. The great successes in this direction—the syllogism, the truth tables of mathematical logic, mathematical induction, and the inspectable structure of proof in mathematics—all indicate that the structural issue is how to control some informational value of the last sentence of a sequence on the basis of the presence of information in initial sentences, by means of constraints on their structure and on connectives between them. It has already been seen in the analysis of discourse (9.3) that any sustained treating of subject matter involves sentence-sequences in which (after transformational alignment) something of the nth sentence is retained in the $n+1^{th}$, aside from intervening sentences of various kinds. The only overt whole-language structure that connects successive sentences is the set of bi-sentential operators, O_{oo}, including the conjunctions. These impose weak word-repetition constraints on their arguments (9.1), but no recognizable transitive or long-range constraints. In the set of all natural language discourses, the only common constraint stated in 9.3 is retention of operator–argument (or of some transforms of it) among words.

In mathematics, the constraints of material implication (or truth tables) and substitution (with no likelihood differences) determine the possibilities of proof. In formulaic science languages, these mathematical constraints are unavailable because of the specific combinations to which the various word classes are restricted, which produce families of sentence types, thus precluding such a general relation as material implication. In natural language this complexity is due directly to the particular likelihood (and, therewith, meaning) properties of each word. In science languages, this complexity is reduced to the combinatorial restrictions of word subclasses; but the reduction is achieved at the cost of having several different sentence types. But this very limitation may open the way to a different constraint, new to natural language, which appears in some successions of science sentences: regularities of change in the successive words of a given word-class (within successive sentences) in respect to changes in the other word classes of their sentences. This word-change is syntactically reminiscent of substitution in successive sentences of a mathematical proof, but is very different in that it is a constrained and successive substitution, specifically in the double array that characterizes discourse (9.3). This word-replacement rests upon the fact that the word classes of science have a common meaning interrelated with the meaning of their environing word classes, so that in successive sentences the words of a given position and the words of their environments (the columns of the double array) change together.

In addition, science languages have another property, lacking in mathematics, which can be used for constraints on sentence-sequences: the fact that sentences of the science can carry many varied metascience and evidential operators acting on them—O_{no} and O_o in contradistinction to the bi-sentential (and largely causal or temporal) O_{oo}. The importance of all this is seen in the fact that the Discussion sections in scientific articles are built not out of new sentence types of their own but out of previous Methods and Results sentence types with more-or-less constrained word-replacement, metascience and evidential operators, and conjunctions.[12]

To this should be added one important consideration: the choice of initial sentences. Whereas in the whole language the choice is arbitrary, and in mathematics it is in principle limited to proved theorems, ultimately to structurally recognizable axioms, the practice in science writing is a selection of Result sentences—made according to the wishes of the writer—with a selection of hypotheses, posed alternatives, and the like. As in mathematics, all these sentences are not generally used in a sequence obeying a single constraint. Rather, initial sentences (premises and results) are stated, and each may be followed by a constrained sequence amplifying or supporting it, whereafter another initial sentence can be brought in. The whole becomes—somewhat like a proof—a partially ordered sequence. Nevertheless it remains a progression, satisfying the essential requirement that from given initial sentences certain constraints be handed down through successive sentences to affect the concluding sentence as a 'consequence' related to the initial ones.[13] The greatest problem in the case of science discourses is in establishing criteria for the initial sentences, in relation to the subject under discussion.[14]

In science languages, the effect of the type of sequence considered above on its concluding sentence is not that the truth of the conclusion is guaranteed. Even in mathematics the effect of proof is only to guarantee that there is no loss in truth value, in respect to that of the initial sentences. Since, in language and in science, we lack a criterion to assure the truth of initial sentences (at least to the extent that axioms are taken as true), and lack the truth-retention of the succession-constraints of mathematical proof, we

[12] More generally, the set of sentences which go beyond data-description to constitute a theory of a subject do not merely have additional words, for constructs and concepts that go beyond the words of the data-description. Rather the new 'abstract' words have positions in the theory sentences related to the positions which corresponding 'concrete' words have in sentences of the data-description. The sentences of the theory are thus structurally related to the sentences of the data, or may contain (as their arguments) classifiers of the latter sentences.

[13] For what little it is worth, it may be noted that in English both *sequence* and *to conclude*, which are contained in *consequence* and *conclusion*, suggest that the latter refer to reaching a point at the end of a sequence.

[14] The structure of argumentation, and of the data-presentation prior to it, is developed not only by the individual scientist's response to the science material, but also by the views and understandings of the science community in general. It thus has something of the public condition of natural language (2.5.1, 12.3).

could not even say how to recognize by its structure the truth of an arbitrary scientific sentence—or of any sentence. But lesser goals may be available, given the constraints in science languages: for example, consistency, retention of possibility or plausibility, even the certainty that something follows from explicitly given initial sentences.

It is of interest here that the syntactic and discourse properties of proof-like argumentation in science are also found in sentences expressing causation. The term 'causality' covers a welter of situations: an agent or a precursor causes an event; an earlier stage of a process or chain 'causes' a later stage; even membership in a set 'causes' a member to have the properties of the set. But in many cases the reporting of causality, i.e. stating that a causal relation holds between events, has the structure of a sequence or partial order of statements culminating in the statement of the caused event, as in the QED of a proof.

In any case, under the constraints, the final sentence is a consequence of the initial ones. Any attempt to specify the constraints on succession more fully, and to formalize scientific argumentation or make it subject to inspection and control, will meet with great difficulties. Although the development of science is certainly affected by unpredictable advances in technology and in methods of observation on the one hand and by social controls and personal interests on the other, it is also affected by its own results and conclusions which in themselves are not arbitrary or faddist or competitive, but rather follow the cumulative direction not only of its increasing data but also of its own argumentational constraints.

10.5.5. *Notation*

The structure of the sentential formulas of a science constitutes a notation for it: that is, as a set of symbols with a syntax (combination-constraints) on them. In mathematics, notation began as shorthand for the objects and operations in current mathematical statements; but today, within the limits of meta-mathematical knowledge, a notation is determined a priori by how a set is closed with respect to operations and properties. In the various sciences, no meta-science knowledge suffices to fix a priori a notation for a science. The alternative proposed here is to derive the notation from what is actually said in the science, by procedures that assure its objective para-phrastic relation to the original, its objective reproducibility, and its coverage of the field. The latter requires that the notation contain inherently the capacity to change with any change in what is written—change of knowledge, of theory, and of attention (i.e. of what is topical) in the science. In 10.4, it was seen that procedures of sublanguage analysis can serve for that purpose.

An adequate notation does much more (as well as less) than express what is present in the science. It also limits what can be expressed, which in the case of always-changing science requires an inherent capacity for notational change. This capacity is in principle possible since change in science is not arbitrary even in the case of unexpected discoveries, but almost always includes continuities in knowledge and even in problematics. The notation also provides certain kinds of critique of the expression of knowledge. In part it does so by reducing into precise and prosaic formulas any metaphoric, rhetorical, or sloganizing presentations which may have been made for communicational effect, as well as any analogic argumentations, and any teleological terminology.

Consider, for example, the case of teleological terms, as when a process is named in terms of its outcome, even though the process must have been occasioned by other factors before its outcome has come into existence. It is not always easy to decide semantically whether a term for an entity or event in a given field is teleological: does the meaning of the word, or its uses outside the field, relate to its end-product rather than to its substrate or composition or mechanism? If it does, it is teleological, as an inverse of reductionism. As an example of the need to avoid end-product teleology: in evolution, every step has to have had a survival value for the species, without taking into consideration the value of what it has led to (e.g. in the development of feathers, before flying with their aid). Another example is the use of the term 'communication' for certain intra-species correlated animal behavior (12.4.1). Thus to use end-product terminology such as *information*, *messenger* in the DNA–RNA literature is somewhat like the inverse of what we would have if we named information-carrying messages in the macroscopic world by a term taken from the biochemical action of 'messenger' RNA. In notation, such problems as teleological and meta-phorical terminology are more manageable: to the extent possible, each zero-level term of the sublanguage should not have other meanings or connotations than are required for the selection of first-level operators with which it occurs in the sublanguage, except for reasonable generalizations or the like; and similarly for first-level operators in respect to their zero-level selection. In mathematical research, one finds little rhetoric, but even there discussion in a natural language shows departures from the limited universe defined in the notation. In various sciences, there are various degrees of loosely stated or uncertain data, observations, and argumentation, as also of metaphoric or end-product talk. This is all the more reason for having by the side of the natural-language presentations also a procedurally obtainable mapping into a formulaic language.

10.6. *Mathematics as a linguistic system*

The question of how language can be described as a mathematical system was discussed in Chapter 6. Here we will consider how mathematics can be described as a linguistic system. Only a sketchy account will be given here, because anything more fundamental would involve broad considerations of the syntax of mathematics, as well as meticulous grammars of the sentence structures and discourse (proof) structures in mathematics and logic. The gross syntax of mathematics and logic is clearly similar to that of language. In all of these, the texts consist of sentences (propositions, formulas), each of which is a sequence of symbols that are readable as words, the occurrence of some types of symbols being dependent on the occurrence next to them of other types (their arguments). Roughly, in the case of mathematics and logic, the tilde of negation ($\sim$) requires one argument, *plus* and *minus* signs require two, and the signs for *equals* and *implies* and *contains* also require two arguments, but in different conditions.[15]

The linguistic status of mathematics is seen in the fact that the sentences of logic and mathematics are within natural language at least under translation.[16] That is, it is possible to read mathematical and logical formulas in a natural language, although in some cases only by language constructions which differ structurally from the mathematical notations. (Hence not all mathematical notation satisfies the conditions for natural-language syntax.) The converse does not hold: not all sentences of a natural language can be written in mathematical or logical formulas. The symbol-lists (vocabularies) and grammars (syntax) of mathematics and logic do not provide stable differences in likelihood of combination, such as language has in its operator –argument selections; nor do they have the ability to borrow this capacity from language. Hence, they cannot capture the meanings expressed in language. Hence, also, not all sentences or constructions of natural language have a 'logical form', i.e. not all can be adequately represented and distinguished by expressions in mathematical logic.

There are crucial similarities and differences between mathematics and natural language in respect to the syntactic status of the symbols (or words). The zero-level arguments of mathematics and logic are the numerals—the

---

[15] The detailed syntax is a bit different if the notation of mathematics is freed from the linearity of the symbol-sequences. Certain properties of operations, for example, can then be stated less redundantly and more essentially. Thus instead of defining commutativity as the equivalence of two orderings ($a+b=b+a$), we can say that when the operation $+$ on a set maps a subset of members onto its image the subset is unordered: the unordered pair a, b is mapped onto its image c. And instead of expressing associativity by grouping—$a+(b+c)=(a+b)+c$— we can say that applications of pairwise mappings of members of a subset onto their image are unordered: i.e. in a, b, c, any ordered pair (a, b or b, c) is mapped onto its image in the set, and then the pair consisting of that image with the remaining member of a, b, c is mapped onto its image in turn.

[16] One might also say that the statements of different fields within mathematics and logic are intertranslatable, in their notation, in the case when the fields are abstractly equivalent. André Lentin has called my attention to the possibilities of translation between two mathematical systems which are not isomorphic. Cf. ch. 6, sect. 11, 'Polymorphisms; Crypto-isomorphisms' in G. Birkhoff, *Lattice Theory*, American Mathematical Society Colloquium Publications 25, 3rd edn. Providence, R.I., 1967, pp. 153 ff. Note here the comment (p. 154) about the possibility of defining the same abstract algebra in several non-polyisomorphic ways.

names of the numbers—and variables defined for each text over a specified domain of values. Signs such as *plus* operate on two zero-level arguments, the result being again an argument of the same level; such an operator does not exist in natural language. The negation sign operates on one sentence and produces one sentence out of it (like the O_o of language). The *equals* and *greater than* signs operate on two arguments, with the result being a sentence (like O_{nn} in language). The implication sign operates on two sentences (propositions), with the result being a sentence (like O_{oo} in language).

To see more specific differences from natural language, note for example that the symbol ϵ for membership of an element in a class cannot be O_{nn}, as it would be in language, since under ϵ the second argument (a class) cannot be of the same sort as the first argument (N: an individual). In contrast, the first argument of *is* in such sentences as *An opossum is a marsupial* can also be its second argument, as in *This is an opossum*: this *is* is closer to *is a case of* than to *is a member of*. In the grammar of logic, we have two choices. We can say that names of individuals and names of classes are two different sets of zero-level arguments (something which can be defined in the external metalanguage of logic). Alternatively, we can say that the names of classes are O_n predicates, with ϵ being an integral part of them (their operator indicator, G37, 54): *is-a-marsupial*. Then *is-a-member-of-class-X* is an operator, as is *is-large* or *sleeps*. The inclusion relation $\supset$ then has to be an O_{oo} operator, since each of its arguments is a class, hence an O. This makes it syntactically possible for the logic of propositions to be the same as the logic of classes: the same notation applies for the ability of a class to include another as for a proposition to be implied by another. In the case of mathematical notation, the variables function somewhat similarly to different indefinite nouns, that is, as zero-level N arguments having no limitation of likelihood in respect to their operators (G70). Operators that map a set A, or its Cartesian product $A \times A$, into the same A, can be defined in mathematics on various sets; but in natural language they can be defined only on sentences, i.e. on operators with their arguments, and not on zero-level arguments (N).

Because of its differences from natural language, the notation of mathematics is a separate linguistic system, a science language, rather than a science sublanguage of natural language. But the grammar of mathematics, its external metalanguage, is a science sublanguage, which has to be stated in natural language. The natural-language discussion in mathematics, e.g. the talk around proofs, or about discovering proofs, would presumably also be a sublanguage. Although mathematics is a discovery of relations which exist on their own, while language would seem rather to be an invention (even if evolving and unintended), the system of mathematics requires the existence of natural language in order to be formulatable; but for it to be exhibitable (in its own notation) does not presuppose natural language. Note however that the exhibited notation, like science languages in general, does not have the structure of a mathematical object (such as might aid in its deciphering in the absence of a natural-language metalanguage).

By the side of the crucial similarities in sentence formation and the differences in argument and operator classes, there are also encapsulated differences in sentence structure. One is the status of variables as indefinite *N*. The meanings of *N* (other than indefinites) in natural language, to the extent that they can be obtained in logic and mathematics, are provided by class-membership predicates (compare *All men are mortal* with *For all x, x is a man implies x is mortal*). The fact that the *N*-variable position in logic and mathematics formulas (sentences) is occupied only by the equivalent of an indefinite noun, which contributes no likelihood (selection) limitations or meaning to the sentence, makes the information in the sentences of logic and mathematics arise entirely from the operators—unary predicates and binary relations. In terms of syntax, it is this that makes mathematics a science of relations, and makes a mathematical object a system defined by its relations 'saying more and more about less and less'. In natural language too, there is a possible grammatical description providing only one *N*, an indefinite, so that, for example, *A book fell* would be from *Something which is a book fell* from *Something fell; that something is a book*, and similarly *John phoned* from *Someone who is called John phoned*. But this complicates the status of the *is* in *Something is a book*, etc.; and the class-membership predicates would then be restricted by the same selection properties as the *N* they replace, though perhaps with compensatory advantages.

Another overt sentence-structure difference is in quantification, which in logic is an operator on sentences (as in the *for all x* above). While the details do not have to be discussed here, the language apparatus is close to that of logic since plural, *all*, *each*, *any*, and the like are analyzed here not as modifiers of nouns (as they appear overtly in language) but as derived from operators (such as *for each*) on sentences (G51, 169, 252, 328, 337).

In reference, too, the difference is not major. Logic and mathematics have conventionalized cross-reference under the scope of a quantifier, and in the identity of symbols for variables within a proof or specified sections thereof. The pronouns of language too serve for cross-reference rather than for reference, with demonstrative pronouns being derivable from cross-reference ones. But in the case of language, the great variety of positions for antecedents of the cross-reference requires a special locating apparatus such as that of the relative clause, which is absent (and not needed) in the syntax of logic.

Aside from the lack of selection differences in nouns (variables), the overriding difference between natural language on the one hand and the set-theoretic systems—logic and mathematics—on the other is not so much in constraints on sentence-structure but rather in constraints on sentence-sequence. Although many things were known in ancient and non-written mathematics without benefit of proof, the great edifice of mathematics and logic rests on the methods of proof; the problems that have accumulated around mathematics in the last hundred years, even in the criticisms made by the intuitionists, have not involved the basic apparatus of proof (other than the marginal method of the excluded middle). In the structure of proof—with slightly different strategies in different mathematical areas—the barest essentials are that in a sequence, certain initial sentences of a fixed kind be evaluated by axiomatic status or by truth tables, with the constraint on the sequence (or partially ordered set of sequences) being such that its last sentence, the QED, is not less true than the initial ones. These essentials can be stated a priori and inspectably only in conditions such as those that obtain in mathematics: where the truth value of certain elementary (axiomatic) sentences, underlying the initial sentences of a proof, can be determined by their syntactic composition, or by listing, or in case of difficult infinite sets, by constructive formulation; and then the truth transmission of the fundamental operators is given explicitly in the truth tables. In natural languages, these conditions are not met, largely because of the likelihood inequalities among members of a word class, tied to the shifting knowledge or perception of the real world.

Even within a science, the likelihood inequalities of operators and arguments in respect to each other are too numerous and too changeable for us to establish any useful set of axiomatic sentences whose truth value can serve non-trivial proof-sequences. However, as noted at the end of 10.5.4, weaker types of sequence-constraints, suggesting consistency and consequence, may be statable for science languages, largely based on regularities of word choice within the co-occurring operators and arguments.

In a different way, the language in which computer programs are written, and the compilers which mediate between them and the binary code of the computer, all constitute science languages or linguistic systems not identical with natural language. Each program is a text in such a language, following constraints which lead—largely in a partially ordered way—from one step to a later step. The flowcharts that were used to design programs, and the line indentations used in writing programs, are examples of partial ordering of the sentences of the text. Despite the similarities of programming languages to natural language, the lack of 'logical form' in natural language (cf. at the beginning of 10.6), and the present lack of any adequate constraints that would relate conjunctions to the particular words under them (10.5.4), exclude any present simulation, representation, or equivalence of computer programs to natural language and to the thinking it presents.

10.7. *Non-linguistic systems; music*

It may be appropriate to consider whether there is any comparability between natural language and other systems of notation and symbols, such as music, the language arts, animal communication, and deaf signing. The major case here is music. As above in respect to mathematics, no attempt is made here to characterize music except as to its difference from (and similarity to) what has been seen here about language. In comparing the two systems, we will see that while both are built out of constraints, the kind of constraints and their relations—and their content—are entirely different. It should be noted from the start that whereas all languages have in some respects a common underlying system, and all fields of mathematics do (in other respects), there is no evidence that there is something systemic common to all music.

First we consider the physical elements. The first difference is that in language the phonemes are essentially discrete objects because they are defined by distinctions (7.1), whereas the notes of music are discrete partly as a matter of convenient notation (except in the case of stopped instruments). In both language and music, the physical elements are sounds, variously combined in largely linear (or multilinear) order through time, and each has attained discrete notation. One might think of comparing the notes of music—written as discrete sounds with fixed lengths—with the phonemes of language.[17] But the phonemes are irreducible distinctions in respect to repeatability and distinguishability of utterances. In contrast, musical notes are direct cuts in what is first of all a continuous scale of musical sound, although the relation to overtones is discrete. Between two successive notes in a scale other notes (or musical sounds) can be recognized even if lacking a notation; the like is meaningless between phonemes.

Notes are not arbitrary components of a musical phrase: they and their relations in the scale are the inherent material of the phrase or chord even though the whole can be shifted to a different key, that is to different notes in a comparable scale relation. In contrast, the phonemes by which words are recognized have no relation to the content of the word, and the same content can be given synonymously or suppletively (or in different languages) by completely unrelated phonemes; also the same phonemes can identify homonymously a completely different word. The similarity of notes to phonemes is slightly greater in the case of onomatopoetic words, but these are necessarily irregular rarities in language (11.1). If the human ear could distinguish as many single sounds as there are words in the language, there would be no need for a separate stage of phonemic composition of words,

[17] Differences among notes are differences in pitch, i.e. in sound-wave frequency. Differences among phonemes, however, are, acoustically, differences in timbre, i.e. in amplitude of the various overtones, as are the differences between musical instruments. Hence the recognizability of, say, the phoneme a, and its difference from e, is constant no matter at what pitch it is pronounced (as in men's and women's voices). The existence of notes as distinct notations, or discrete structures in musical instruments, and even the availability of more than one musical instrument and of different timbres, are all limited to particular musical cultures and traditions. Both the scope of the present book and the experience of the present writer unfortunately limit the discussion here to the relatively recent culture of the 'Western world'.

but a comparable situation would not reduce musical phrases to a single note. Also, the meaning of words and sentences can be given fully in writing (including in ideograms) which can be understood by a reader who does not know the sounds of the spoken form. In music, however, a score can indeed be read without being heard, but this is done necessarily with respect to the known sounds.

Aside from its sound elements, music differs from language in being able to have a space of so many standardized note-lengths per bar, in which these sounds are located; this is actualized primarily by differences in stress, attack, etc. The distinction between this systemic timing and physical time is seen in the fact that a *ritardando* can stretch the final phrases over more time without altering their relations to the bars or (in many respects) to the composition. This distinction is also seen more generally in the fact that the same music can be played slower or faster than the indicated time, usually without losing essential properties of the composition.

We next consider the functional entities of the system (i.e. the entities in respect to further combination), which are formed from these elements. In all languages, we have two segmentations: words as fixed sequences of phonemes, and sentences as regular sequences of words—regular in respect to the combinability of those words in all sentences in which they appear. In music, short musical phrases can be considered comparable to words (with the relevant chords serving as classifiers of a sort); this especially if the composition uses them as explicit elements, e.g. by structured repetition or variation of them. However, if we do not limit the consideration to particular styles or compositional forms, music does not show two levels of segment-ation such as word and sentence, each recognized by systemic regularities in the level below it. All this entails the lack in music of anything like predication, since the basic regularity of sentences in respect to words, the operator–argument dependence which creates the sentences, is unavailable in music. Many varied relations exist between a longer musical line and its subsegments, but not the single fixed dependence of words to sentence. Furthermore, where in language expanded sentences are made out of elementary sentences in much the same way that elementary sentences are made out of words, there is no regular such relation of larger to shorter stretches of musical composition.

As to effect, or meaning, of all these constructions: in language, there is a fundamental difference, ultimately a single one (predication), between the meaning of a sentence and that of the words which compose it. In music, there may be no one essential difference between the effect of a composition, short or long, and that of the musical phrases or lines of which it is composed.

Another structural consideration is the relation of the sentences, paragraphs, and chapters to the discourse, or in the case of music, the relation of the larger musical segments to the whole composition. Both in linguistic systems and in much music, there are initial or underlying sentences or themes which are repeated and modified in making the discourse or composition; and in both there are subsections with internal regularities different from how the subsections relate to the whole. In language we saw the double-array requirement of repeating the word-relations (9.1, 3); in science languages, a few stronger similarities and progressions; in mathematics, the sharp constraints that characterize proof. In music, on the other hand, a great variety of complex modifications and developments of the themes and chords enter, together with other elements, into the total musical work. These 'discourse' structures are obvious for example in rounds, fugues, the sonata form, and differently in the demand to return to the tonic. One vague and second-order similarity between compositions and texts is that, in both of them, the structural relations among successive short segments (musical phrases; sentences) are more stringent and 'convincing' than those among long segments (movements; paragraphs or chapters).

In addition to the specific differences such as those noted above, the overall structural difference is of two kinds. One is that linguistic systems—including mathematics—are built by a few and permanent constraints, whereas in music the constraints are more variegated, and differ from one culture and period to another.[18] The other difference is that in linguistic systems most of the complexity of sentences and discourses arises from non-informational elements—the transformations in shape. The meaning-effect that this complexity contributes is quite different from that of the initial informational sentence-making constraints. These last are simple in themselves. In contrast, in music complexity can arise from the intricacy and interweaving of the initial constraints themselves. The effect of the complexity is not necessarily different in kind from that of the smaller compositional elements on which the complexity is carried out. The nearest thing to grammatical transformations is in the musical form of variations upon a theme; but the compositional status of this form is not different from other compositional complexities.

A major issue at this point is the relation of the structure to the content, to what is being said. In the case of language, it seems that the content is achieved by the real-world associations of both the word-choice and the initial constraints (11.4, 6). We cannot say that something comparable holds for music, but we know that a composition has an effect, which may be related to the affective-experience association of its sounds and sound-sequences, its harmonic structure, and the way its compositional constraints are developed. We may also suspect that just as certain grammatical features are unrelated to the information of the sentence (11.3.3), so

---

[18] The permanence of basic constraints in natural language and in mathematics is what makes translation possible between languages that have in detail different grammars. In the case of mathematics, intertranslatability is what is called 'abstract equivalence' between mathematical fields. In music, there is no intertranslatability.

certain compositional features may be unrelated to the content or effect of the music. Nevertheless, having seen the contribution that the sentential word-dependence and the further constraints make to the informational capacity of language (11.3.2), we can understand that the structure of a musical composition—as much or more than the themes—can be a direct bearer of what is expressed in the composition. Indeed, some musical themes may have little expressive effect ('meaning') until the composition works on them, as may be the case for the *aria da capo* in a set of variations. The relevance of combinatorial constraints (rules of composition) to content can be seen in the need to 'solve problems' in composing, and to change styles in music and art in order to express fresh things: not to compose today in the style of Mozart or to paint in the style of Rembrandt, no matter how much the originals may still be appreciated today. Given this we can understand how a composition or painting can express to us meanings, by virtue of the structure we see in it, that the maker may not have felt explicitly. And we can understand that new developments in the formalism— the 'language'—of music and the other arts can contribute to what is sayable in those arts. New devices, local or global, in the language or style of an art can carry or facilitate new meanings. With all this, the rules of composition no more suffice to make interesting music than the rules of grammar suffice to make interesting discourses. But we can see how popular art and 'high' art can be housed in the same system, as ordinary conversations and great literature can all be housed in natural language.[19]

We now turn to the content of all these systems. We have seen that the specific content of natural language is transmissible information, a capacity provided by its essential discreteness, its word meanings, its dependence (predication) structure, and its likelihood inequalities. Comparably, the content of mathematics is preservation of truth values through a sentence-sequence, a capacity provided by its having only the indefinite N (variables) with no likelihood variation, its structurally characterized truth tables, and its constraints on proof-sequence.

Although the universe to which language can refer is roughly the whole perceived universe—language can talk 'about everything' including itself— what language expresses structurally is just 'about' this universe, i.e. just the information of it (ch. 11). Furthermore, the discreteness of phonemes and the conventional pre-set finite stock of words at any one time assures the largely error-free transmissibility of language, so that what the final hearer receives is in principle the same substantive information that the original speaker issued. Thus language cannot indicate quite everything that is

[19] The contentual power of a language in art, corresponding to that of a grammar in natural language, may perhaps be seen in the need to change the language of music after Bach, and in the inability to paint purely cubist paintings after cubism reached its peak. And the difference between a language of art and an expression made in that language may perhaps be seen in the similarities and differences between corresponding cubist paintings of the otherwise very different Braque and Picasso.

indicated, say, in a picture. It cannot express in a direct and full manner the whole range of human meaning—what a person 'means' to himself, what an experience means to him, what precisely is the sensation of pain or whatever that he has. In particular, language cannot give by its structure any direct expression to emotion, except by bringing in extra-grammatical material, i.e. items which have no regularities of co-occurrence in respect to other items of the language: intonations such as of sarcasm or anger, or marginal constructions and words such as *How wonderful*, *Gee!*, *Ouch*. Feelings are carried in the language only by informational sentences, possibly false, which state that the speaker has the feeling (*I am happy*), or else by art-activity (on which more below) of manipulating language, in ways, even non-grammatical, that carry emotional or aesthetic impact.

In 11.6 it is suggested that the capacity of language to have an effect or to 'mean' is by virtue of the association of its elementary parts with something in the world of experience, and by the constraints on combination of these parts—or rather, by the association of these constraints with aspects of these experiences. Further, the effect that language structure has is not the whole range of meaning but only information, and more specifically the kind of public information that can be transmitted with little or no error. Music too has parts, and constraints on their combination: notes as compared with phonemes, musical phrases as compared with words, methods of composition compared with syntax. Despite some physical similarities between the parts and constraints of language and of music, they differ deeply in their structural properties and in their content—in the aspects of experience with which they may be associated. However, the very fact that meaning—one might say subject matter—in language is limited to information leaves room for important areas of meaning or expression to be covered elsewhere. Indeed, if language is specifically an information system (to express or communicate) rather than generally an expression system (for all content), then all types of expressing other than informational ones must be lodged in other modes of expression and other types of activity.[20]

As to music and the arts, the content which they indubitably have must be related to human experience, since art is made by people and responded to by people. Many features in the constraints of art and the response to it, for example the difference in response to art as against the response to decoration, suggests that here, as in language, the effect (meaning) may be in response to the kind and complexity of constraint (or stabilities, instabilities, and organizings) which is met with in the course of human experience—different from the constraints relevant to language (11.6), but nevertheless constraints

---

[20] How different are informational expression and non-informational expression is seen in the fact that most language expression is communicated only once, after which the recipient hearer can use it accretively with other received information. In contrast, the expressions of music and other arts are often accepted many times by the same recipient, and are not combinable with other expressions in any regular way to comprise some joint result.

on experience. These meanings in art do not have to be of the same kind as those talked about in language, for there may be feelings or sensings which are too private or subterranean or non-informational to have been identified in the fixed, publicly recognizable, and informational meanings of words.[21]

The difference between the meanings of music and of language can be seen most sharply when language and music collaborate, as in songs, arias, and recitatives. We may think that much of the music in Bach's St Matthew's Passion fits the words marvellously well, or rather fits what Bach undoubtedly felt about those words; but people of the modern world can respond perhaps as fully to the music without responding comparably to the words. A given piece of music can support a variety of texts expressing quite different information (e.g. in recitatives), and different specific emotions (as when revolutionary poems replace the words of folk-songs). Similarly, a given text can be set to music in quite different ways (for example, the Mass). Indeed, in order to increase the aptness of music to words, the *Lieder* tradition uses items of theater—word-phrasings, facial expression, etc.—which no matter how they enrich the performance are additions to, rather than parts of, the musical composition.

These rather obvious observations are brought here only in order to point out that pieces of music cannot be identified with those things that language can identify—certainly not with information, but even not with those feelings that are connected with specific information or specific situations discernible in language. This is all the more obvious in the 'pure' instrumental compositions, which can have great emotional and aesthetic impact without our being able to characterize it in terms of language or specific content. This holds even though many musical compositions are felt to be 'saying' (or rather, expressing) something (e.g. the late quartets and sonatas of Beethoven), and some compositions more deeply than others, as some language discourses are deeper than others. The inadequacy of language for music (and the converse) holds even when what music is 'saying' may seem to be indicated by the words of its title or by epigraphs. Music does not carry or express information in the sense that language does, and neither music nor language can be translated—even roughly—into the other (as mathematical notation can be translated into language).

The fact that the informational specialization of language leaves room for other things to be otherwise expressed does not open the door for the non-rational to intervene in the semantic domain of language. Music can indeed say things, but not things that can be put into words. Its essential divorce from information suggests that music—and art in general—does not provide us with some 'immediate knowledge' of the world such as could be added to our language-borne knowledge. Language and reasoning are effected by con-

---

[21] The fact that to this day we cannot say—or agree—what it is that music expresses serves as a cautionary tale about the otherwise invisible limits of what can be said in language, or at least about the difference in content between language and music.

straints on the form and on the sequence of predications, none of which are available to art or to feeling. It is not that these last cannot complement what we have in language, but they do not supplement it within its own universe by supplying any immanent material that can then enter into, and correct, the specific predicational and sequential-argument structure (10.5.4) which constitutes the informational power of language. Indeed, as language is important to us in giving us some structuring of the mass of objects and events which we experience, music and art may be important to us in giving us some structured expression of the sensings and feelings which we experience both internally and externally.

In this sense, and given the difference in structure between language and the arts, Pascal's 'The heart has reasons of its own, unknown to reason' is questionable: any extra-rational status or validity ascribed to 'the heart' (for Pascal, faith) cannot add to or correct what has been reached by the rational structure of language. Whatever it is that man finds in art, or for that matter in art-like response to nature, it cannot contribute to what man finds through language; it cannot make him wiser in respect to his public problems, whether material or social. Conversely, art and feelings cannot be said to be by their nature 'true'. In the matter of validation, the rationality of language may be considered to be a way of dealing with the world, with successful (effective) outcome; but the arts may be considered to be affective responses to the world, with 'success' being either undefined or highly indirect. It is not the case that what is said may be erroneous or false whereas what is felt is *ipso facto* authentic and valid. There may be 'false' emotions, and cliché emotions, which a person may indeed feel but which may be an artefact of inculcation, or of institutional and public pressures which led him to feel so. What is more, while the validity of ideas and statements in language can be tested by reasonably explicit, direct, and public methods of critique (10.5.4–5), the validity of feelings can be tested only by such complex considerations as those introduced by Freud and by Marx. Somewhat as one may consider a discourse superficial, one may consider a composition superficial. And as one may question an idea stated in language, one might question a feeling, for example awe, which one might consider no longer befitting man's relation to nature and society. More generally, as language cannot directly express emotions, art cannot directly express ideas. Artists may be influenced by cultural winds and social movements, but that content appears in their work only in secondary and often amorphous ways.

The difference between information and feeling may also clarify some of the difficulties in discussions of mystical experiences. The latter may consist of real emotional experience associated with some particular informational (factual) content, but where no publicly relevant connection can be made between the experience and the informational content (whereas in language such connection exists, 11.6). The difficulties may not be so great if we understood that the experiences exist as such, due to whatever conditions, but that the associated factual content which triggered the emotions in that experience may not be contentually related to that experience, any more than the words of a song are contentually related to its music.

A similarity between language and music is that in both of these the creator of the discourse expresses a meaning and the hearer receives a meaning, an activity quite different from, say, play or work. Music however is unlike language in the matching of the author's act and the recipient's perception. In language, by virtue of its publicly established phonemic distinctions and pre-set vocabulary, the hearer is expected to get out of a sentence precisely what the speaker put into it; any appreciable difference constitutes a failure in language use—in information transmitting. An important failure, for example, is the inability of the speaker to indicate, in a general way within the grammatical apparatus, which word is the antecedent for a particular occurrence of *he* or the like. There are also types of mismatches which are not due to grammar inadequacy. One type is when the speaker uses circumlocutions or euphemisms because of custom, or of personal or social discomfort with what he has to say, in which case the hearer is expected to recognize the shift and to correct for it, understanding the statement to mean what is intended rather than what is actually said. Another is when the speaker speaks with indirection, hoping to mislead his hearer into accepting what he actually says rather than perceiving his intent. But aside from such special cases, a sentence is correctly understood only when the hearer receives as closely as possible what the speaker meant or intended.

In contrast, the arts have no apparatus to ensure that what the maker meant is what the recipient receives. It is accepted in music and the arts that the recipient—viewer, hearer, performer—may perceive something different from (though presumably related to) what the maker meant or intended. It is acceptable (at least to many) that different historical periods have different ways of playing Bach, and that music written for Baroque instruments may be played on modern ones, and certainly that different performers may interpret a composition differently—that is, play it differently, see different things in it—even when the composer's own performance is known, as for example in Stravinsky's conducting of his own works. The importance of performing tradition and individual performance over and above the score is not just a matter of incomplete notation.

This difference between reception in language and reception in the arts is understandable when we keep in mind that language transmits more or less explicit information from person to person, in a system based on full repeatability of pre-set material, whereas in the arts, the maker expresses an emotion or a sense of something in ways that call out feelings or sensings, not necessarily identical, in the recipient; the recipient's respondent feelings depend in part on his own emotional nature and experience. Indeed, the maker can manipulate materials and symbols as a technical or random activity, without an intended expression of feeling, and yet the recipient may react with feeling, as he may react to unintended nature. By the same token, when the work of art contains expression of feeling or intentional structure from the maker, it is not necessarily the case that the feelings it evokes, or the structure that is perceived, in the recipient are precisely the same as the maker's, or the same for all recipients. Thus expression and communication are not identical in art, and transmission is only a secondary use of it. In the absence of systemic control over repetition and sameness, such as phonemic distinctions provide for language, authentication of what the maker did or meant, and the communicational problem of maker–recipient differences, are not systematically resolvable.[22]

Several activities other than music should be mentioned here in order to distinguish them from linguistic systems. One is the use of language itself for purposes of art. This is not a second function of language, but a re-use or manipulation of language. The language remains in its original and only definition: words with meanings and operator–argument statuses, sentences as informational operator–argument combinations. The art use of language adds to this—and also alters it—by introducing not only additional features of discourse structure but also non-grammatical features: sound similarities (onomatopoeia, alliteration, rhyme), time-intervals (syllable count, stress, rhythm, rhyme), play on word meanings (including allusion, nonce extension of selection). Modern literary devices also violate some of the grammatical requirements, from word combinations that are not within selection to ones that clash with the operator–argument structure or linearization. All this may be interconnected with the meanings of language material, that is with its ordinary use; or it may be virtually independent of that. In any case, it does not constitute an independent system, but rather a set of structural modifications whose purpose is tangential to the 'function' of language or totally independent of it. These modifications may, of course, follow a systematic or even conventionalized way of manipulating language or playing upon it, with emotional or aesthetic effect. The property of acting within and against rules of a system is more pronounced in music and the language arts, as it is in puzzles and games, than it is in such arts as painting and sculpture where the artist creates his independent arrangements of objects or symbols.

Another activity that has to be mentioned is animal communication, whether cries or standardized behavior (postures, etc.). There is little question that some of these are expressive for the individual, and some are in effect communicative for the group. The only thing that has to be said here is that these activities differ from human speech not only in degree—the vocabulary being minute—but also in kind, because there is no operator–argument relation, no predication, no organized likelihood-inequalities, or

[22] On authenticity in art, and also on notations in art (10.5.5 above), cf. Nelson Goodman, *Languages of Art*, Hackett, Indianapolis, 1976.

anything approaching the structure that makes sentences (12.4.1). It should be noted that a two or three orders of magnitude increase in vocabulary is in itself a difference in kind, because the whole vocabulary can then no longer be efficiently utilized without classification, and because its possible combinations are then so many as to require constraints and regularities.

Finally, it seems that the sign language of the deaf does not have an explicit operator–argument partial ordering, nor an internal metalanguage, but rests upon a direct juxtaposition of the relevant referents. This applies to autonomous sign languages, developed by the deaf without instruction from people who know spoken language.

V

Interpretation

11

Information

11.0. *Introduction*

Language is clearly and above all a bearer of meaning. Not, however, of all meaning. Many human activities and states have meaning for us, and only some of this can be expressed in language: feelings and vague sensings can be referred to only indirectly (10.7); non-public information, such as proprioceptive sensations, can be named only with difficulty; certain kinds of non-language information can be translated directly into language (e.g. graphs and charts); but other kinds (e.g. photographs) can be represented in language only loosely and selectively. Meaning itself is a concept of no clear definition, and the parts of it that can be expressed in language are not an otherwise characterized subset of meaning. Indeed, since language-borne meaning cannot be organized or structured except in respect to words and sentences of language (2.4), we cannot correlate the structure and meanings of language with any independently known catalogue or structure of meaning. In each language, we do not know a priori which specific aspects of meaning will be referred to by words, and how much will be included in the meaning of a single word. Even when the meanings are well-defined, it is not always possible for words of a language to mirror in their structural relations the relation among the referents. For example, the relation among the integers is given in Peano's axioms, but language cannot thereupon name them as *one*, *successor of one*, *successor of successor of one*, etc. What languages do in this case is to give arbitrary names to the first ten (or six, or whatever), and then to use these and other names to indicate successive multiples of the initial set: the integers are thus named modulo that initial set, by a language-based decimal or other expansion. Such cases suggest that we have to study the specific words and structures of a language if we wish to see what meanings they cover, and how.

As has been seen (5.4), the phonemes are irrelevant to meaning though they underlie communicability: the phoneme-sequences that constitute words are not in general constrained in any way that is relevant to the meanings of the words. The words are thus a fresh set of primitive elements, which can be identified without phonemes (spoken or written) as in ideographic writing, and which are associated with meanings, in a manner to be discussed in 11.1 and 11.6. The meanings of words are modified by their likelihood relations to their operators and arguments, and in some cases differ according to their operators and arguments (11.1). And given words with their meanings, it will be seen in 11.2–3 that the constraints on word co-occurrence create further, 'grammatical', meanings. By separating all occurrences of language into occurrences of words and constraints on their co-occurrence (rather than syntax being simply a classification of word combinations), we can show (1) that the meaning of a sentence or discourse is the meaning of its words plus the meanings of its ordered constraints; and (2) that each of these meanings is constant for all occurrences of the given word or constraint in all sentences (even under transformations and zeroings, 3.3); that is, no meaning beyond these is carried by the structure of grammatical transmissible language. Within a sentence, these constraints are all included in the partial order of words in respect to sentences (3.1–2). For discourses and sublanguages, the constraints have been described in Chapters 9 and 10 and their meanings will be described in 11.3.2.

All of the above yields structural locatings of specific contributions to the meaning of a sentence (11.4). In addition, it will be seen that language is limited to particular kinds of meaning (11.6), and that information has in certain respects a language-like structure (11.7). The relevance of information to language is seen also in the fact that non-contributing of information is the grounds for zeroing (3.2.6).

11.1. *How words carry meaning*

The first question is: what are the least parts (forms) in language to which meaning adheres? Not phonemes: there are no regular ways in which the meaning of the words of a language can be obtained as combinations of meanings assigned to the phonemes of those words. Indeed, phonemes do not in general have any linguistic meanings; and the pair test which establishes phonemic distinctions is related not to the meanings of the tested words but only to the recognizability of their repetitions (7.1).[1] If in all the occurrences of a word the phonemes were replaced by others, we would simply have a variant form of the same word. But if we replaced some or all of the occurrences of a word by a word which had different selection, i.e. whose normal occurrences were different, we would have a different meaning. Hence onomatopoeia, in which the sounds of a word suggest its meaning, is a rarity in language. By the same token, the cases in which different morphemes that are similar in sound are also similar in meaning must be haphazard and

---

[1] As noted at the end of 7.4, the phonemic composition of words is needed because there are not enough pronounceable and audible distinctions among fixed single sounds to distinguish all the vocabulary of a language. The structural and semantic properties of words can be carried even by having a single sign for each word, as in ideographic writing.

not regular; and these sound–meaning correlations are then not usable in any regular way in the grammar.

However, in each language there are certain listable phoneme-sequences, called words or morphemes, which carry fixed meanings. It has been seen that these sequences can be isolated out of the flow of speech even without any knowledge that they are the words of the language, or what their meaning is (7.3, 11.3). A sequence of phonemes cannot be established as a word simply on grounds of being judged to have meaning of its own: it must also have regularities of occurrence relative to established sets of other words. There are cases of a phoneme-sequence which seems to have a characteristic meaning but whose immediate environments do not them-selves combine in a sufficiently regular way, as morphemes. The morphemic status of such phonemic sequences may therefore be border-line. Such are, in English, the *sl-* of *slick, slip, slither, slide, slimy, slink, sling, slog, slosh, slouch, slow*, etc., and the *gl-* of *gleam, glow, gloom, glisten, glitter, glare, glide*, and the *fl-* of *flick, flip, flit, float, flash, flare, flap, flop, flutter, fling, flow, flee, fly* (7.4). Such is also the case for *-le* in scores of words: *handle, dazzle, nozzle, nuzzle, muzzle*, etc., even though in some words the *-le* had been a more regularly combining English morpheme. On the other hand, a phoneme-sequence may have to be accepted as a morpheme even when it has no assignable meaning, just on grounds of its regularity of combination: e.g. the *re-* of *recommend*.

In view of all this it is necessary to consider what is the source and status of word meanings, and how these relate to the combinational regularities of the words: how do words carry meaning? First, it is clear that most words have some association with meaning independently of their occurrence in sentences (i.e. in combination). Since a large vocabulary was presumably in use before sentences developed (12.3.1), these words must have had meanings in isolation. In addition, many words that enter the language at various times get their initial meanings from the experiences that gave rise to them rather than from any grammatical combinations. This applies to newly made words such as *gas* and *boysenberry*, to borrowings whose meanings may be adapted or specialized from the meanings in the source language (e.g. *opera, piano*), and also to word combinations that are not made on a syntactic basis (e.g. *flip-flop, wishy-washy, wheeler-dealer*). In addition, many words change or specialize their meanings in the course of time on the basis of cultural and historical developments, without apparent regard to their etymology or to the syntactic combinations into which they otherwise enter: examples are countless, e.g. *Quaker, verb, plane*. And of course very many syntactic word combinations have a specialized meaning beyond their composition: whereas *school-books* are any books specifically for school, *snow-shoes* are not merely any shoes specifically for snow but a particular type of object.

How words are associated with their non-syntax-based meaning will be discussed in 11.6. Before reaching that question, however, it is important to note that this association of a word is in general not to a unique object, event, state, or experience, but to a set of these—a set which outside of a few special conditions (such as in science) is not fully delimited and defined either extensionally or intensionally.

As to uniqueness of referent: there is nothing to prevent proper names from being pluralized (hence able to refer to more than one object), or from being used about more than one referent. A pet can be called *Zeus* or *Napoleon*; and while the names thus used contain allusions to their original bearers, it is hard to say that they retain the original bearer as referent. Similarly, geographic names can be used for more than one location or formation. As to the extension of referents, it may be impossible to determine the complete set of individuals which would constitute the denotation of a given word. It thus seems that words cannot be limited to a single referent or a closed and unextendable set of referents. They cannot be defined by their extension.

As to the intension of a word, for most words it is difficult to state a definition, or a set of alternative definitions, which would precisely characterize all cases in which the word can be used; this even though one might be able to state definitions that would suffice to distinguish the meaning of the given word from that of all other words (a much more delimited task). As will be seen, words in general refer to properties and classifications (of objects or events) which would not necessarily be otherwise distinguished—i.e. except for their being named by the same word. Words thus select properties, and classify their bearers (referents).

There are also various problems in fitting together these two aspects of word meaning: the set of referents for which a word is used—its denotation or extension—and the sense which it conveys—its intension or dictionary definitions. Not only can a word used in a single meaning refer to different individuals who may not otherwise be identical or equivalent, but also two words or phrases with different senses can unbeknown to speaker and hearer or known to one or both refer to the same individual: Sir Walter Scott and *the author of the Waverley novels*, *New Amsterdam* and *New York*, *Franklin D. Roosevelt* and *That Man* (as he was called by opponents), *the evening star* and *the morning star*.[2]

For some words there is just one specific meaning as in *boysenberry*, or closely related meanings as in *time*. For many words the meaning is a set of referents with a fuzzy boundary: e.g. as to what is included in *chair* (a *chaise longue*?, a one-legged spectator seat?, a rock hollowed out for sitting?). A word may have a single more or less continuous range of meanings, as for *chair*; or more than one range of meanings, as for *station* (rail and bus, radio and television, but not telephone; meteorology, and perhaps not astronomy, etc.), or for *field* (agricultural, sports, region, range of interest, and as

---

[2] This difference is involved also in words, such as *unicorn* (11.6.2), whose set of referents in the real world is null, but whose sense or meaning is not empty.

defined in physics and in mathematics). The distinctions and boundaries of meaning-range are complicated and vague. It will be seen that meaning is more easily stated as a property of word combinations (or of words in combination) than of words by themselves.

Although meanings exist for most words in isolation, it can readily be seen that this is not all there is to say about word meanings. In the first place, a phonemic sequence may have different meanings, and be considered different words, when in the environment of different words, e.g. two words *pen* in *Fill the pen with pigs* and *Fill the pen with ink*, in contrast to one word *spill* in *The pigs spilled out of the pen* and *The ink spilled out of the pen*. This occurs also for many words which may be considered the same in their different occurrences despite their different meanings: e.g. for *spur* in *The spur on the riding boots was too short* and *The spur inside the flower was too short* and *a railroad spur*, or *policy* in *Fire losses* (or: *Foreign adventures*) are *not included in our policy* and *Foreign adventures (or: Industry bail-outs) are not included in our policy*. Also, two words which are very different in meaning may be almost synonymous in a particular environment, although each keeps its own meaning throughout: e.g. *the cells divide* and *the cells multiply*. The environment is thus a factor in the meaning of words.[3]

The dependence of meaning on environment clarifies somewhat the problem of meaning-range. Once we see that the meaning of a word in a particular occurrence depends on its environment, the categorization of meaning-ranges can be replaced by a categorization of the environing words. Note that when we judge the meaning of a word-occurrence by what can replace it in the text, we are really judging the meaning by textual environment. We need not ask the virtually uninvestigable question: is the meaning-range of all occurrences of a given word a single continuous one, or a composite of several ranges? Instead we ask: do the environing words, of the various occurrences of our given word, fall into distinct sets? The environments of *spring* of metal and *spring* of gushing water are readily distinguishable, even if we can find a few common to both (e.g. *The force of the spring is surprising*); this even though these words share an etymology ('are the same word etymologically'). Such is the case also for *table* as a piece of furniture and *table* as presentation of data, although the awareness of a metaphor-like relation is stronger here. In contrast, a clear separation may not be possible among the environments of all occurrences of *field* (above). Environment distinctions, which are more measurable than pure meaning-distinctions, can make it possible to distinguish homonyms from separate meanings of one word, and the latter from an indivisible, coherent meaning-range.

[3] Even for the many words whose meaning is constant the semantic effect may differ in different environments. When said of an unknown object, *small, large* may be used relative to the size of speakers and hearers; but said of a known object, these are relative to its class of objects: *small elephant, large flea.*

Observation of the data makes it clear that the relevant environment for meaning-specification of a word-occurrence is its immediate operator or argument, more rarely its farther operators or arguments and its co-argument; modifiers of a word are counted as operators on it in a secondary sentence (3.4.4). Because the environment of a word delimits its meaning, the meaning of a syntactic construction is sharper than the meaning-ranges of the component words taken separately. Thus the correlation of meaning with syntactic constructions of words is more precise than it is with words alone; and indeed, a dictionary is frequently forced to refer to grammatical environments if it is to present word-meanings with any precision. The meaning is in the words, but the precision for each occurrence often comes with the combination. Cases in which environment differences do not match meaning-differences are only apparently such: for example, there is much similarity in the environments of certain words whose meanings are opposites (*left*/*right*, *small*/*large*, etc.); it is relevant here that many opposites (*good*/*bad*, *young*/*old*, *mother*/*daughter*) are known to have important semantic features in common. Indeed, certain aphasias and slips of tongue can replace a word by its opposite.

New extendings of environment and meaning are constantly being made. They are not determined merely by the speaker's intent. In many cases there are alternative possibilities for the environment, and the extending is affected by various factors. Thus when a new operator was needed for the motion of a plane in the air, *fly* was not the only possible choice (the Zeppelins *sailed*): similarly for *fly* as the moving of a flag. When words are needed for incapacitating a tank, or for eliminating a bill in Congress, *kill* is not the only word with a somewhat appropriate meaning. And steamships and motor-ships *sail*, but motor-boats do not. In the many such cases of new word combinations, the meaning-range of the chosen word (*fly*, *kill*, *sail*, etc.) changes to accommodate the new use.

A case of this is seen when metaphors, originating as a simile, yield a new meaning of a word, as in the case of *see* in the sense of 'understand' (*I see what you're saying*), or in the meaning of *understand* itself as it developed out of its etymological source. Metaphor is an extreme case of an apparent, or *de facto*, change of meaning in a word. It arises in a simile when indefinite nouns and verbs, and also the *as* of the simile itself, have been zeroed, leaving the remaining word to carry the meaning of the zeroed material in addition to its own. If the metaphoric *see* above is derived from some source such as *I treat what you're saying as one's seeing things* (through an unsaid *I treat-as-one's-seeing-things what you're saying*, reduced to *I treat-as-seeing what you're saying*, G405), then the metaphoric *see* has the meaning of the original physical *see* plus the meaning of *as* (while the verb-classifier *treat* and the indefinites *one*, *things*, carry little meaning), 'I as-though-see what you're saying'. Other cases of words which are left to carry the meanings of zeroed operators on them are seen for example in the 'zero causative' of *He walked the dog* from roughly *He took the dog walking* (through an unsaid but grammatically constructible *He took walking the dog*). Such situations can even lead to words having apparently opposite meanings, as in *He dusted the furniture* from *He treated the furniture for (its) dust* as against *He dusted the crops* from *He treated the crops with (insecticidal) dust*. A different case is that of abstract nouns, as nominalized operators, which arise when the indefinite arguments of an operator are zeroed in a nominalized sentence (i.e. a sentence under a further operator): e.g. *hope* from *people's hoping things* (in *Hope springs eternal*), *vehemence* from *one's being vehement* (as in *Vehemence is uncalled for here*, 11.2). 'Lexical shift' (i.e. meaning-shift in words) is thus simply the effect of environment extension plus zeroing.

The most general factor in the varied and changing meanings of words is simply the constant though small change in likelihoods—what words are chosen as operators and arguments of other words, and how frequently they are thus used. Beyond the argument-requirement dependence, the grammar does not limit each word to particular other words as its operator or arguments: any O_n operator can make a sentence—even if a nonsensical one—with any N word, and so on. Given this, the particular choice (or grading of likelihoods) of words depends largely on what makes sense (or intended nonsense), given the current meanings of the words; but it depends also on marginal distinctions that the speaker wants to make, on attention-getting surprise, on euphemism, on the analogy of how related words combine, and on other factors.

The changing frequency and acceptance of certain combinations affect what comes to be seen as the main meaning of a word, as do increasing specializations (i.e. frequencies of particular meanings) of word use. Over time one sees many changes in combination which produce recognizable differences in meaning, e.g. as between *I was flying in a 747* (an extension in the meaning of *fly*) and *A moth was flying in the 747*. There are words in which successively extending the frequency of the environment-choices has resulted in large differences in meaning: for example, within the history of English, *fond* from 'tasteless' to 'affectionate', *like* from 'to please' to 'to be pleased by, to favor'; in *wear*, by the side of the continuing meaning 'to be clothed in' there was also a development to 'to deteriorate' and 'to pass' (*The night wore on*).[4]

Although the meaning of any word at a particular time is a factor in contemporary choices of operators or arguments for it, the meaning is clearly not sufficient to determine those choices. But those choices in turn affect what is understood as the meaning of that word. After new choices of operator or argument have been made for a word, we can often find an

---

[4] Changes may be so natural, given the changing culture, as not to be noticeable. Thus, *the old verities* (literally 'the old truths') once referred to what users of the language must have understood as, indeed, 'the old truths'. By now our suspicions as to the once-accepted truths are such that when we say *the old verities* we are close to meaning 'the old falsehoods'.

extended core-meaning for that word, such as would fit both its old and its new environments (e.g. *discharge an employee, an obligation, an electric potential difference*). But we would not necessarily have come to this meaning-core before the new environments had come to be used: consider *to charge the enemy*, or *the accused*, or *the purchase*, and to *charge a battery*. That is, knowing the meaning of a word does not suffice for us to predict what new environments would or would not be used with that word. Different resolutions in this respect are found in different languages: thus although English and French have rather similar systems of kinship nomenclature, seemingly opposite choices appear in *grand-daughter* as against *petite-fille*, each being a reasonable way of adding 'one stage farther' to *daughter*, but *grandmother* is paralleled by *grand'mère* (and *aïeule*). (Note that *great-granddaughter* is *arrière-petite-fille*, and for *great-grandmother* one can say not only *arrière-grand'mère* but also *bisaïeule*.) Extending to new environments is all the more complex when we consider nonce forms,[5] metaphor, analogy (especially that), and literary turns of phrase. All these many-faceted similarities and differences in meanings of words can create regularities and patterns of meaning-relation among words, which are a source of interest in lexicography and literary criticism.

Attempts to categorize and to find regularities in extending and changing meanings of words have not fared well. But what is important is that such attempts have succeeded far better when directed to the choices of environment rather to any inherent dynamics (such as narrowing or widening) in meanings as such. That is, a word is extended from one environment to another on environmental grounds, e.g. on the basis of the environments in which that word already occurs in comparison with the environments in which variously related words occur.[6]

The meaning that a word has with a particular operator or argument is constant for all its occurrences with that operator or argument. Thus not only is this local meaning more sharply defined than the general meaning-range of the word, but also it is this meaning that is preserved under further operators. This meaning is also generally preserved under transformations. However, a word may seem to change meaning when it incorporates the meaning of attached words which have been zeroed, as in *expect John* in the sense of *expect his arrival*, and in *She's expecting* for being pregnant. In contrast, when words come to be used in greatly broadened selections, their meaning becomes less specific (and in some cases their form is reduced) when they are in the broadened selection. For example, *of* is derived from *off*, and *have* is weakened from 'possess' to 'be in (a situation)' when it

---

[5] In nonce forms and certain jokes a word is used in a new environment with the understanding that this is not a precedent, that the meaning of the word should not be modified to fit the new environment.

[6] H. M. Hoenigswald, *Language Change and Linguistic Reconstruction*, Univ. of Chicago Press, Chicago, 1960.

occurs with all verbs to make the perfect tense. (Note that in the French perfect tense, e.g. *J'ai fini hier*, the *ai* is shown by the *hier* to be no longer present tense, but in English *I have finished by now*, where there is no *I have finished yesterday*, the *have* is largely still present tense.) Furthermore, in some cases a word combination changes its meaning so that it no longer follows in a simple way from the meanings of the component words: these are the idioms (e.g. *kick the bucket*, *fix his feet*). This may take place even if the non-idiomatic meaning of the combination continues to exist, as in *take care of him* both in the sense of caring for him and in the sense of neutralizing him.

It would seem to follow from all this that meaning has to be stated as a property of word-occurrences (i.e. of a word 'token') rather than of word (i.e. of its 'type'). But this is wasteful, since very many word-occurrences have the same meaning. The regularity of meaning is in terms not of word as such nor of word-occurrence, but of word combination: the word has fixed different meanings when with different subsets of its selection (immediate, or somewhat farther).

In sum: first, the meaning of a word in isolation is in most cases the starting-point, and its meaning affects its current selection of operators and arguments, though only in a loose way. Second, the operator–argument selection of a word, while affected by its current meaning, can be changed by the pattern of the selections of variously related words (especially in analogic extension), which in turn affects its ensuing meaning. But it must be kept in mind here that the selection of a word is not simply the set of operators and arguments with which it occurs, but the set it occurs with in above-average frequency (hence e.g. not as jokes); we include here cases when the frequent co-occurrent is somewhat distant in the dependence chain. It is thus that selection can be considered an indicator, and indeed a measure, of meaning. Its approximate conformity to meaning is seen in that we can expect that for any three words, if two of them are closer in meaning to each other than they are to the third, they will also be closer in their selection of operators and arguments.

By speaking of a word's selection instead of a word's meaning in isolation we leave room, first, for different meanings (taken from the range in isolation, or extended out from it) in the presence of different operators or arguments, and second, for meanings to change and to change differently in particular environments. The latter has been shown to occur not only on the basis of processes of meaning change but also on the basis of environmental similarities and differences. For example, a word can change meaning in suffix position while it does not change in its free-word occurrences (compare *-ly* with *like*).

Selection is objectively investigable and explicitly statable and subdividable in a way that is not possible for meanings—whether as extension and referents or as sense and definition. Indeed, one can take a reasonably large sample of sentences or short discourses and list the complete selections for various words, although in many cases there may be uncertainty as to whether a particular operator or argument has selectional frequency for the given word or is a rarer co-occurrent which does not affect its meaning. Characterizing words by their selection allows for considering the kind and degree of overlap, inclusion, and difference between words in respect to their selection sets—something which might lead to syntax-based semantic graphs (e.g. in kinship terms), and even to possibilities of decomposition (factoring) for particular sets of words. Such structurings in the set of words are possible because in most cases the selection of a word includes one or more coherent ranges of selection (3.2). The effect of the coherent ranges is that there is a clustering of particular operators around clusterings of particular arguments, somewhat as in the sociometric clusterings of acquaintance and status (e.g. in charting who visits whom within a community).

Selection sidesteps certain boundary difficulties that are met with in considering meaning. For example, words whose referents have at all times null extension (*unicorn*, *centaur*, spiritualists' *ectoplasm*, or for that matter *laissez-faire capitalism*) present no problem here, because they each have characteristic and stable selections, and ones which adequately distinguish, say, *unicorn* from *centaur*. Such words can also be characterized by their definitions. Indeed, the selection of a word can be used to generate the set of its partial definitions: a *chair* is something on which one can sit, which in most cases has four legs, etc. As to different words which are not really synonyms but have the same referent (e.g. *Sir Walter Scott* and *the author of Waverley*), these have or can acceptably have the same selection—but not in sentences where their difference is involved (e.g. not in *The king did not know that Sir Walter Scott was the author of the Waverley novels*). Words whose definitions are not readily statable, or which may have no definition or meaning, such as the spiritualists' *ectoplasm* or *there* in such sentences as *There's a man here*, can nevertheless be characterized by their syntactic position and their selection.

Selection also treats directly the cases where meaning and phonemic composition do not correlate: in suppletion, different phoneme-sequences have the same meaning; here we note that they have the same environments except for being complementary in respect to a particular part of the environment (e.g. for *is*, *am*, *are*)—a situation which has a distant similarity to the issue of *Sir Walter Scott* above. In free variation, partially different phoneme-sequences have the same meaning; here we note that they have the same environments (e.g. the two pronunciations of *economics*, or the *not* and *-n't*, and *will* and *wo-*, of *will not*, *won't*). In science languages (10.5) a given word symbol, with fixed sentence-type positions, has a sequence of English phonemes in English articles and a different sequence of French phonemes in French articles. In homonymy, the same phoneme-sequence can have different meanings; here the phoneme-sequence occurs in two or more unrelated selections—environments of kinds which do not otherwise have the same phoneme-sequence selecting them coherently (3.2.1, e.g. *see*, *sea*, *Holy See*). In synonymy, to the extent that it exists, different phoneme-sequences have the same meaning and the same selection.

More generally, words which have certain kinds of selection have (or get) certain semantic properties. For example, words which occur under virtually all operators with little likelihood difference have the meaning of indefinites (e.g. the set *someone*, *something*), and, interestingly enough, do not have different meanings under different operators. Also, a small set of words or affixes which have this indefinite-like property but are approximately complementary in respect to some set of further operators usually constitutes a grammatical category or paradigm: e.g. the tenses in respect to adjoined time words (*Someone will act tomorrow*, *Someone acted yesterday*), or the plural and singular affixes in respect to adjoined number words (*Three things seemed important*, *One thing seemed important*). Words that come to be used with a large variety of arguments lose some of their individual meaning: e.g. *take*, *give*, *do*, *have* in *take a walk* or *a look*, *give a look* or *thought*, *do work* or *a jig*, *have looked* or *have a look*, etc. More closely, bi-sentential operators (words or phrases) which become frequent on a wide variety of sentence-pairs lose some of their specific meanings: c.g. *I will go provided* (or: *providing*) *he doesn't*, as against the specific meaning in *His setting up a trust fund provided for her going to college*; similarly for *is due to* and other conjunctional phrases (*My writing you is due to her*, or . . . *to her insistence*, as against *There is still some money which is due to her*). A sharper case of this loss of specific meaning is in the prepositions, where the largely locative original meanings (*in the box*, *on the table*) are almost lost in their much-widened range of environments: *interest in a problem*, *people in charge*, *thoughts on a subject*, *bets on a horse*.

We see then that meaning is specified not by a word but by a word with its choice of operator or argument (usually its immediate ones) (cf. n. 9 below). Indeed the vocabulary of a language is not large enough for the variety of experience, and this is partially remedied by the meaning-range of words in different environments (which include the secondary sentences that become modifiers of a word and so can specify submeanings of it, as in *steamship* from something like *ship using steam*, *medical examination* which is quite different from *doctoral examination*, etc.). In general, the continuous nature of much of experience does not always lend itself to discrete representation by words. Words do divide up the world of meanings, and much of the division is much the same in many languages, greatly facilitating translation. But there are given words in one language which cannot be translated, or not precisely, into another—if there are no words in the other language having similar coherent selection to the given words in the first one; there are cases of events and actions for which the other language has no word. We cannot make words mean what we want, because we cannot restrict them to the particular selection which would support that meaning. Nor can we reach words in any organized way by starting from their meanings, quite apart from the fact that there is no known way in which we can objectively and consensually organize the world of meanings. But for the most part, the argument-requirement of a word is preserved under translation: that is, O_n words (e.g. *walk*) usually translate into O_n words, O_o words (e.g. *occur*) into O_o, and so on. And for each word, what the learner or analyzer of a language does is not to think deeply about what the word 'really means' but to see how it is used, keeping in mind the fact that the various life-situations in which it is used are largely reflected in the various word combinations in which it occurs. The grammar cannot be built out of meanings; but it can be built out of word combinations.

11.2. *Information in sentences: Predication*

One can readily see that the meaning of a sentence is more than a selection of the meanings of its words: *John saw Bill* does not mean the same as *Bill saw John*, and the meanings which we can garner from collecting the words *John*, *Mary*, *Tom*, *Smith*, *call*, *tell*, *will*, *to*, *and* are not the explicit, and varied, information that we get from *Smith will tell John to call Mary and Tom* and *Mary will call John Smith and tell Tom to*. The sentence information over and above the meaning of its words is given by the meaning of the operator–argument relation. To see what this meaning is we note that the structural property is not merely co-occurrence, or even frequent co-occurrence, but rather dependence of a word on a set: an operator word does not appear in a sentence unless a word—one or another—of its argument set is there (or has been zeroed there). When that relation is satisfied in a word-sequence, the words constitute a sentence. This statement does not exclude other words or combinations, especially zero-level words, from constituting exceptional types of 'sentence', as perhaps in *Fire!*

The operator is thus not something one says by itself, but rather something one says with—about—the argument, a predication on it.[7] A word that requires the presence of one or another of certain other words means—in addition to its own lexical meaning—that it waits on those other words, i.e. that it is said about them. Hence each occurrence of the dependence relation contributes a predication meaning (between its participants) to the sentence. Thus the meaning or information of a sentence is not just a sum or juxtaposition of the meanings of its words. And since the predication relations in a sentence are a partial ordering of its words, we see that the meaning of each partial ordering of words (i.e. each sentence) is that same

---

[7] Predication is a relation between two meanings, and is the meaning of a relation between two word-occurrences. It is not the same as assertion (5.5). In *John's accepting is uncertain*, the *accept* is predicate on *John*, but not asserted, as also in *John will accept if he is invited*.

partial ordering (i.e. predication relation) of the meanings of the words. Thus not only the words but also their partial order in a sentence have a semantic interpretation.

More generally, all grammatical relations among words in a sentence contribute their particular meanings to the total meaning of the sentence, over and above the meanings of the words (which are identified by their phoneme-sequences). This holds for the various 'modifier' meanings of relative clauses, adjectives, adverbs, secondary clauses, all as modifiers of their hosts; and the meanings of pronouns as second occurrences of nearby-occurring arguments.

Similarly, -*ness*, which makes a noun out of an adjective, is not just a suffix meaning 'abstract'; it has that meaning because in changing a predicated property (e.g. *serious*) into an argument of further predication, the suffix is attached to the adjective alone, independent of its subject, so that it seems to name (nominalize) the property by itself: *John is serious, John's seriousness impresses people*, or (1) *Seriousness impresses people* (from (2) *Anyone's seriousness impresses people*). In the present theory, all relations other than predication can be derived from the predication relation (e.g. for English in G, *passim*). Their semantic contribution, however, is observable (or felt) independently of this derivation: (1) is seen as speaking of *seriousness* in the abstract, while in (2) it is speaking of the property of a specified subject, which is the indefinite and hence zeroable, yielding (1).

Significantly, every grammatical relation except predication can be paraphrased by word combinations that do not use the relation in question: e.g., for the modifier relation: *A new book arrived* paraphrased by *A book arrived; it was new*, or more precisely by *A book—the book was new—arrived*. And what is most translatable between languages is the partially ordered predication relations in a sentence: which words are predications on which other words: the ordered predications are a constant of the two paraphrastic (intertranslatable) sentences. In contrast, many meanings which are carried by other grammatical relations or constructions in one language are carried (and translated) in another language by some combination of words without that special construction. It is thus impossible to express the meaning of predication, in a given language or by translation into another, except by word combinations that are constructed with predication. This is because the dependence that creates the operator–argument relation does not just carry a meaning like all word meanings. Its meaning produces something new, which we may call language-information (11.7)—the difference between the meaning of words and the meaning of sentences. As will be seen in 11.6, we can say that a given word, say *tree*, has its meaning by virtue of its association with something in the world of experience and by its range of combination with other meaning-bearing words in sentences. Comparably in the case of predication, we can say that its characteristic dependence has a meaning over and above the meaning of its words by virtue of its association with specific conditions in the world of experience, so that *Trees fell* does not mean just any juxtaposition of these two meanings—e.g. that the speaker saw or thought of trees and that something then fell—but specifically that falling is a condition of the trees. This aboutness is what creates an item of information out of a collection of meanings. Information differs from meaning not only in that it reflects a situation, a fact, rather than an object or a property or event, but also in that sentences may be felt to 'give' information whereas words 'have' meaning: information characteristically records and communicates, where meaning indicates.

11.3. *How language carries information*

11.3.0. *Introduction*

We have seen that when the constraints on form, i.e. on word combination, are isolated we can distinguish between constraints that carry information and a (mostly reductional) residue that does not. The constraints on combination constitute a linguistic system by themselves, the base sublanguage (4.1), but the residue does not. Thus the information in language is the information in a structurally distinguished sublanguage, which consists of base forms of each sentence in the whole language.

11.3.1. *Where the information is located*

In the base the relation of form to information is much sharper than in the whole language, and here we can ask precisely what kinds of form carry what kinds of information. We can take a base sentence, ultimately an elementary sentence no proper part of which is a sentence, and adjoin to it various word-sequences which make it into a longer sentence, and then ask what information has been added by each additional part. (We disregard here the removal of parts of a sentence, by zeroing, which does not remove any of its information—up to ambiguity.) A more organized account of such accretion of information is given by the semilattice structure of the sentence. Here every word that enters into the making of the sentence has a fixed point in the partial order, at which its form and its meaning are contributed. Here, then, every leg of the partial order contributes its own grammatical (ultimately predicational) meaning at that location; and here it is found that each type of meaning has its characteristic location. Typically, these contributions include: zero-level nouns; first-level operators, as their properties and relations; the linear order of the first ('subject') and second ('object') arguments of a relation; the difference between things (nouns) and states or events (sentences); second-level operators, as properties and relations of sentences. The further contributions to information are noted in 11.3.2.

Translation, which seeks in a second language the same (paraphrastic) information as in given sentences of a first one, shows that some features of form are not universally specific to information. Chief among these are: the phonemes of words in one language or another; the meaning-ranges of particular words; even, up to some limit, the size of the vocabulary (the number of nouns in one language can be smaller than in the other, since existing nouns can be given modifiers (secondary predicates) to make up for a semantically lacking noun); morphology (the affixes of one language can be translated into free-word operators of another); linearization of the partial order.

In contrast, some features are preserved under translation and hence are features of information. Thus the operators in the original sentence are generally translated into operators in the other language. Specifically, for the most part, the various argument-requiring classes are preserved under translation, so that N translates into N, O_n into O_n, O_{no} into O_{no}, etc. And the semilattice—that is, the partial ordering of the words—in the translation is generally found to be very similar to that of the original sentence.[8]

11.3.2. Successive stages of information-bearing constraints

It should be clear first of all that every vocabulary or grammar choice made in constructing a message (sentence or discourse) is an application of a constraint defined as being available at the given point in the construction. Each chosen constraint makes a fixed contribution, at the point of its application, to the substantive information carried in the message. For a given type of choice, either the contribution is always null (e.g. trans-formations, 11.3.3), or it is accretively meaningful. Furthermore, the amount of choice at a given point may be graded: the more expected a given choice-decision is at that point, the less information is contributed by the expected choice. And if what has been chosen can be determined (reconstructed) from the environing choices, no free choice has been exercised, and no meaning contributed, at that point.

The meanings of all words and grammatical entities can be indicated via the constraints listed below, which may be grouped as follows:

(1) words in isolation (with their meanings), viewed as a non-syntactic constraint on phonemes;

(2–5) the constraints involved in environment-based specifications of word meaning;

[8] Cf. the English and Korean examples in Z. Harris, *Mathematical Structures of Language*, Wiley-Interscience, New York, 1968, pp. 110, 112.

(6–8)   the constraints that underlie the meanings of sentences over and
        above the meanings of their words;

(8–10) the constraints that yield the further meanings of sequences of
        sentences;

(11)   the constraints that yield the meanings of word subclasses and
        sentence types in sublanguages.

Loosely, the element of meaning common to all the constraints is that in
each sentence the entities between which an ordered constraint holds are
relevant to each other at that stage in the construction of that sentence. It
will be seen that each additional constraint adds informational capacity to
the language, and that this capacity is constant for all occurrences of the
constraint.

(1) As an initial approximation, we note that words have a range of
meaning when said in isolation or in irregular combinations (such as might
have existed before syntax became established, 12.3.1). These meanings in
isolation cannot be due to regular constraints on word combination. They
can however be described as the meaning associated with individual constraints on phoneme combination, although there is little regularity in this
constraint. Such a constraint would be defined on the resultants of the
phoneme-establishing constraints (7.1, 2).

(2) In limited cases, when a given word occurs in a particular subset of
environments, it has a different meaning than it has elsewhere. That the
difference is due at least in part to the environmental constraint can be seen
from the relation between the meaning-difference of the given word-
occurrences and the meaning of the environing words in these occurrences
(*The field was ploughed for the new planting*; *A field is defined as having two
operations*). The ability of words to have different meanings under different
operators is a special case of word meanings being characterized by their
selections, (4) below. Such ability would be inexplicable without (4), i.e. if
the meaning of words is assigned only as an inherent property of words in
isolation, as it is to a first approximation in (1). The extreme case is the
idiom, in which a word-sequence (construction) has a meaning not currently
related to the meanings of its component words.

(3) The ability of words to refer to words has been seen (ch. 5) to
create several interrelated informational capacities: the metalanguage, cross-
reference (for word-repetition), and the performative *I say* for sentences as
said. Each of these capacities involves the constraint of unusual environments: for the metalanguage, the zero-level positions of quoted ('mentioned')
words; for the performatives, the ability of *I say* to occur on every sentence;
and for cross-reference the fact that fixed-antecedent cross-reference is
limited to positions where repetition has high likelihood (a constraint).

(4) The fact that different words have different selections, i.e. have
different operators or arguments with which they occur more frequently

than average, is a constraint: that constraint is the existence of relatively stable frequency differences for word co-occurrences.

(5) Various regularities in the frequency yield the meanings of special kinds of likelihood, such as: the likelihood of a word's being repeated, which makes referential pronouns; of having roughly equal likelihood under all operators, which makes indefinite nouns (*something*); of a word's having the same operator likelihood as its modifying noun has, which makes fragment-nouns and set-nouns (*piece*, *crowd* G204).

(6) The dependence that creates the operator–argument structure has been seen in 11.2 to bear the meaning of predication.

(7) There is also a common meaning to each argument-requirement set of operators: O_n as 'intransitive' states or properties of something, O_{nn} as relations and grammatically 'transitive' acts, O_o as aspects of events (sentences), even O_{no} as largely purposive ('human') relations to situations or events, and so on. Furthermore, there is a common meaning-difference between being the subject and object of an operator, which is given by the linear ordering of the arguments under an operator. This last is a dependence not between word classes but between word-occurrences. These meanings are clearly related to the constraint that creates the class or structure.

(8) Certain operators on sentence-pairs impose a constraint favoring that some word or other in the second sentence under them be the same as some word in the first. This gives these operators a meaning of 'connective' (e.g. *but*, *because*, *if*), and makes the two sentences under the connective more relevant to each other (9.1).

(9) Discourse has been seen (9.3) to be characterized by a constraint, whereby operator–argument combinations (or the equivalent) of word subclasses special to the discourse repeat in its successive sentences. This repetition of sentence types expresses a relevance among the sentences, thus advancing from the recording of solitary observations to the focusing of various items of information around a topic in such a way as to constitute a discussion.

(10) There is a further constraint within parts of discourses, governing the relation of certain successive sentences, so as to create argumentation and, in the extreme case, proof. This constraint is formulated explicitly over the main discourses of mathematics and formal logic. In natural language as a whole it exists only as an inexplicit and often tenuous constraint, appearing episodically in various sections of various discourses, more particularly in science. There it may be a constraint in the making. It creates a relation between the information of the initial sentences of the sequence and its last sentence: the last is a 'conclusion' from the initial ones, i.e. not less true, or not less plausible, than they are. The fixing of syntactic constraints which ensure that the last sentence is no less true or plausible than the initial ones is an addition to the meaning-bearing constraints in language.

(11) In natural language as a whole, the constraint that creates syntax, i.e.
sentences, is the operator–argument partition of words or word-occurrences
into argument-requirement classes, while the likelihood constraint on word
combination in (4) above is simply a not-well-defined variation within this
partition. However, in science sublanguages word combination is so sharply
limited that the place of likelihoods is taken by definite constraints which
further partition word-occurrences into subclasses of operators and arguments. This seems a novelty, changing the non-grammatically-differentiating
likelihoods of (4) into grammatically-differentiating subclassifications. However, it is actually no more than an extension of the conventionalization of
use which had produced the original operator–argument partition; cf. (5)
above. In the whole language the use of the dependent words was distinct
and conventionally fixed enough to create distinct word classes, but likelihoods were not; in science sublanguages the likelihoods, i.e. the use of
specific words, are distinct and conventionally fixed enough to create
combinatorially and semantically distinct subclasses of words within the
whole-language classes. This is the major information-bearing constraint of
sublanguages, especially of the languages of sciences.

The result is a characteristic structure for the semilattices of sentences in
each science language, with a strong connection to the information in the
science, and with the various word subclasses occupying particular positions
within the semilattices—the meta-science material, if present, being at the
top of the lattice of each sentence as secondary sentence. In addition, a
science language may have new argument-requirement classes which do not
appear as such in the whole language, especially in respect to quantity
relations (e.g. *ratio*). In all, the word classes, sentence structures, and
discourse structures of science languages provide not only the specific
information of the science but also its characteristic overall information
structure, much as the structure of the whole language provides both the
information reported in the language and also—in the operator–argument
system and its secondary grammatical relations—the structure of language
information in general.

11.3.3. *Non-informational constraints*

Of the various constraints, on sounds and on morphemes, some have no
regular effect on which words combine with which, and hence no effect on
the further structure of language or on its information. Such are the specific
phonemes of a language, and the phonemic composition of each word,
morphophonemics (the change of phonemes when a word is in particular
operator–argument and reductional environments), assimilation and dissimilation (the change of sounds when a word is in particular linear phonetic
environments). Such also are the automatic morphemes, such as English -*s*,

-*ing*, which have been analyzed for the present theory as being merely indicators of operator status or of argument status, rather than being reduced words such as enter a sentence by virtue of their argument-requirement status (G37, 40).

Also non-informational are the reductions, as noted above (3.3). Since these are changes only in the phonemic shape of words, or in their position, we can say that the word remains in the sentence even if in altered or zero shape, and that its grammatical relations are unaffected, so that reduction when so analyzed makes no change in word combination or in information. Even if we consider the overt appearance of sentences, where reduction of a word to zero would be taken as absence of it, the loss of the word would not entail a loss of information because words that are reduced in this way are reconstructible from the environing words, and thus had contributed no new information to their sentence. The non-loss of information is also seen in the fact that the bulk of transformations, and especially reductions, are optional, so that both the original and the transformed shapes exist, often in the same discourse-environments, and the transformed are clearly paraphrastic to the original. In morphophonemic reductions, too, the paraphrastic character is clear: *I will* and *I'll, going to* and *gonna*, etc.

Nevertheless, reductions can indicate meanings, without contributing new word meanings. This happens because reductions indicate high likelihood of a combination and even, as in the case of the compound form or the adjective order, the relative stability of a combination (G231, 233). When operators which have high likelihood of undergoing tense variation get the tense affixed to them, and so become 'verbs' (*He aged*), while for the other operators the tense is given in a separate word, leaving them as 'adjectives' (*He was old*), the different localization of the tense creates a fixed distinction between less and more stative operators (G188).

Some reductions affect or indicate the speaker's or hearer's relation to the information in a sentence, without adding to the information or subtracting from it. The effect is primarily in the way the information is presented. One such effect is in the amount of information which comes through at a given point in the sentence. This results even from some morphophonemics: given the *am, is, are* forms of the operator *be*, the person who hears *Am I responsible?* knows already from the first word that the argument will be *I*. (But for the continued presence of *I*, one could have analyzed *am* as *is* plus *I*.) Another effect on presentation of information is the degeneracy of certain reductions, whereby different reductions in different sentences produce identical word-sequences, which are thereupon ambiguous (homonymous): *He drove the senator from Ohio to Washington* is produced by one reduction from *He drove to Washington the senator who was from Ohio*, and also by another from *His driving of the senator was from Ohio to Washington*. The ambiguity did not exist in the sources, before the reductions.

The alternative linearizations of the partially ordered words of a sentence can bring words of particular grammatical statuses in the sentence to its beginning, as its 'topic', without altering the grammatical relations: e.g. *This I deny* as against *I deny this*. There is also the interrupting of a sentence so as to insert a secondary sentence as a modifier (relative clause) within the first (3.4.4).

The capacity for paraphrase in a language is relevant here, because paraphrase, like translation, is intended to preserve information. We can therefore ask what features of form are preserved under paraphrase, thus correlating with information. Paraphrase is of several kinds. One is reduction, as in *I expect him* from *I expect him to come*, where the zeroed word contributed little or no new information. Another is the transformational recasting of a sentence, as in *John is the author* as against *The author is John*, or *Its wording was mine* as against *I worded it*, where the relation of subject to verb is seemingly reversed. A third paraphrase capacity is synonymy or addition of modifiers, as in *The plane almost crashed* and *The aircraft almost crashed*, or in *The igloo was hardly visible* and *The snow house was hardly visible*, where words or phrases with similar definitions and selections are interchanged.

Reductions can also contribute to certain episodic classification of information, which may not enter into the perception of the users of the language. One example is when a reduction applies to a particular domain, e.g. the *be here*, *come*, etc., which are zeroable under *expect*, or in English the fuzzy domain of the male humans and higher organisms which are pronounced by *he*, and the same (female) plus traditional other (pleasurable, etc.) objects such as ships which are pronounced by *she*, all in contrast to *it*. Such categorization is seen also in the case of information which is conveyed in specially reduced or patterned forms, such as that covered by affixes in morphological languages, or in the time-order and plural covered by such paradigms as conjugations and declensions, or in the fuzzy set of English auxiliary verbs (*can*, *may*), or the set of 'count' nouns in English which require *a* in the singular (*a tree*, *a man*, etc., as against *water*, or *man* in the sense of 'mankind', etc.). Such categorization is stronger when there is a paradigmatic requirement, i.e. when words of a given set (e.g. nouns) must occur with one or another morpheme of the category (e.g. gender, or the distinctions made in the Bantu noun classes), or when absence of a form means not just absence of its meaning but presence of an opposite meaning (e.g. when absence of the plural suffix means not unstated number but specifically singular, as in English). Although such structures are very prominent in use, not all languages have them.

There remain scattered additional similarities and structures which have some meaning-properties but no explicit information. Some of these are in grammar (which includes all regularities) but not in syntax (which specifies word-class combination); such are occasional onomatopoeia (e.g. *teeny*), or cases of phonemic and semantic similarities among words (e.g. *mother/*

father/brother/sister, or the *sl-* and *fl-* words). More important, but still not
explicitly informational, are such literary devices as poetic meter, rhyme,
and alliteration, all of which relate one word to a particular other in the text
(whatever else they do); or allusion to absent words by phonemic similarity
to them in words that are present in the sentence (in literature, jokes, swear-
words), and various kinds of literary and popular word-play.

11.3.4. *The meanings of word relations*

In sum: if we now survey the meaning-effects listed in 11.3.2–3 we see
that combinatorial constraints (11.3.2) affect the meaning of the combinations, chiefly by suggesting relevance between their parts, while the other
relations, largely reductions, rearrangements, and phonetic similarities,
which do not change the words' co-occurrences, do not affect the substantive
information, but only indicate the speaker's attitude to it (e.g. by the topic
linearization, 3.4.2) or affect the hearer's receipt of the information. In
particular: in (2) of 11.3.2, the occurrence of a word under a particular
operator selects a relevant meaning for that word.[9] In (3), the ability of
words to refer to words comes from extending the selection property of (2);
and it creates new metalinguistic meanings appropriate to the new kinds of
co-occurrence (as in *Mary is a word. The word which is Mary has four
letters*). In (4), the above-average frequency of each operator is constrained
to particular argument words; and, except for nonce forms, whenever a
word is used with some other operator or argument its meaning is adjusted
to be relevant to that new environment. In (5), similarities and patternings
among the co-occurrence likelihoods of various words correlate with similarities and patternings in their types of meaning. In (6), the partition in respect
to co-occurrence that creates the operator–argument relation among the
words of a sentence yields the meaning of predication which connects the
meaning of the operator to that of the argument. In (7), the similarities and
differences in respect to argument types make similarities and differences in
the predicational scope. In (8), the likelihood of word-repetition in the two
arguments of an O_{oo} operator brings out the expectation that the two
argument-sentences be relevant to each other in meaning. In (9), the
likelihood of word-class operator–argument repetition in connected discourse creates the relevance of the sentences to each other. In (10), the
conventionalized requirements of sequence and repetition in the sentences
of an argumentation creates the meaning of 'leading to a consequence'.

---

[9] One might say that the environing words select this relevant meaning from among the
meanings that the word has. However, the only evidence that the word has the given meaning is
that it has it in the given environment or in reference to such an environment. Hence we cannot
say that the word in and of itself has a stock of meanings from which a given environment
chooses, but only that the word is accorded a given meaning in a given word-environment.

In all these, the meaning of syntactic ('institutionalized', 12.3.5) word co-occurrences is that the occurrence of each word has relevance to the occurrence of the others.

In contrast, the word relations of 11.3.3 do not add or subtract word co-occurrences, as is shown by the reconstructability of the base (unreduced) form; and they do not yield new information. Some phonemic changes and reductions in words change the point in the sentence at which the hearer gets a particular bit of information, or create degeneracies for him when two different sentences receive the same form. Some permutations call attention to particular words in the sentence (thus creating a secondary kind of relevance). Many transformations give a sentence alternative shapes which may yield stylistic benefits at a cost of complicating the decoding of the information and the correlation of form with content. In some cases transformations that are based on broad likelihoods of word-occurrences create 'grammatical' categories such as quantity, time, gender, or the verb–adjective distinction (again creating kinds of relevance). But these are just collectings of words with phonetic or other similarities into sets having slightly semantic properties. Finally, some relations hold between a word in the sentence and words which may not be in the same sentence or may not be present in the same discourse. This includes partial phonetic similarity among semantically related words (e.g. *father*, *mother*, etc.), or various kinds of allusion. Relations within the text include alliteration, and meter and rhyme. In all of these the sounds or the position of one word call attention to some other word or some other meaning, and so create new semantic relations and relevances among words.[10] All these meaning-effects, and the forms that engender them, are distinct from the informational ones seen in 11.3.2.

11.4. *Form and content*

We can now attempt a summary of the relation between form and content in language. As in so many issues which are clouded by being conducted in catch-phrases, the resolution here is not one or the other of form and content or even both form and content together, but a particular relation between certain forms (produced by certain constraints) and certain content. Specifically, certain types of content-bearing elements enter into certain types of meaning-affecting constraints on their combinings, and vice versa. At this level of generality, the situation is not too different from the relation between form and expression in music and other arts.

The relation between form and content cannot be a correlation in the ordinary sense of the word—a covarying of members in two independently

[10] Differently, contrastive or emphatic stress does not relate to any particular other word. It functions as an additional word in the sentence, a synonym of *emphatically* or of *rather than anything else*, or the like.

known sets—because content as we know it from language, i.e. as a system of information, is not known without reference to the form of language. Some meaning and information—even some that are expressed in speech— exist without grammatical language. But there are certain kinds of meaning and information that exist only in words and sentences. It has been seen above that much of this meaning is allied to particular features of form. This alliance is seen even in the case of the different kinds of word meanings: words having the same argument-requirements generally have some aspects of meaning in commmon. Zero-level words (N) largely refer to concrete objects; operators with one-place arguments (O_n) are intransitive; and so on. Nevertheless, there are major cases of different forms for the same meaning: synonyms (and definitional phrases replacing a word), trans-formational paraphrase, morphophonemics, and for that matter the different phonemic forms of words which have approximately the same meaning in different languages. And there are major cases of the same form for different meanings: homonyms, transformationally ambiguous sentences, different meanings of a word under different operators, and the idiomatic versus literal meanings of idioms.

More generally, the universes of language and of meaning are not quite identical. On the one hand it is not the case that, in any particular language, everything can be said, or said within reasonable limits of directness (10.3.4). Not only are there extra-grammatical meanings in speaking, such as those of gesture, glances, and intonations of emotion, but also there are various language-type meanings which are expressed in no language, or not in one language or another. For example, if something would involve complex reference to various parts of a sentence, requiring the addressability of arbitrary phonemes and morphemes in it, it cannot be said; for even though such complete addressability is definable within the reference capabilities (sameness-statements and relative location) of language, it is not provided by the pronoun systems of actual languages.

Individual languages may also have particular types of meaning that can be said only by circumlocution, either for lack of vocabulary or because of grammatical restrictions. There are also form–content mismatches, as in *He caught the glass before it hit the floor* (or *He died before he got the chance to see his grandchild*) as against the less common *He caught the glass before its hitting the floor*. The former seems to include the false assertion that the glass hit the floor or that the man got the chance to see his grandchild. We may explain away such a form by appropriate semantic interpretation of tense or of assertion; but such adjustment may be *ad hoc*, due precisely to such 'wrong' forms.

An indirect type of form–content correspondence is the following: by the present theory, grammar has to describe not merely the not-well-defined set of actually said sentences, i.e. the popularly acceptable sentences, but the set of all possible sentences within a stated structure. This condition on grammar is not just an artifice for convenience of formulation. It has linguistic support, and it has a specific semantic effect: it accounts for a certain part—the imaginative and nonsensical part—of the content of language. For recognizing the ability of any zero-level word (N) to occur as argument of any first-level operator, and the like, makes it possible to create word combinations that are beyond any experienced meaning. This is found not only in some schizophrenic talk, but also in imagination, in fairy-tales and science fiction, in literary effects, nonsense verse, and word-play—and in various irresponsible discussions.[11] Indeed, one can make a sentence without quite knowing what its meaning is.

Despite all these reservations, the great bulk of information that is given in language is associated with specifically relevant features of vocabulary and syntactic form. Even something which is considered a semantic datum about sentences, namely how a sentence is characterized by its consequences, can be largely stated in syntactic and discourse-structure terms. For this last, we recall that the capacity of a discourse to lead to a logically acceptable conclusion from given initial sentences is obtained by particular proof-like constraints on sentence-sequences in the discourse (10.5.4). We recall also that the meaning of a word can be syntactically approximated by the set of its more frequent than average environments. We can then say that the meaning of a sentence is syntactically approximated not only internally by its construction, but also externally by the set of its successor sentences under the proof-like constraint, i.e. by the set of its consequences.[12]

The association of form with meaning can be studied in detail if we start with sentences of any size (initially, zero) and ask what partially ordered accretions of words would enable them to carry specified accretions of information.

In sum: when the methods of analysis in Chapters 7 and 8 are applied to arbitrary samples of natural language we obtain, for reasons discussed in 11.5, a structure that conforms to our general experience of the world (11.6). When the same methods are applied to samples of a relatively closed subject matter, we obtain a structure that reflects the objects, relations, and events of that subject matter (ch. 10). When they are applied to a single discourse by itself, we obtain a structure which in many cases shows what the discourse had set out to do (9.3). In each case, the structure reveals something about the burden of its universe of data.

---

[11] In general, situations which call not for information but for some other kind of expression may occasion sentences whose overt information is not what is really being communicated. For example, if people have encountered each other several times in one day, they may say 'It's a small world' or simply, with a smile, 'We meet again', where what is being communicated is not the facts that are said, but the recognition of the unlikelihood of so many encounters—the content of the smile.

[12] H. Hiż, 'On the Rules of Consequence for a Natural Language', *The Monist*, 57 (1973), 312–27; and 'Information Semantics and Antinomies', in G. Dorn and P. Weingartner, eds., *Foundations of Logic and Linguistics*, Plenum, New York, 1985.

More particularly: the numbered form–meaning features of sentences in
11.3.2, 4 are each a constraint which acts on the resultants of prior constraints,
and which brings a further informational capacity not available in its
absence. Since such features and those noted in 11.3.3 cover the structural
devices of the language, and since each one adds its own kind of informational capacity to the language, we see that indeed the structure of the
language is made out of an accretion of specific constraints which produce a
stepwise accretion of informational capacity and properties.

This accretional analysis of structure with its informational capacity
underlies the result noted above that the meaning of the sentence as an
ordered set of constraints (including the word-choice as a constraint) is that
same ordering of the meaning of the constraints (including the chosen
words) in the sentence (9.1).[13] Such a statement is the more understandable
if we see that information is not just a correlation or interpretation of the
constraints, but is an intrinsic effect of them. This view of the step by step
correspondence between form and content in language (where both form
and content are more explicitly analyzable than, say, in many arts) has
various applications. For one, it is possible to make critiques of the language
of a science to see if its vocabulary is more meaningful or teleological than
the specific entities and relations or events of the science require. For
another, it is possible to see why in the case of an art, the introduction
of a fruitful new 'language' or apparatus may be not less important than
the production of new works. In a time-slice, the step by step correspondence
shows form effecting ('informing') content, but developmentally we have
form arising not only from its own dynamics but also from the conventionalization of its largely content-oriented use (12.3.5).

The situation of form effecting content and of content as a factor in
favoring the development of form has some similarities—for all the essential
differences—with the relation of morphology to physiological function and
to behavior in biological systems and their evolution.[14]

11.5. *A theory of language or a theory of information?*

Given all this, we can consider the question: is the theory presented here a
theory of language, or of information? It is in any case a theory of language
structure, since it attempts a reasonably complete survey of the constraints
in language, some of which are essential to it (and must be defined as
independently of each other as possible) because of the lack of any external
metalanguage in which the structure of language could be stated. And it is a

[13] Even idioms are included here, their constraint being a special case of a word having
different meanings under different operators.

[14] It also has an echo in the use of the term 'semantics' as 'what a system can do' in analyzing
computer programming languages.

theory in which meaning was not used in making the analysis. It also produces a structure of language-borne information because a distinguished subset of these constraints—ultimately the operator–argument dependence and the word likelihood within it—are characteristically associated with an informational interpretation; they each have the effect of information. These constraints thus create the essential nucleus of language, and also create its information. The relevant question is whether in choosing the methods that yielded this theory, other methods might have been available that would not have yielded this close informational interpretation. But the search for a least grammar was justified in 2.2 without any consideration of meaning. That the constraints on equiprobability which yield the least grammar carry informational interpretation is due to the nature of information: both word combination and information are characterizable by departures from randomness. Features which do not give new information, such as the linearity of language, or the time and quantity categories that are created by paradigms, are distinguishable structurally from the informational ones.

There follows from this, what is in any case borne out by experience, that we cannot in general impose our own categories of information upon language. In a limited way we can arrive at certain categories of information independently of language, but we cannot locate these types of information in any regular way in a language unless the language has a structural feature whose interpretation this information is. We cannot determine in an a priori way the 'logical form' of all sentences. We cannot determine their truth except to the degree that we can locate them in a proof-like constraint on sentence-sequences, in respect to known prior sentences or later consequences. We certainly cannot map them in any regular and non-subjective way into any informational framework independently and arbitrarily chosen by us.

The other side of the coin is that we cannot disregard any information that is carried by a structural feature of a particular language. Various languages express time-order and quantity with various distinctions and relations. We need not go only by the most overt expression in the vocabulary but can also seek second-order evidence, in more subtle restrictions on the environments of this vocabulary, to see what time relations are expressed in the language. But we are bound to the regularities of word combination in the given language. For example, some languages have overt forms (suffixes and the like) for the 'aspect' of verbs (mostly features of duration), while others mark this only by regular differences in word combination, especially in respect to time words, as against yet other languages which have no regular indication of aspect (G274). An informational category of 'aspect' either must be or must not be recognized for the language, accordingly. And the non-informational constraints (11.3.3) too remain important for each language.

If in spite of the informational dependence upon the demonstrable structure of the particular language, excellent approximate translation is so freely possible among all languages, it is because the basic partial-ordering relation is identical in all languages, and the meaning (equivalently, likelihood) ranges of words, or words with suitable modifiers, are sufficiently close to permit the translation.

To sum up: both language and information are necessarily structured by departures from equiprobability acting upon departures from equiprobability (which is why languages are informationally intertranslatable), although many kinds of information are not expressible in language, and some items of language structure do not express information. In addition, language universally develops informational facilities that do not arise directly from dealing with the real world. One such is the metalanguage; another is the capacity for saying nonsense. The similarities and differences between form and content surveyed in 11.4 show in what ways the theory of language structure can be or cannot be a theory of information (or of the language-borne subset of information).

11.6. *Whence the power to 'mean'?*

11.6.0. *Introduction*

What gives language the power to mean something? We have seen that the meaning of a sentence is not a primitive unanalyzable entity, but rather the product of accretions from the fixed meanings of its words and the constraints on their combination. We therefore have to ask what gives the words and the constraints the property of having these accretive meanings. The usual view is that words are associated (as symbols) with some selectively perceived aspects of experience (as referents); a less selective representation, such as a photograph, would not be considered a symbol of what it represents. As to the meanings of grammatical relations, they have been little studied and are presumed to be a new primitive element created by language.

In what follows we will see: first, that the dependence relation of words reflects what one might consider dependences within man's perceivable world (11.6.1); and second, that word meanings and co-occurrence selections express a categorizing of perceptions of the world (11.6.2).

11.6.1. *Co-occurrence and the perceived world*

Rather than begin the discussion from word meanings in language as a whole, we start from the analysis of word combinations in science languages. Here, especially in sub-sciences, we see something new: words classified together by their having similar word-choice environments, especially in their immediate operators or arguments, sometimes despite differences in their meanings in the science as a whole (10.3–4). When we consider these word classes, we see that they reflect the actual objects and relations of the science. If antigen names, antibody names, and cell names constitute different classes in immunology language, so are antigens, antibodies, and cells different classes of objects in the immunology laboratory. If there are sentence types in which antibody names are co-arguments of cell names under a specific class of operators (*contain*, *produce*, *secrete*), so are there in the laboratory such specific relations and events which hold between antibodies and cells (*Cells contain antibodies*). To a large extent, a given sublanguage class of zero-level arguments (N) has as referent a class of objects which can indeed be put into one class in that they have much the same status in respect to the activities and observations in the immunology laboratory; and the sublanguage operators on, say, two N classes indeed refer to the relations between the objects named in one class and the objects named in the other class, in the real world of the laboratory. Thus each class has a class meaning (e.g. 'cell', of whatever kind, or 'contain') which is characterized by its total set of sublanguage operators or arguments, and also by its laboratory referents. As to each word within a class, it has a meaning identified by all the laboratory experience with the relevant object or event—all of which can be expressed by the various operators and arguments co-occurring with that word in the science language.

What we see here is not only a relation between the meaning of a word and its co-occurrences with other words but also something of a correspondence between the co-occurrences of words and the perceived co-occurrences of objects and events about which those words speak. This correspondence is sufficiently consistent, so that the types of word co-occurrence can be considered to correspond to types of co-occurrence among objects, properties, relations and events in the world described. In the real world, properties or events 'co-occur' with the objects which participate in them.[15] This is reflected in the basic co-occurrence types of language: the dependence of an operator on its arguments.[16] The co-occurrence or connection (in an appropriate 'action space', not necessarily in time or locale) of the objects

[15] When we recognize (above) the meaning common to a class of words as being related to the meanings of the word-environment which is common to those words, we can also recognize that the different meanings that a word has in different environments are related to the different meanings of those environments. Thus (a) word meanings that depend on word-environment are not different in kind from (b) word meanings that depend on association with the real world (11.1 and n. 9 above); in case (a) the co-occurrence of a word and its environment reflects the real-world co-occurrence of the here-selected referent of the word with the referent of its environing words.

[16] Given that *cat* is a zero-level word and *smile* a first-level word (operator), one can see a Cheshire cat with no smile or other distinguishable expression, but whether one can see a smile without a Cheshire cat or some other bearer is questionable, Lewis Carroll's claim to the contrary notwithstanding. Similarly, when an abstract noun is taken as a nominalization of sentences having indefinite arguments (11.2, 11.6.2), it presupposes the existence of such sentences, even if their arguments are unknown or fictitious.

participating in a relation is reflected in the linear ordering of the arguments (co-arguments) under an operator. The connections between properties, relations, and events between which causality or any other relation holds in the real world are reflected in the second-level operators whose arguments are operators, hence sentences.

The correspondence can even be seen in more global relations. For example, just as a prior science must exist before one can base oneself on it in another science, so the sentences of a prior science appear as already-present zero-level arguments under the operators of the other science, i.e. as descriptively earlier in the partial order of the sentences in the other science (10.3). And the grammatical fact that the meta-science material, which describes the scientists' actions and thoughts, appears as the top operator over the separately existing science sentences conforms to the real-world fact that the scientist is an external factor in respect to the separately existing—again generally prior—events of the science.

The correspondence, specifically in respect to co-occurrence, between language and the world it describes is not surprising, since each of these can be described within the same universe—that of departures from random-ness.[17] Each co-occurrence, or joint involvement of entities in a relation, can be viewed as a departure from randomness. On the one hand, each contribution to the meaning-bearing structure of language has been seen to consist in a constraint on combination of phonemes and of words. On the other hand, each object, property, relation, and event in the world can be considered an unstated constraint or departure from randomness in the physical world. It is not surprising that some of the constraints which constitute the perceived world should be reflected in constraints on the word-occurrences with which we speak about it. It is this correspondence that makes information about the world carryable by a sentence, and in a manner that is objective rather than subjective and private. The whole correspondence is reminiscent of the relation between art and experience, that colors our response to works of art.

The explanation above of the language-world correspondence in science languages shows that the correspondence is based on properties of language in general rather than on any special properties of science sublanguages. And indeed, more loosely and less demonstrably, the correspondence holds for the basic types of co-occurrence in language in general. The fact that these basic types of constraint are rooted in properties of the physical world, which are experienced in common by all humans, explains why the partial ordering (and to a lesser extent some of its secondary constraints) is apparently universal for language, and while different from one language culture to another is nevertheless far more similar among languages than the choice of what phoneme-sequences make words and carry particular meanings.

[17] The necessary caveat is given in 11.6.2.

Most generally: the mathematical object created by the dependence-on-dependence relation can be considered to reflect—to 'mean'—the co-occurrence relations of objects, situations, properties, in the perceived world. Beyond this, we find various types of co-occurrences of word-dependence: repetition of a word in sentences under certain conjunctions, repetition of grammatical relations between particular word subclasses in discourse and sublanguages, constraints on sentence-sequence in proof. These can be looked upon as reflecting subtle dependences among the co-occurrences in the real world, or else as building something new—'information'—about the real world. Information is thus not an entity, but a relation of symbol to the world: the recording and communicating of the occurrence of perceived objects and events. And the structure of language, and of information, reflects not the structure of the world but only this much about it. It should be noted that if we want to build an alternative or a further representation of the real world, which will 'mean' something further about the world, we cannot do it by simply defining new word classes, but must define new grammatical relations such as third-level operators (that depend on operators that depend on operators) to house more complex concepts such as *ratio* or *co-vary*.

What we have here is not merely that meanings can be expressed by grammar, i.e. by arrangements of words as well as by words themselves, but rather that when we get down to the elementary constraints we see that each constraint has a fixed meaning (8.2.2) as each lexical element has a fixed meaning, although a different kind of meaning: the words reflect (or image) objects and situations in the world, while the constraints reflect the co-occurrence of these in the world.

Finally, language and information, because of their limited relation to the world, make possible certain structures about the world that could not be made directly with the vastly more complex and never fully known actual events of the world. Such informational structures are: generalizations, imaginative extrapolation, questioning, discourse, and controlled argumentation. Somewhat similarly, mathematics and logic, because of their yet more limited relation to the real world, make possible the controlled structure of proof, which is true about the world but could hardly be obtained directly from observable relations of events.

11.6.2. *Language and the perceived world*

Given what has been seen in 11.6.1 of the correspondence of language constraints with constraints in the perceived world, we now ask what is the case for the whole vocabulary and structure of language. As has been noted, words are associated first of all with objects, properties, and events of the perceived world. We may call these the referents of the words, but this does not have to be restricted to their extension in the logical sense. *Unicorn* is associated with certain more or less specific perceptions of a horned horse, and co-occurs with a particular selection of operators, even if its extension in the real world is null. Most language-expressed meanings which do not consist in directly perceived aspects of the world are found to be grammatical derivations of words with more concrete meanings. Simple cases of this are the pronouns, and metaphoric meanings. More complex cases are: the meaning common to a class of words that have similar word-environments (11.6.1 and n. 15 above); the fact that most 'abstract' nouns are nominalizations of operators whose arguments have been zeroed as being indefinites (i.e. a disjunction over all members of the argument class, G225)—thus *coercion* is derived from *anyone's coercing someone*; and even the meaning 'not-less-true-than' which is produced by the sentence-sequence constraints of proof structure.

However, even for the many words which seem directly associated with objects and events in the world, the association is a complicated relation. For a word to mean something in the world involves isolating and choosing a particular (not always delimitable) part of the world or a (not always specifiable) property in the world. It further involves the *de facto* classifying of various such parts or properties as are referred to by the same word, since words do not have unique referents (9.1). Such distinctions and classifications may not have counterparts in the real world, and may be fuzzy within one language and may differ as among languages.[18] They arise because languages—their speakers—have no other way of usefully associating symbols (the words and their combinings) with the perceived world. To this we must add the compositional properties of words: that they and their components, the phonemes, are discrete, distinguishably repeatable, finite in number at any one time and finite in length, and pre-set within a language community. All these create a stable, public, and transmissible structure for the words. In this way Procrustean symbols are formed onto which aspects of the world are projected, although the word-symbols can then be extended into less symbolizable (e.g. abstract, relational, or private) uses.

The meaning of a word is known not only as an association with aspects of an experience, but simultaneously as a particular selection of operators on it or arguments under it with which it occurs in higher than average frequency. If a word has several distinct meanings, it has them in corresponding distinct ('coherent') subsets of selection. To use a word in a different meaning is to use it in a different selection, so that to establish a new meaning for a word is not merely to announce the new meaning but to get a new selectional subset for the word. In accord with an observation of Henry Hoenigswald, a new definition for a word is in itself a new operator–argument for it, and it entails

[18] There are nevertheless many similarities in the meaning ranges of words, as among unrelated languages. Many unrelated cultures have quite similar species distinctions in their words for flora and fauna.

the occurrence of the word in other new environments; which are coherent with those in the new definition. It follows that not only word meanings, but also word likelihoods in various combinations, are associated with aspects of the perceived world, except for those word combinations that are fostered by social and institutional factors (fashion, fiat, etc.) or by intra-linguistic factors such as analogy; or communicational efficiency. The relation to people's experience of the world holds also for a word's extra-high likelihood in particular combinations, that permits reduction of the word.

That meanings of a word are for the most part determinable both by association with the real world and also by the 'selection' of the word becomes explicable if we say that it is the sentence, or word combination, that has meaning by association with experience; and the parts of the sentence and also their arrangement have meaning by association with the perceived parts of that experience and with the way these parts co-occur in it.

On the basic and universal language structure thus formed, with its component meanings thus rooted in the perceived world, there appear various further structures, not all of them universal and not all related in a direct way to properties of the world's phenomena.

Some of these are local developments based on non-informational regularities such as phonemic similarities among words: e.g. morphophonemics, or the differing declensions and conjugations in some languages. Here we must include the use of a single word for various meanings, differently in different languages. Others are conveniences in presenting information. This includes reductions and other grammatical transformations, which clearly do not reflect anything in the referents of the words and sentences. It also includes grammatical developments that bring distinguished kinds of information into common form: e.g. the cross-reference pronouns (5.3.5), or the paradigmatic arrangements of tense, person, and number. These last categories, for example, are certainly involved in the physical world, but not in the same relation to objects and events in the world as they have to nouns and verbs in the paradigms of many languages. A slightly different case is that of transformations that are based on word likelihoods but that create new grammatical distinctions and partitions out of graded differences in experience-reflecting likelihoods: e.g. the partitioning of operators into verbs and adjectives (as in English) by establishing two distances for the tense on an operator, possibly on the basis of how frequent is tense-change within a discourse for the different operators (5.5.3, G265).

Still other further structures consist of new constraints on co-occurrences, expressing new information, but not based on any immediately recognizable constraints in the physical world. These—all of them constraints on sentence-sequence—are the word-repetition around conjunctions (9.1), the operator–argument class-repetition discourses (9.3), and the proof-structure sequences in mathematics and some science writing (10.5.4). One might say that these constraints do not reflect or record perceptions of the world, but rather make possible our useful manipulation of that information. But one might also wish to argue that this usefulness in language constraints reflects corresponding regularities among events in the real world.

Finally, there are further structures and capacities which arise not from representing the world or from manipulating that representation, but from the developmental possibilities of the representational structure. Such is the constraining of the dependence relation into a dependence-on-dependence and into a mathematical object (3.1.2). This dependence-on-dependence has the further effect of divorcing the predicational constraint (the partial order of 3.1) from the likelihood constraint (the selection of 3.2), with the result that the real-world predicational relation can act on non-real-world (non-selectional) pairings of words. This admits as-though-experiential meanings: imaginative, jocular, nonsensical, and other novel statements, 'sensibly' grammatical yet freed from the need to reflect experience. Another systemic factor going beyond real-world experience is the development of metalinguistic capacities, including the metalanguage and the performative *I say*. And the development of metalinguistic and systemic concepts such as the 'relevant to each other' above.

Given all this, it is clear that the correspondence of language with the world—in the co-occurrence of 11.6.1 differently from in the association of 11.6.2—is a frame for thinking about meaning rather than an explicit relation between language and the world. For one thing, the departures from randomness in the case of language are stated constraints acting in a stated hierarchy; nothing so specific can be said for the world referred to. For another, many words refer to particular selected aspects or classifications of objects and their relations, rather than to referents which would necessarily be distinguished or selected in the absence of language—this even though many languages make much the same referent-choices in the world, as can be seen in the structural simplicity of much of their intertranslatability. Finally, an untold amount of the world is not referred to in any language— not only because it is not public, or not discernible at the given time, or not of interest to record and communicate, but also because it would be more than human beings—finite and of a particular size—can note or respond to. As a result, we can not relate the structure of the referred-to world to the structure of the referring language; this was already noted in 2.4, where the comparison was with meaning rather than with the world referred to. Nevertheless, we know that whatever structure the world has is a matter of departures from randomness. It is only in this vague sense that we can suggest that when an occurrence of language refers to a situation in the world, the departures from equiprobability which constitute that occurrence might be said to have as their referent something of whatever departures from randomness constituted the situation in the world. The correspondence mentioned here differs therefore from any correspondence theory within philosophic realism. Indeed, rather than a general correspondence with the world, what we have in language is, first, a correspondence between the proper property-object co-occurrence in perceived nature and the operator–argument non-zero probability in language; and, second, a correspondence between the relative likelihoods of property-object associations and operator–argument co-occurrences. Thus rather than reflecting the structure of the world, the co-occurrence structure that creates language creates the concept of 'information' about the world.

We have seen that in the absence of any independent knowledge of what information is, the information of language can be considered to be the meanings of words and of the constraints on their co-occurrence. We have also seen that precisely these parts of language are related to aspects of the perceived world, as symbols (in part) in the case of words, and as correspondence (in part) in the case of the constraints. In addition to this, there is information which cannot be expressed directly in language or in its grammar; and there are structures in language which add no information. More importantly, the known structure of the world is far beyond what is directly expressed in the structure of language. Also, many structures in language are only indirect expressions of aspects of the world: for example, the modern French perfect has taken over meanings and selection of the past tense while retaining the structure of a present tense, while the English perfect has done so only in special environments.

Hence if one says that form follows upon content in language and also content follows upon form (11.4), this is by virtue of two meanings of 'content'. In 'form follows upon content', the word-choices and co-occurrence constraints are based, in two different ways, upon the objects, properties, and events of the perceived world and upon their co-occurrences. In 'content follows upon form', the information or meaning of a sentence is given by the grammatical relations of its word-choices. The statable information about the world is thus mediated through the isolating and classifying properties of words as symbols, and the devices of co-occurrence constraints in syntax.

The commonplace that while mathematics is a discovery (of something about the world), language is an invention or convention of man is thus not quite so. Language is indeed a convention, but one that associates sound-combinations with perceived aspects—objects and their relations—of the world, and is therefore constrained by what constitutes successful perceptions in dealing with the world—thus a discovery.

11.7. *The structure of information*

We begin with information as an undefined term, with no known structure. In language, we distinguish certain accretional steps in constructing a

sentence and discourse, ultimately the partial ordering of words and constraints thereon, with words themselves characterizable by their status and
likelihoods in the partial ordering. These steps are found to yield in a fixed
way a distinguished part—the lion's share—of the information carried in
the given sentence and discourse. Thus the information itself is structured
by such accretional steps. Indeed it has been seen that given a system
or message with certain constraints and certain information, the defining
of additional constraints in it makes possible corresponding additional
information (11.3.2, 12.3.3).

First, the choice of a particular word at a given point in constructing a
sentence is a departure from equiprobability of phonemes and words there,
and yields a meaning.

Then, the relation of departures from equiprobability to informational
contribution is seen throughout the grammar. For example, words lessen
their information as they broaden their selection (i.e. as choosing them
constitutes less of a constraint on their argument or operator, 11.1), e.g. in
the process of *provided*, *providing* becoming conjunctions (10.3, and n. 2
there). Words with broadest selection add least information, as in the case of
the indefinites (3.2). And words which are reconstructible from the environment, i.e. add no information, are zeroable (3.3.2). In addition to word-
choice, each constraint that creates the partial order of words in the sentence
is also a departure from randomness in this language universe, and yields a
meaning. The information in a sentence or a discourse is thus formed by
departures from equiprobability acting upon departures from equiprobability.
The information is not simply an unordered collection of departures from
equiprobability, but a specific structure consisting of one step acting on
another.[19]

In the mathematical theory of communication (Information Theory), the
total departures from equiprobability (the redundancy) of a message or a
channel is discussed as its informational capacity, indicating the amount of
information it can carry.[20] The analysis of amount does not identify or

[19] The analysis of language information as successive steps of constraint makes possible
comparison with the production of meaning-controlling sequences of steps for computability, in
Post Productions (E. L. Post, 'Finite Combinatory Processes: Formulation I', *Journal of
Symbolic Logic* 1 (1936) 103–5), and in the Turing Machine (A. M. Turing, 'On Computable
Numbers, with an Application to the Entscheidungsproblem', *Proceedings of the London
Mathematical Society²* 42 (1936–7) 230–65, 43 (1937) 544–6; R. Herken, ed., *The Universal
Turing Machine: A Half-Century Survey*, Oxford Univ. Press, Oxford, 1988). However,
computability cannot be fruitfully compared with externally imposed arbitrary semantic representations of language, but only with the structural steps that produce the meanings of a
sentence. As was seen in the comparison of natural language and mathematics (10.6), the
essential difference between the two is in the variety of likelihoods within a set of symbols in
respect to the next step in the constructions.

[20] C. E. Shannon and W. Weaver, *The Mathematical Theory of Communication*, Univ. of
Illinois Press, Urbana, 1949; A. I. Khinchin, *Mathematical Foundations of Information Theory*,
Dover, New York, 1957; A. M. Fano, *Transmission of Information*, MIT/Wiley, New York,

distinguish the contributions to the total information, or the structure of their combination if any. It is clear that a random set can carry or express no information except via an external code-book. A set organized by one repeated system of departures from randomness, e.g. a pure crystal, can carry only very limited information. This can also be seen in precisely repeated simple decoration, e.g. a geometrical frieze, as distinguished from art (where the issue is paucity of expression rather than paucity of information). When we decompose the form, and with it the meaning, of sentences and discourses into the accretional departures from equiprobability, we are exhibiting that same structure for the information therein. In this way we are able to speak of the structure of information, at least of the kind of information expressed by language. We see that information is not produced, except in a trivial way, by a single departure from equiprobability or an unstructured aggregate of such departures. Rather, in a system which embodies or produces, and not merely carries, information, the total amount of information is obtained out of a specifiable structure of certain departures from equiprobability acting upon other departures from equiprobability.

1969; A. Kolmogorov, 'Logical Basis for Information Theory and Probability Theory', *IEEE Transactions on Information Theory*, IT-14 (1968) 662–4; C. J. Chaitin, *Algorithmic Information Theory*, Cambridge Univ. Press, Cambridge, 1987.

12

The Nature and Development
of Language

12.0. *Introduction*

The syntactic theory presented here enables us to arrive at some under-standing of the nature of language. To be able to draw conclusions from the structural properties of language, it is necessary to have an organized and reasonably complete view of the whole structure of a language; otherwise the discussion can be supported only by episodic examples. It will be seen that certain properties are apparently common to the known languages, others belong to speech as distinguished from language structure, and still others are limited to the grammar of particular languages or language families. There are also certain kinds of phenomena which do not combine in any regular way with the main body of material in a language, and which therefore have to be left out of any precise language structure: for example, the voice of each speaker, coughs of hesitation, and intonations of anger or surprise.

12.1. *Overview of language structure*

As for the phenomena which show regularities, they are describable, for each language, by the following relations and processes:

(1) **Words**. Sequences of phonemes (the latter defined by phonemic distinctions, 7.1), bearing (in the great bulk of cases) meanings which are associated with aspects of the perceived world, and are characterized by their likelihood of combining with particular other words in the dependence relation. The elements of communicational distinction are phonemes, but the elements of meaning-carrying are words and their relations. The words, however, are not simply symbols assigned by their users. Their ability to have a particular meaning (i.e. to be such a symbol) is identical with their having particular clustered (coherent) selections of their co-occurring operators or arguments, (3) below.

(2) **Dependence**. A relation of 'dependence on dependence' (3.1.2), which partitions the set of word occurrences in utterances into dependence (argument–requirement) classes of operators and their arguments. A sentence, or minimal discourse, can be defined as the locus of the satisfactions of the dependence relation: when every word in a sequence has its argument satisfied therein, the sequence is a sentence. At each stage of sentence-making, the words between which a direct dependence holds are next to each other, except for the effect of prior sentence-making steps, and of later transformations. All other grammatical relations are either paraphrastic rearrangements of this dependence, or else further constraints on the resultants of this dependence.

As was seen in Chapter 3, the dependence relation can best be recognized as a constraint, which states what is excluded from the language. In a less essential way the other basic constructions, (1) and (3)–(4) here, can also be recognized as constraints.

(3) **Likelihood**. The dependence-on-dependence says nothing about how likely the occurring of a particular word in an operator class is in respect to particular words of its argument class, and how constant such likelihood is. An additional constraint in language provides that each word indeed has a particular, though fuzzy, likelihood (even down to nearly zero) of appearing as operator on each individual word of its argument class, and that this likelihood is roughly constant within particular communities and periods.

In the syntactic theory, words have no intrinsic properties other than this dependence status, and, within that, their likelihood of particular combinations. To these two constraints in sentence-making, the meaning of the sentence is closely related. A word's phonemic composition identifies it, but not intrinsically. This is because other words (with other combinatorial likelihoods) may homonymously have the same phonemic composition (as in *hart* vs. *heart*); and because different phoneme-sequences may have the same meanings and the same combinatorial likelihoods (as in the two pronunciations of *economics*), or the same likelihoods with a particular complementary difference (as in *I am*, *he is*).

(4) **Reduction**. In stable operator–argument environments, certain words which have highest likelihood of occurring there may be reduced in their phonemic composition, even down to zero; also, the distance between words that are grammatically relevant to each other may be reduced (as in modifier position and in affixation). The situations in question include both being the most likely in respect to the particular neighbors in the given sentence, and being normally likely in respect to exceptionally many possible neighbors in any sentence.

(5) **Miscellaneous**. A few independent processes are found in particular situations: pre-syntactic morphology, other than that which can be considered to be produced by reduction (12.3.1); asyntactic sentence-making as in *Fire!*

or *Hey you!*; constructions of one sentence on the analogy of another, as in *We had a quick cup of coffee*, built like *We had a quick bath* (from *We had a bit of quick bathing* from *We bathed quickly*, G377); sentence structures which are paraphrastic ('transforms') to another structure without being reductions of it (8.4, 11.3.3); and occasional foreign borrowings, euphemistic word-replacements, or the preservation of now non-grammatical idioms or proverbs (e.g. *The more the merrier*).

These last constructions differ from the preceding processes above in that these last may have no recognizable relation to an operator–argument source: we may derive *have a quick bath* from *bathe quickly*, but have no such source for *have a quick cup of coffee*. This lack of a now-recognized relation may hold even for forms which have historically been derived from operator–argument sources, as in the case of *goodbye* from *God be with you*.

Conversely, sentences which arose historically by analogy may be interpretable as though directly reduced from operator–argument sources. This is evidenced by the possibility of finding for most of these 'miscellaneous' exceptional sentences an as-though source formed in accordance with operator–argument relations, from which the exceptional form could have been derived by suitable reductions of known types (e.g. *have a quick coffee* from an artificial *have a quick coffee-drinking* from *drink coffee quickly*). In such cases analogy or pre-syntactic morphology, or the like, is the process that created the sentence in question (mostly on the basis of some existing operator–argument sentences), but the operator–argument structure is the framework in respect to which the word-sequence in question is understood as a sentence. Only a reasonably small number of word-sequences are used without being referable to the operator–argument framework. Other than these last, the 'miscellaneous' cases which are indeed referable to operator–argument examples are systemically secondary to the operator–argument process which has created the framework for them.

There is another respect in which the operator–argument process is central to sentence-making. This is that the sentences of the grammar, i.e. the metalanguage, which constitute a subset of the sentences of the language itself, cannot themselves be formed without the operator– argument process of (1)–(3) above; but they can be formed without any of the other sentence-making processes (5.2). Thus all the structure needed to form every sentence of the language can be defined in terms of dependence-on-dependence and combinatorial likelihoods alone, or through metalinguistic sentences that are so formed. The lack of a non-language-like metalanguage (one which itself would not have (1)–(3) above) leaves these basic (1)–(3) processes to be characterizable only by exhibition of their constraints.

In addition, the processes of (1)–(3) above each embody a particular kind of information; and in constituting a sentence, each application of one of these processes contributes a fixed meaning to the sentence. Furthermore, these processes suffice for all the information in language. Every sentence in a language is paraphrastically derivable from a sentence found only by (1)–(3), or is paraphrasable by such a sentence.

(6) **Conditions of speech**. There are certain additional properties of language, which are due not to the processes of making sentences out of words, but to the conditions of speaking. One is the arbitrariness of phonemes, i.e. their lack of regular relation to the meaning of words or the structure of sentences.[1] Another is the contiguity of constructions (2.5.4). A third is the linearity of constructions (2.5.3). Yet another property is the ability of words to refer not just to extra-linguistic objects and events but also to words in sentences: this makes a metalanguage possible as a subset of language. And nothing prevents a word in a sentence from referring to the occurrence of a word in its neighborhood: this makes cross-reference (pronouns with antecedent) possible.

(7) Finally, **language changes** (12.2). Different parts of the structure—sounds, vocabulary, likelihoods, reductions—change in different ways at different rates. Languages change somewhat differently in different—especially in separated—communities, thus creating dialects and 'daughter' languages and related languages.

Given the list above, we may note that the processes of sentence-making, in keeping with the conditions of speech above, not only produce the sentences of a language but also yield the particular constructions and grammatical relations of each language, as regularities of those sentences. These include: major word classes such as conjunctions, adjectives, adverbs; special small word classes such as quantifiers or the English auxiliaries (where characterizing them as special combinations of general constraints circumvents the problem of their peculiar character and fuzzy membership); modifiers (from secondary sentences); morphology (that this can be derived via reductions explains how it can be a major yet not universal apparatus); paradigms and their special categories (person, time, number); special rules, and exceptions to rules, and in general the special domains of certain reductions.

In spite of the specificity of the processes and conditions, the set of sentences in a language is not well defined: there are many marginal sentences, largely because the normal likelihoods of certain grammatically important words (e.g. *take a* —, *give a* —, 12.2), and the domains of certain reductions, are only roughly statable. There may also be cases of constructions and regularities that do not apply to particular words in their expected domain. Overtly, this situation resembles the 'exceptions' so common in the traditional rules of grammar. However, not only do far fewer 'exceptions' remain after reduction analysis than in traditional grammar, but those that

[1] The fact that speech is in general continuous plays no role here, because language is built out of distinctions in speech, which constitute discrete entities.

remain have more assignable causes or systemic properties. Partly, they are due to lack of regularization, by uneven applications of analogy or by retention of old, frequently used forms (as in the morphophonemics of *is–am–are*). And partly they are due to language being constantly in process of change at one point or another, so that when the description is sufficiently detailed one is bound to come upon reductions and analogies that are in the process of extending their domain or of becoming reinterpreted into the operator framework—and perhaps differently among different speakers. They may also be due to arrested syntactic change, as in the vagaries of *do* in the course of the history of English.[2] And they may be due in complex ways to the overlapping domains of different reductions.

Although non-regular constructions, and exceptions to the regularities, thus remain, it is necessary to analyze each language in respect to its operator structure as far as this analysis will go. Only then can we see for what forms, if any, the operator–argument reconstructions are impossible, or too costly in complicated use of the basic apparatus, so that we may prefer to accept the form as part of the finite material that is outside the operator–argument system. We can also see how numerous these residual forms are— far fewer than the exceptions and inexplicable forms found in traditional grammar—and what characteristics and sources they have. And above all, as is seen below, for the rest of the language we obtain a simple and unified and procedural—even computable—structure, with everything referrable to first principles, something which could not have been obtained by eclectic and episodic analyses or by purely semantic considerations.

In contrast to the constraints of (1)–(4) above, which delimit what can be and what cannot be in language, the processes described thereafter, including the special constructions in particular languages, indicate specific structures in the language without in general saying what is excluded.

In contrast to the derivational theory developed in the preceding chapters, it is possible to think of a maximally organized accounting of the similarities among sentences, somewhat along the lines of cladistic classification in evolutionary theory. Such an approach would not assume the primacy of initial sentence-formation over analogic back-formation. Rather, all similarities would be treated on a par. Nevertheless, the fact that most analogic formations are on the model of existing syntactic word combinations makes the pre-analogic, derivational, constructions the key syntax. This derivational syntax has been found to be best described as generated by a set of constraints, if for no other reason than that the set of word combinations of a language is too large and fuzzy to be listable, whereas the constraints that generate them are interrelatable and listable even if sometimes with fuzzy domains. However, the constraints with their domains do not quite match the set of word combinations in the sentences of a language, a large

---

[2] Otto Jespersen, *Growth and Structure of the English Language*, Macmillan, London, 1948, pp. 218 ff.

part of the mismatch being due to analogic processes over parts of the domains. We are thus back to recognizing the constraints as primary processes and analogy as secondary. As to the derivations which describe the relations among the sentences created by the constraints, they are largely accretive and recursive: one sentence is derived from another by the accretion of a constraint on that other, even if the constraint itself consisted of removing something from the sentence rather than adding to it.

12.2. *Structural considerations from language change*

In modern linguistics it has been a tenet that synchronic descriptions are separate from diachronic ones, because many systemic relations are relevant only among forms that are used together, in various sentences that are available to the speaker and hearer. However, since at any one time various forms and uses in language may be in process of change (2.5.7), a sufficiently detailed description of a language may be expected to have various points at which something is not quite in accord with the rest of the system. In purely synchronic description, such irregularities-in-process would be fitted, as one kind of deviation or another, into the regular material of the language, just as we would try to do with permanent irregularities. In many of these cases history shows a more regular source, or suggests a more regular future form which is being approached. Such past or future regularities may reveal the present irregularity to be a transition between states that are regular in the given respect, and thus may locate it among the constraints that create the regular structure of the language. In some of these cases, historical considerations show how forms that appear to be syntactically peculiar can be restated as peculiar morphophonemics (or peculiar transformational domains) on syntactically regular forms. English examples are: the auxiliary *can*, etc. (G298), the 'pro-verb' *do* (as in *I do not want it*, G297), the periphrastic tenses (*He has gone, He is going*, G290, 294), suffixes like *-hood* or *-ful* which are close to being compound nouns (G173). At any one time in a language there may be peculiar forms which may be cases of words moving between one regular status and another: for example, the partially tense-like and partially auxiliary-like *used to*, as in *I used to go*, where there is no *I use to go*, *I will use to go*, *I have used to go*.

In order to bring earlier or later historical stages to bear upon the analysis of a given form, we must know how to recognize which forms constitute the earlier or later stages. This requires knowledge of the mechanisms of change in phonemes, meaning ranges, and syntax, and also of the possible sources and directions of the largely unpredictable changes. Such knowledge is much too vast to be summarized here, but there are two properties of change in word use and in syntax which are particularly relevant. These are, following Manu Leumann and more recently Henry Hoenigswald:[3] first, that word-change is not directly word-replacement but a matter of change in use of sentences, and second, that new sentences do not suddenly replace old sentences, but rather the 'new' sentences come to be used for the same meaning as other sentences and so come to compete with those others for a particular informational niche (i.e. for the same word likelihoods around them), sometimes with one winning out and the other dropping from use (cf. in 12.4.4).

Some changes yield no appreciable change in the structure of the language: such are the effects of most sound-changes on the syntax of the language, and the effects of most changes of word likelihoods, or the introduction of new vocabulary (by borrowing or word-coinage), or certain replacements of one word by another (as of *inwit* by *conscience*). Other changes have greater effects: large extendings of a word's selection (i.e. of the list of those words with which it is likely to occur) can significantly alter its meaning (e.g. *provided*, 11.1), and can even create a new (secondary) grammatical relation and meaning, as in the case of the adverbial suffix *-ly* which is a reduced form (with greatly expanded selection and modified meaning) of earlier *-lice* 'with the body or form of' (G191, 307). Many other suffixes, too, show phonemic, likelihood, and semantic changes in contrast to their known or suspected sources. As has been noted, some words show large shifts in meaning and in grammatical environment, as in *fond*, which earlier meant 'foolish'; or *like*, which changed from 'pleases' to 'is pleased by' (the earlier form of *I like it* was *It likes me*). Even smaller but regular changes in selection can create new semantic or grammatical effects, as in the well-known replacement of English *cow*, *calf* in matters of eating by Norman-English *beef*, *veal*, which created a limited distinction between animal words and food words; or the use of *meat* for *flesh* in matters of eating, so that *meat* ceased to be used for 'food' in general except in *meat and drink* and in *sweetmeats* (in this case the selections of *flesh*, *meat*, *food* were redistributed, leaving irregular boundaries).

In many of these changes the earlier structure is still relevant, though less visible. When *give* and *take* have a nominalized verb as object (*give a good look at it*, *take a good look at it*), both their selection and their meaning has been extended, but their original character can be traced not only in their precise meaning here but also in the existence of *give it a good look* (like *give him a new idea*) as against the unavailability of *take it a good look* (corresponding to the unacceptability or marginality of *take him a good idea*). More importantly, as noted, the 'perfect' tense in English (*I have written*) can still be analyzed to a large extent—with historical justification—as a

---

[3] Manu Leumann, 'Zum Mechanismus des Bedeutungswandels', *Indogermanische Forschungen* 45 (1927) 105–18; H. M. Hoenigswald, 'Ancestry and Descent: An overview of Historical Linguistics', also 'Obsolescence, Neologism, and Replacement', both in *The Oxford International Encyclopedia of Linguistics*, 1990.

present tense *have* with *state, condition* (as presumed 'source' of *-en*) for object (G290): one can hardly say (1) *I have written three letters yesterday evening* (as compared with *I wrote three letters yesterday evening*; but in French one says *Il a écrit trois lettres hier soir*); however, one can say *I have written three letters by now*, like *I have considerable experience by now*; and *I have written three letters since yesterday* and *I have this condition since yesterday*. Differently from English, the French perfect has already beome a past tense. Note that in English, the inability of nominalized sentences to carry tense is circumvented by using *have*, which is structurally tenseless ('present'), to carry adverbs of past time: *My having written three letters yesterday evening left me with fewer chores today*, where *have* must have been added to the nominalized *My writing three letters yesterday*, since one does not say (1).

The English auxiliary verbs also show various traces of their being the ordinary verbs that they once were (G298).

Nevertheless, some constructions and distinctions may get lost from language: in languages which have lost their case-endings on the arguments of a verb, not all the distinctions have been taken over by prepositions or by relative positions of the arguments. In addition, some new grammatical situations may be formed. One example is the English auxiliary verbs which could once apparently accept on the verb under them a different subject than their own (as though one said *I can for John to win*); this is no longer possible, and we have only *I can win* (from a required non-extant or no-longer-extant source *I can for me to win*). Another example is the definite article, which is apparently a relatively recent development, and which has different grammatical requirements in different languages. Also, the English *-ing* is a conflation of two very different earlier morphemes (a verbal adjective and a verbal noun), but the single modern phonemic form has been broadening its selection so as to become increasingly a single construction (G41, 294).

We see that any theory of language, even just of language structure, has to provide for change in language (2.5.7). It is also clear that change is partly related to the existing structure, though not determined by it, and that it in turn affects the structure in various ways and degrees. In the present theory, there are intrinsic relations between structure and development (12.3.0). Formulating the structure in terms of constraints is different from formulating it in terms of objects or constructions that are defined by lists or by external properties. The constraints leave room for particular elements to be unclear as to their satisfying (i.e. being in the domain of) one constraint or another. They even leave room for particular elements to satisfy a constraint in more than one way (depending on how it is analyzed)—either because it is in process of changing, or simply because it is irregular or not fully domesticated (reinterpreted) into the system.

Furthermore, the history of sentences, words, and constructions is the history of their use, that is, of their combinations in speech. But it will be seen below (12.3.5) that it is the very use of language that becomes conventionalized and is finally standardized into the constraints of language structure, so that the structure of language is an end-product of the history of language, which indeed could hardly be otherwise. The relation between structure and history makes it possible to evaluate and compare the change in words and in grammatical combinations of words (i.e. constructions): for example, to say wherein a word or construction is or is not the same as a past word, or the same as a cognate word (i.e. a historically related word in another language).

There are certain things which apparently do not change. These are: the dependence-on-dependence relation as the formative relation for sentences, the existence of stable though changeable likelihood differences among words in respect to co-occurrence, the ability of words to refer to words, and apparently the existence of low-information reductions. One indication of their permanence is that these are apparently universal, and so must have gone unchanged since their coming to be. A consideration for this permanence is that if these relations are what created sentences as we see them now, then they cannot be replaced by any other process except through an innovation as basic as the ongoing structure of language—something which is not likely to happen as long as language proves roughly adequate.

The essential adequacy and stability of these basic relations depends in part on their being hospitable to changes and differences in words and constructions. If the dependence relation that makes sentences held just among fixed classes of words, defined by lists or external properties, and if likelihoods of co-occurrence were fixed, then new words—entering by borrowing or cultural change, or altered in their interrelations by phonemic change—might be unworkable within the existing syntax. A pressure might then grow to develop other ways of forming sentences, or informational expressions, with these new words. But the fact that the relation is a dependence-on-dependence of words, that the fuzzy likelihoods are changeable and not mutually exclusive, that reductions are made by low-information properties rather than by a fixed listing, enables new words to enter and function in the system. It is this that makes the system stable, and unchanging in its fundamentals.

12.3. *Formation and development of language*

12.3.0. *From descriptive order to time order*

The origin of language, more specifically of sentences, is not normally associated with a theory of language structure, and is a subject diligently avoided by modern scientists. However, the present structural theory is

intrinsically relevant to issues concerning primitive—or in any case, early—
formation of sentences, and this in two respects. First, language structure is
obtained here not in terms of composition but in terms of processes—the
processes that constrain. These processes rest on the standardization of use
in speech (12.3.5), and are thus of the same kind as the processes of language
change (12.2). The latter we can follow back historically for a few thousand
years. But these processes have inherently the potential not only of regularizing an already existing (and possibly differently structured) speech, but
also of creating sentences out of an unstructured stock of separate or
arbitrarily combined words. Hence it is natural to push the investigation of
these processes backward, and ask how the relations which they create
might have begun.

Second, the structure ascribes a descriptive order to its main constituents:
first-level operators are defined as operating on zero-level arguments to
create sentences, and perhaps conversely; second-level operators are defined
as operating on first- or second-level ones to expand the sentence; reductions
are carried out on existing sentences. When we are dealing with a least
description (2.2), covering almost all the structures of a language in a single
system of constraints, and especially when the core of the system carries the
responsibility of being a mathematical object, then the descriptive order of
entities bids fair to be an essential definitional order, not an artefact of a
particular approach. True, it is possible that the description in terms of
operators and arguments may well be some kind of large restructuring
of previously existing otherwise-formed sentence-like word combinations,
with the operator–argument form being attained by disuse and modification
of any word-sequences that did not conform to it. The language we know
may have been preceded by very different kinds of speech, possibly different
kinds in different human communities. But in any case at some period
speech took on a form which can be characterized by the above descriptive
order of processes, which is their definitional order: $B>A$ in descriptive
order if B is defined in terms of A, and A is not defined in terms of B.

In respect to time order, the descriptive order suggests two possibilities.

The descriptively later may have existed first, with the descriptively
earlier being a regularization of its structure. Such a time order is reasonable
in the case of phonemes: although words are defined as sequences of
phonemes, we can hardly think that the phonemes existed without words;
rather, we would suppose that a stock of words existed without benefit of
phonemic explicitness, and that the distinctions developed as a way of
increasing their transmissibility.

Somewhat differently, defining likelihood inequalities on the resultants of
the dependence relation (i.e. on operator–argument pairs) is an artefact of
the theory: the datum is the likelihood of combination, and the dependence
is a name for all likelihoods which the theory considers to be positive, no
matter how small.

In the case of sentences, however, it is more reasonable to think that the descriptive order is also the time order, with elementary sentences being formed first, by a preferred dependence in the co-occurrence of words. Even if there also existed pre-vocabulary utterances in which similar segments were gradually distinguished as separate words, the definitional order suggests that, rather than a systematization creating language, there may have been temporally different developments of successive vocabulary classifications: first zero-level and first-level (forming elementary sentences), then second-level words. At the very least it suggests how language could have grown, by a recursive repetition of a simple dependence relation that created the elementary sentences. And the considerations of 12.3.5, about how such a relation could come to be used, suggests that the successive processes proposed in the present theory would not have been momentaneous inventions, but the gradual conventionalization, through time, of communicationally successful habits.

12.3.1. Pre-syntactic speech

The descriptive order of operators and arguments refers to their status in sentences. Both types of word may have existed pre-syntactically, i.e. before the development of sentences, but not with these classificatory statuses. There are certain things that most likely existed in speech before sentences came to exist. There must have been words, for sentences to be formed of. This means in the first place conventional sound-sequences associated with particular referent meanings, i.e. with particular types of objects and experiences. It may well be that the sounds were already then subject to phonemic distinction, because otherwise distinction and transmission of a large vocabulary could be inefficient (7.1); phonemic distinction can be characterized simply as those sound-features which are preserved under transmission. From the explanation of the early development of sentences (12.3.2), it would seem to follow that word combinations must have already been in use before syntax, with different likelihoods for a given word to combine with different other words. Or very many distinct utterances must have been in use, which were or became sufficiently similar in some of their parts to become decomposable into segments which we now call words. Such explanations suggest that a large vocabulary was already in existence, so that the standardization implicit in sentence origin would have served to bring some order into an existing stock of word combinations.

The lack of fully structured sentences in this presumed developmental stage does not preclude that many speech communities may already have had morphology, that is, particular phoneme-sequences with particular meanings which could be affixed to certain words, e.g. for plural, time, masculine–feminine. If all speech communities did, then the morphology must have been lost in various languages, since not all languages now have morphology; forgetting is not unknown in language history, as in culture history and in politics. It is also reasonable to suppose that pre-syntactic speech had intonations or special words or affixes for command and question and the like.

12.3.2. *Early development of sentences*

If words are said in combination, certain combinations will readily add up to some fuller interrelated meaning, over and above that of the words alone (*boy sleeps* or *sleeps boy*), and certain others will not (*walks sleeps*). This arises from the actual relations of the words' referents in the speakers' experience. There are different likelihoods for different conbinations in what people would normally say. When a great many word combinations are said in life situations, each word will be found to occur mostly with one or another of a particular set of other words: *sleeps* and *walks* perhaps with *boy*, *dog*, etc.; *flows* perhaps with *water*, *milk*, *sand*, etc.; *see* with the pair *I* and any of the above. Such combinations may occur far more frequently than *sleeps* with just *walks* or *sees*. It is reasonable to suppose that the less useful word combinations were less said, as not constituting effective communication. The meaningfulness of word combination is especially important in the case of transmission: in face-to-face communication the hearer can judge the meaning of a word combination with the aid of some of the extra-linguistic situational information that the speaker has. But in transmission beyond the immediate situation, the words are on their own. If speakers tended to have certain words occur with any of a set of other words but not alone, then that preference could be conventionalized into those words not being said except with one of the other set. This standardization would then constitute a constraint on how those words combine.

We thus reach a particular constraint, a dependence among words. The dependence may impose relative positions in the linear order of speech: the dependent word may appear usually in particular positions in respect to the words on which it depends. Within this dependence, the likelihood of a particular word occurring with particular words of the other set remains as it was before the standardization into a constraint. This dependence, which is found in the grammar of individual languages, is thus not an invention or a plan; it could be neither, since the concept of such a relation could not have preceded the existence of the relation itself. Rather it is a conventionalization of natural limitations on the use of word combinations, which are imposed by the need for effectiveness in communication.

Since the limitation on use was due to the lack of usefulness of particular word combinations, the dependence which was created thereby carries, and expresses, a meaningfulness in the very co-occurrence of the participating words, over and above the meanings of the individual words. The meaning of combining is that the parts are relevant to each other. And since the form of the dependence is that the dependent words are said only in the presence of the others, the interpretation is that the dependent word is said about those others, that it predicates something about them. The dependent word is then the predicator or operator on the words on whose presence it depends; the latter are its arguments. This is not to say that the idea of predication underlies the dependence relation. Remembering Piaget, one does not have to understand in order to do. People do not need an awareness of predication in order to speak or understand language. They need only to know that certain words are not said except in the presence of certain others—aside from later reductions—and that this means that the dependent word (recognized by its position as well as by its likelihoods) is said 'about' those others. The awareness of the relation could only arise from experience with its use in language, and becomes explicit much later, in grammar and in logic.

The dependence relation creates sentencehood. Every combining of a word with words on which it depends constitutes a sentence, and saying it has the meaning of asserting—either asserting the words themselves, or asserting that the speaker wishes, wonders about, demands, or asks the contents of those words. It has thus created, out of a particular dependence relation of co-occurrence among words, a new object (sentence), with a new kind of meaning (predication), and with new relations (of requirement) within its set. Non-predicational utterances exist in languages, chiefly one-word ones such as *Fire!*, *Ouch*, *Hello*, *John* (the latter may be a case of calling, or surprise, or pointing and the like). Whether these one-word utterances should be called sentences or non-sentential utterances of the grammar is a matter of terminology. It should also be pointed out that non-dependent (i.e. non-sentential) word combinations may or may not have some rough meaning, according to the semantic possibilities of the words: more in *John Bill Mary saw with* than in *Up vacuum light eats*. But a sentence, i.e. a satisfaction of the dependence relation, always has meaning and has it explicitly even if nonsensical: *John saw Mary with Bill, Light eats up vacuum.*

The evolving of this dependence relation is the major step in the formation of language structure as we see it now. Not only does it make sentences, all on the basis of a single relation among words, but it also brings into existence structural criteria: criteria for recognizing sentences and relations among sentences, and criteria for distinguishing n^{th}-level word classes. It establishes a structural constraint whose meaning is relevance or aboutness. In addition, it establishes the concept of information as a particular (predicational) relation among word meanings.

The separation of pre-syntactic vocabulary (12.3.1) from syntax (12.3.2) makes it possible to think that different parts and functions of the brain were involved in the development of these two contributions to language, and that the development of the two in the brain took place at different times. It also specifies the difference between these two contributions to the existence of language: a large set of symbols, as simply sounds associated with experience, and a relation of dependence in their use (their co-occurrence in speech), in which the symbols become modified in meaning and in effect redefined as cogs in respect to a system.

Beyond the formation of the elementary sentences come the complex sentences, formed by words which depend on words that themselves depend on something: e.g. *continue* on *sleep* which in turn depends on *boy* or *dog*, etc. Such words as *continue* therefore in effect depend on sentences; and with them they construct an expanded sentence. In the same way, *know* depends on the pair of a zero-level argument and a sentence (*I know the boy likes the dog*); and *if* depends on a pair of sentences (*The dog barks if it sees a cat*); and so on. In principle, the possibility of defining further dependences for words is unlimited.

Another further event or constraint, which can be looked upon either as making other (derived) sentences or as making new forms of existing sentences, is the reduction in phonemic content (or the change in position) of certain words in certain dependence and likelihood relations to other words in the same sentence. As we know, the conditions for reduction are mostly that the word in question adds little or no information to what is carried by the other words in their given dependence relation; these other words can be looked upon as being present in the sentence (as it is being built up) prior to being joined by the new word. Each reduction has a particular domain, i.e. a particular set or type of words on which it can be carried out; and each language has only certain reductions and conditions for reduction which are special to it. Most reductions are optional, so that the unreduced sentences continue to exist in the language, or at least to satisfy any reasonable grammar that can be made of the language, even if some unreduced sentences are wildly complicated and never naturally said. For the reductions, their basis in use is the convenience of word and sentence abbreviation and of identifying the antecedent for cross-reference words (5.3.2), weighed against the ambiguities that many reductions create for the hearer. The effects of reductions create various special structures that come to characterize certain portions of sentences in one language or another, and that may package the presentation of information to the hearer in various useful ways. An example is the creation of modifiers (relative clauses, adjectives, adverbs) out of a secondary sentence which has become attached to something in the primary that has been repeated in the secondary (3.4.4).

12.3.3. *Further development of syntax*

The arising of structural effects from specializations of language use does not cease with the initial formation of syntax. Structural development continues, in long-range trends of change which can be characterized only vaguely. There is some overall evidence of a trend to syntactic regularization, yielding simpler grammars. Such simplification may be withstood by some of the most frequently used entities (e.g. the variants of *be*: *is*, *am*, *are*, *was*, *were*); this is to be attributed to frequency, not to any special semantic importance of *be*. There is also evidence that structural irregularities, usually encapsulated, are created on various occasions as by-products of historical changes.

Types of irregularities include the disappearance of cumbersome or stylistically disfavored sentences that are the unreduced source of existing reduced sentences; thereupon the base set of sentences lacks the material of the reduced sentences (and so no longer contains all information of the language) unless artificial source sentences are reintroduced, with the necessary reductions, in formulating a grammar of the language. Another type of new irregularity is the restricting of the arguments of particular operator words. For example, it has been seen that the auxiliary verbs (which are O_{no}) restrict the first argument (subject) of the operator under them to be the same as their own subject: in *I can solve it*, the subject of *solve* is *I* and the reconstructed source has to be *I can for me to solve it*. This is not a restriction on the argument domain itself, which would violate the mathematically important condition that the dependence of an operator is only in respect to the dependence properties of the argument words and not in respect to any external property or listing. And indeed, *solve* can have as subject any zero-level argument (i.e. any word which itself depends on zero), as can *can*. The only new requirement is that under *can* both subjects must be the same. This more specific restriction is still an irregularity, but a lesser one: rather than restricting the arguments of *solve* or *can*, it restricts the arguments that *solve* or any other operator has when under *can*—and not to a particular list but to being the same as for that occurrence of *can*.

There are also various other irregularities, as when what is obviously the same word satisfies the conditions for two different classes (without current grammatical derivability of one from the other) and has to be multiply classified: e.g. *that* as a zero-level argument (in *I know that*), and *that* as argument-indicator (in *I hope that he's here*) which shows that a sentence (*He's here*) is appearing as argument (of *hope*). An irregularity of a different type, in the complex conditions for quantity words, is seen in the alternative forms *A small number of people is coming tonight* and *A small number of people are coming tonight* (in the latter, *small number of* is a hard-to-derive adjective on *people*).

There is another development which could take place only after orderly sentences came to exist, but which modified in detail the nature of sentence structure. This is the spread of analogic formation and analogic word combinations, and other extensions of word use, which made word combination and local structures within sentences (e.g. phrases, paradigms) more rule-based and less a matter of the semantic appropriateness which was assumed in 12.3.2 to have underlain the development of the dependence relation.

In contrast with such slow or small changes, there are great specializations and further structures in language when the things being talked of (the subject matter) have specialized relations among their perceived objects. This is seen in concentrated form in sciences and above all in mathematics, as noted in 10.2–4. The greater precision and intellectual complexity of scientific writing is creating minor grammatical innovations, as in the use of *respectively* to make cross-connections, e.g. in *A and B produce Y and Z respectively* (from *A produces Y and B produces Z*); such a criss-cross of the source sentences would be exceptional in natural-language grammar, and the use of *respectively* in colloquial English is not restricted to this environment.

Other innovations are larger but still can be housed within a generalized language structure. One innovation arises out of sentence-sequences, i.e. out of dependence among words of neighboring sentences; these also have inherently a limiting feature of having a common subject matter. The simplest case of this is seen under certain (conjunctional) O_{oo}, which connect two sentences by operating on the pair of them: here the high likelihood is that there is some word that occurs in both sentences (9.1). A larger development is that of a double array of word-class dependences in the sentence-sequences that constitute discourse (9.3). Another development is the singling out of certain constraints on sentence-sequences, whereby the last sentence has the semantic status of a conclusion from the others: proof in mathematics (sayable in English if we wish), and—in a far less specific way—well-constructed argumentation in science. This creates a new semantic power for language: a distinguishability of non-loss of truth value or of plausibility (in science sublanguage).

More fundamental innovations in language structure arising from subject-matter limitation involve the overall properties of the operator–argument structure, differently in mathematics and in science languages. In mathematics and mathematical logic, likelihood differences within an argument or operator domain have been eliminated. In argument status, the differences are eliminated by having only an indefinite argument, a 'variable', meaning in effect only 'something' or 'a given thing' (within a domain), it being understood that within an indicated sequence of sentences each occurrence of the same variable-symbol is a cross-reference to the first (while each occurrence of another variable-symbol is not specified as to sameness with the first, unless stated to be the same or different). And the meaning of operators there is related not to likelihood differences but to truth tables that specify the truth value of the resultant sentence from the truth values of its arguments; likelihood then has no semantic or structural relevance.

In science languages, as was seen in 10.3, the consistent restriction of subject matter, and the attention in each set of texts to particular properties or relations within it, is reflected in particular operators being used only on particular words out of their possible natural-language argument domain. What is new here to language structure is the 'only'. As noted, in English as a whole, the O_{nn} *eat* may never have occurred on the *N*, *N* pair *vacuum*, *light*; nevertheless, *Vacuum eats light* is a sentence and we even know what it means, though the meaning may be nonsensical if *eat* is taken as 'nourishes itself on', or simply untrue in a general sense if *eat* is taken as 'effects diminution of'. In contrast, in immunology the O_{nn} *contain* cannot occur on the ordered *N*, *N* pair *antibody*, *cell* (though it occurs on the pair *cell*, *antibody*): the possible though nonsensical English sentence *Antibody contains cells* is not considered a possible sentence of immunological language. This creates subclasses, largely superseding likelihood gradation: a novel language-structure situation in which there are several sentence types (operator–argument structures) at the same level (e.g. the O_{nn} level, or the O_{oo} level). And it creates a novel semantic power for language: a distinguishing and eliminating of irrelevance and nonsense.

The syntactic systems of science and of mathematics have lost the mathematical-object status (because their dependence is defined on particular lists of argument words), and they have to be described in an external metalanguage, that is, in some natural language. Because of the semantic differences expressed by these various structural innovations, the formal notations of mathematics and science languages (as in 10.5) can be translated into natural language, but natural language as a whole cannot be translated into them.

Both in the regularizations, irregularizations, and changes in the language as a whole, and also in the larger innovations of science languages and mathematics, we see that change and furtherance in structure can go on, that the development of language is not at an end. We also see that at least the larger changes and innovations arise from the different external conditions (e.g. subject matter) and internal linguistic pressures (e.g. analogy) under which language is used, and from the conventionalization of such usage. The conventionalization is seen, for example, not only in the development of explicit rules of mathematical well-formedness and proof, but also in the peer-pressure criticism by colleagues that keeps scientists' arguments within inspectable grounds.[4] Both the conjectured initial formation of syntax and its further development are thus institutionalizations of ways of effective recording and communicating of information about the world.

[4] Cf. P. Latour and S. Woolgan, *Laboratory Life*, Russell Sage, New York, 1979; K. D. Knorr-Cetina, *The Manufacture of Knowledge*, Oxford Univ. Press, Oxford, 1981.

12.3.4. *The limits of language*

The further, and presumably continuing, development of grammar does not mean that everything needed for saying anything can be reached, or that anything that the syntax makes possible (and that would be helpful for speaking) will some day be reached. There are many inadequacies in one language or another, and in all languages.[5] In many languages some complex applications of grammatical relations cannot be expressed by direct use of the relevant grammatical devices. In the English comparative, one cannot repeat the comparative form in order to compare a comparative with a fixed amount: one cannot say *He is taller than me than me* (or: *He is more taller than me than me*), but only *He is taller than me by more than my own height* or the like. Also, direct nominalization of a sentence in English, by argument indicators placed on the operator of that sentence, leaves no room for the tense that the sentence was carrying: e.g. in *John came yesterday* (in past tense), which is nominalized in *John's coming yesterday surprised us* (with no tense on *come*). The past can nevertheless be expressed here by circumvention, as in *John's having come yesterday* (where *have* is not in the past tense, and *have come* is not really in the perfect tense since the perfect would not accept *yesterday* here); much more rarely, futurity can be indirectly expressed in nominalization, as in the rare *John's being to come soon*, from *John is to come soon* which is present tense but indicates a plan for the future.[6]

More generally, some features which we would consider of grammatical relevance may not be distinguished in language. For example, the use–mention distinction (5.4) is not indicated by any general grammatical device, although quotation marks are often used for it in writing. Another example is seen in the case of assertion. In English, assertion is not

---

[5] Of least interest here is the fact that the specific forms of reductions in a particular language may make it impossible to express certain things in it. For example, a language in which absence of a plural suffix means 'singular' makes it impossible to leave the difference between singular and plural unstated. Furthermore, in such a case it becomes unclear whether one can say, for example, *Either John or his co-workers is responsible* or *Either John or his co-workers are responsible*. Another type of situation is seen when we consider *That he failed is clear*, formed by *is clear* operating on *He failed*. If now we want another second-level operator (e.g. *is not universally accepted* to act on *That he failed is clear* we find that the grammatically 'correct' form, as in *That that he failed is clear is not universally accepted*, is generally avoided in favor of *That he failed is clear is not universally accepted* (which can only be described by an *ad hoc* reduction of the two occurrences of *that* to one).

[6] Another inadequacy involves the antecedents for pronouns. Consider for example the following sentence: *That night he wrote to Tanya telling her that the gipsy woman was in despair and that he could no longer marry her, Tanya* (A. N. Wilson, *Tolstoy*, Norton, New York, 1988). Here *marry her* is ambiguous, and its ability to refer to *Tanya* would have been even weaker if the author had written *saying that* instead of *telling her that*. The alternative, of not pronouning here, by writing *marry Tanya*, is virtually unavailable: *He wrote to Tanya that he could no longer marry Tanya* would seem to refer to another Tanya. The easiest solution is the stylistically forced *marry her, Tanya*.

structurally defined. Whether a sentence is asserted can make a grammatical, not only a semantic, difference, as is seen in certain structural phenomena: for example, as noted, we can say *That he left is false* (or *improbable* or *uncertain*), but not *He falsely* (or *improbably* or *uncertainly*) *left*, because the latter can only be a reduction from *He left; that he left is false* (or *improbable* or *uncertain*), where the negation is too unlikely a sequel to the assertion to support the reduction (3.2.2, end). Nevertheless, it is hard to find what structural features indicate assertion. Many higher operators assert or non-assert their arguments: *I assert that he left* in contrast to *I deny* (or *think*) *that he left*. But in the absence of these, the situation is not always clear. *It hit the floor* is not asserted in *I caught the glass before it hit the floor* (but it is asserted in *I timed the glass before it hit the floor*), and it is only fuzzily asserted in *It hit the floor, so they say*. In *I'm here, looking at a house which exactly fits my needs*, the existence of such a house is presumably asserted, but in *I'm here looking for a house that exactly fits my needs*, it is not asserted: there may be no such house.

More pervasive are certain lacks which are structurally characteristic for natural language, but may be overcome in specialized linguistic systems such as science languages.

One of these is the lack of a complete addressing system (3.4.4, end). An effect of this lack is that when a speaker uses free pronouns (*he*, *she*, *it*, etc.), he knows which neighboring word is the antecedent for it, but cannot communicate this to the hearer within that sentence (see n. 6 above). Furthermore, the partial addressing apparatus which is indeed used consists of words which refer to the positions of other words in the sentence (e.g. *latter*; or *wh*- generally referring to the preceding host noun). The grammatical analysis shows that referential pronouns are not directly cross-referencing devices, but simply reduced repetitions of a (usually) preceding word. Since the word has then been reduced by virtue of its repeating the antecedent, the fact of reduction constitutes a reference to the antecedent. Since no word or word-sequence can be a repetition of its own occurrence, no word can refer to its own occurrence: in *This sentence is false*, the *this* refers grammatically to some preceding sentence (as in *He says 'I didn't know.' But I assure you, this sentence is false*). Thus this form of the impredicative paradox is unavailable in natural language (when the grammar is considered in detail); other forms are sayable, but their impredicativeness is indirect.

Another lack which may be unavoidable is the lack of total regularity in the vocabulary. Some words may be factorable into a sequence of two or more words in some grammatical relation to each other, and this on grammar-simplifying grounds.[7] However, in natural language, the openness

---

[7] However, factoring on purely semantic grounds (e.g. *kill* into *cause to die*) yields little or no advantage, and will virtually always run foul of differing combinatorial restrictions or likelihoods: from *The Queen's command caused his words to die unspoken*, we cannot obtain

of the vocabulary, and the fluidity and multiple determinants of co-occurrence likelihoods, exclude the possibility of a complete and regular factoring of the vocabulary into a least set of words which are maximally independent in respect to co-occurrence. In a science language, where at any one time there are fixed subclasses (even for new incoming words), the combinatorial potentialities of the words may be sufficiently regular to permit factoring the existing words into an organized substrate of vocabulary components.

Also, natural languages have apparently not developepd a set of third-level operators—ones whose arguments are operators whose arguments in turn are operators. Science languages, or the scientifically precise use of certain words in natural language, may as noted above develop such a dependence: e.g. *rate*, as in the rate of change of some property or relation of things.

In spite of the various open features of language, there are certain limits to the direction of its development. Some obvious limits are due to its discreteness and linearity. The basic partial ordering, the repetition positions under O_{no} and O_{oo} (G139, 146, 148), and the status of modifiers (3.4.4) are all simply not indicated in the linearity, though one could imagine, as offshoots of language, systems of writing or graphing that would be linguistic in retaining the dependence and likelihood properties but would be freed of linearity (e.g. the semilattice representation of sentences).

The possible content of language is limited by its associating pre-set words with objects and properties of the publicly perceived world, and the immutable partial order of dependence which creates sentences and which is essential to science languages also. The meaning in languages is thus restricted to what can be meant by these words and their combinations.[8] As a result, language is limited in what it can express, especially in what it can express structurally, and directly (rather than just by talking about it). In effect, by its grammar, language can do directly little more than predicate (which includes, for example, taking questions and commands as statements that one is asking or requesting something). Language is thus primarily a system for publicly communicable information, as seen in Chapter 11. There are many inexplicit

---

. . . *killed his words unspoken*; and from *I'm just killing time here* we cannot obtain *I'm just causing time to die here*.

[8] The particular meanings of extant words are in part accidental, although there are many meanings that are found with little difference in a great many languages. Some types of meanings are harder for words to come by. For example, many abstract concepts such as (a) *understanding*, *important*, or (b) *truth*, *magnitude*, are available in single words either: (a) only as a result of metaphoric specialization of words of concrete meaning (G405), especially of morphologically complex words; or (b) only as a result of operator nominalization (from nominalizing a sentence with zeroable indefinite arguments: *truth* from *things' being true*). In addition, the meaning of a word can be greatly modified by adding modifiers to it, which changes the selection since the further operators on the given word are then determined by the word as modified (as in *The snow house melted*). (It should be pointed out that, as in all derivations in the present book, 'from' above means specifically 'derivable by established reductions or other transformations in the grammar'.)

and private meanings that language cannot express directly. As noted in 10.7, language cannot communicate that which is communicated by music and other arts. Feelings are indicated either by the possibly false statements that one has these feelings, which is hardly an expression of them, or by phonetic activities (intonations, ways of pronouncing) which have no regularities of combination with anything in grammatical structure and so are not part of a grammar. Literary, poetic, and even jocular effects in language are obtained not primarily by special grammatical features (such as allusion) but by special deviation from standard grammar, or by exceptional utilization of grammatical features.

Finally, natural languages all have the property of having different relatively stable yet changeable likelihoods for individual operator words in respect to their individual argument words and vice versa. This makes it difficult for words to be classifiers of other words in any definite manner. It also makes it impossible for language to exclude nonsense—that is, for nonsensical sentences to be ungrammatical. And it makes it impossible to identify single sentences by their structure as true or false: even *and* on a sentence together with its negation is not necessarily false, as in *A politician? Well, he is one and he isn't*. However, in science languages, where likelihood properties are replaced by special word subclasses organized into special sentence types, both of these limitations can be overcome.

12.3.5. *From use to constraints: Institutionalization*

The conventionalization and institutionalization of language use to the point of its becoming a constraint on how one speaks one's language (noted at the end of 12.3.3) may arise from several sources. For one, the limitations on the experience that people express or communicate makes it more appropriate and expected to use certain words together rather than others. Second, speakers and hearers need to understand each other efficiently, and so come into accord on what sounds and word occurrences are considered repetitions of each other, where to expect particular kinds of words (if they are being said), and in what way to understand the relation of words in a combination, and finally, when to understand something as being a reduction of another word. In later specialized constraints, there may also be conscious imitation and agreement on how words are used, and peer pressure for particular uses; this may hold for the development of mathematical terminology and notation, and of styles of argumentation in science reports.

In all this conventionalization, and even in the institutionalization that makes some forms grammatically 'wrong', there is little or none of the indirection and misleading that make so much of culture an institution of control. In this respect language differs from much of culture. The conventionalization here will be seen to have something of the character of biological evolution rather than of power-preserving social structure. It is also important that individual idiosyncrasies in language, as well as cultural and literary styles or social and regional divergences, can all be described in terms of special constraints, even though in a rather different way than the constraints of language structure, of language-like systems, and of language change. For this means that in various respects and degrees all of these can be investigated in similar terms, and therefore can be compared.

The development of constraints from word-specific dependence (i.e. of each word on a selection of certain others, as in *eat* on the pair *boy*, *berry*) to the general relation of dependence-on-dependence was crucial to reaching a stable system. As has been noted, such a system could be generalized so as to be adequate for any new situation, whether this situation consisted of new words in existing classes (e.g. other pairs than *boy*, *berry* under *eat*), or whole new classes (e.g. words, such as *continue*, that might depend on *eat* in the way *eat* depends on *boy*, *berry*). Indeed, defining word classes by their dependence on the dependence properties of other words creates a self-contained and self-organizing framework for language and enables it to expand without limit, but only in terms of this same relation. For example, an O_o class that operates on O to make a sentence does it in the same way that O_n operated on N to make elementary sentences (as in *The boy's eating of berries continued*); and so does an O_{oo} class on a pair O,O to make one sentence out of two (e.g. conjunctions).

It is clear that the constraints developed where they did not exist before, ultimately out of a situation without constraints. However, they did not—and could not—develop out of nothing, but rather out of tendencies to use certain kinds of word combination in preference to others. The initial constraint created the set of orderly elementary sentences where no such thing had existed before as a fixed structure. But this does not mean a sudden innovation or even a recognizable innovation—only that types of combination which were widely in use, in contrast to the non-use of other combinations, came to be used exclusively. This is in effect the same as what happens in language change (12.2). A somewhat different situation of a new structure arising without apparent invention or discovery was seen in the case of the alphabet (7.2.2).

The initial conventionalizations developed partitions of the set of words out of fuzzy differences. To give an artificial example: if, say, *sleep* was used with *child*, *dog*, etc., and perhaps rarely with *tree*, and virtually never with *berry* or *walk*, a comparison with, say, *fall* which occurs with all these words except *walk*, suggests that *berry* be included as possible argument of *sleep* (with only nonsensical meaning), while *walk* is excluded as argument of *sleep* or *fall* and is left to be itself an operator. The partition of the set of words into N, O_n, etc. is thus a reassignment of virtual non-occurrence into rare or nonsensical occurrence in the case of *berry* under O_n, and exclusion in the case of *walk* under O_n (any such occurrence being then called 'not in language' or 'wrong').

Changes of usage do not only conventionalize tighter structuring. They can also create word classes that at least in some respects cut across the O_n, O_o, etc. partition, as in the case of prepositions, which occur in many operator statuses, and which we can attribute to just one operator class only by calling upon transformations for deriving all other uses. Indeed, changes in use can reduce some of the tight structuring of grammar. For example, in various languages the gradual phonetic subsidence of case-endings, which had been a structured and required set of operators, left prepositions, which constitute a less structured set, to fill their syntactic place.

A prominent situation of usage-change-creating tight structure is seen in the special constraints of special word classes; such processes have some similarity to what is called phonemicization, wherein changes in the pronunciation of certain sounds can make them phonemically distinct from other sounds in ways that they were not before. One syntactic case is the loss of the second subject under English auxiliary verbs such as *can*. As has been seen, those verbs may have had a full sentence as their object (second argument), as though one could say *I can (for) him to do it* meaning something like 'I know how for him to do it.' In time these verbs ceased to be used in cases where the subject of the verb under them was different from their own subject. As the sameness of subject for *can* and the verb under it became a requirement, a new constraint arose in the language. These verbs also have an unremovable tense, and they zero the *to* under them. Since several verbs share these requirements, there rises a small subclass of 'auxiliary' verbs different from other verbs. This subclass is thus the product of a set of constraints acting on the same words. The tight structure here is lessened because the domain of this set of constraints is a bit fuzzy: some verbs have only some, not all, of the constraints of the auxiliaries (e.g. *I ought not do it* by the side of *I ought to do it*.)

Another important constraint which presumably grew out of use preference is the verb–adjective distinction. Both of those word subclasses are operators, but, as has been seen, those operators that had higher likelihood of tense-change within a discourse had the tense attached directly to them (e.g. *walked, aged*) and are called verbs, while those that did not had the tense attached to a separate 'carrier' *be* (e.g. *was brown, was old*) and are called adjectives (G188). While operators are graded in respect to tense-changeability, they are partitioned in respect to tense-attachment: *sleep* is a verb even for a hibernating bear; *ill* an adjective even for an intermittent illness. Note that the same distinction can be made without creating a grammatical requirement: Chinese has only optional suffixes for 'past time', but these occur more commonly on momentaneous and active predicates than on stative ones.

In all these cases we see much the same kind of structure-creating process.

The particular accumulation of pronunciational change which makes new phonemic distinctions ('phonemicization'), such historical changes in grammar as the formation of compound (periphrastic) tenses, Manu Leumann's formulation of historical change in terms of use-competition among sentences (n. 3 above), the formulation of word-use change in terms of combinatorial patterns as in the work of Henry Hoenigswald (n. 3 above)—all of these in different ways exhibit grammatical innovation as a result of changing use becoming institutionalized. The same process can also yield the initial formation of grammar out of freely occurring, grammar-less word combinations.

It has been noted that some sentences do not conform to the operator–argument structure. The general and self-contained framework of all these constraints makes it possible for almost all non-conforming constructions to be reinterpreted as though they did conform. They are thus domesticated into the system. This applies both to pre-syntactic forms that may have existed before dependence and sentences evolved, and to many forms that arose later by non-dependence processes such as analogy and change. Reinterpretation of forms as though they were produced by the dependence relation on known words, with reductions of known types, is possible only because the predicational meaning of the dependence suffices for almost all meanings of word combinations, and because known types of reductions suffice in almost all cases to yield existing forms out of suitable dependence-structured sources. For some items, such as the perhaps pre-syntactic tenses and plurals, it is indeed difficult to find reasonable sources (though dependence-structured paraphrases are always available); but it is easy for others, including the recent periphrastic tenses (such as the perfect tense, *has caught*) as well as the perhaps pre-syntactic interrogative and imperative.

As an example of how the arising of a fixed sentential structure can lead to reconsidering the grammatical status of linguistic forms, consider the domestication of word meanings into syntax. Before syntax came to be, the many words that must have already been in use must have had some range of meanings as associations with referents, i.e. with objects, events, states in the perceived world: *rock*, say, would have meant a stone of big or medium size, and *stone* medium or small, and *pebble* small. Once words are used in operator–argument combinations, within which the likelihoods are relatively established, we find a new situation, local synonymy: *rock* can have *stone* as synonym in *I dug up the rock*, but in *They climbed the rock and built houses on top of it*, one would be much less likely to admit *stone* in place of *rock*. Thus the meaning of a word-occurrence in a fixed structure may be only a segment of the meaning-range of that word, and may be the same as the meaning, in that position, of some other word which is not otherwise synonymous with it. This situation makes the meaning of a word depend on its grammatical co-occurrence, and in a way that is more explicit in detail, hence more useful, than presenting the full meaning-range as a single entity.

In a dictionary, a word is given a meaning or a range of meaning, although the set of referents is in many cases not an otherwise existing or useful class. In a more careful dictionary, a word is given a list of meanings, without the implication that these form a whole, a continuous range, or even that they are always interrelated in content. In the occurrence of words in sentences, however, the situation we find is not that words appear in a whole meaning with just part of that meaning used in each occurrence, but that each word-occurrence in a sentential (operator–argument) environment appears in just a specific meaning which may not be the same as that of another occurrence of that word in another environment. Starting from meaning by referent, we have thus reached meaning by operator–argument co-occurrence. The difference between these two is thus a difference between two stages of vocabulary: pre-syntactic (or in isolation, independently of sentences), and in respect to sentences (i.e. when occurring in sentences).

Another way in which syntax readjusts the meaning of words is seen in the pronouns. It is reasonable to assume that pre-syntactic vocabulary contained demonstrative words such as *he* or *this*, whose individual referent varied with the situation of speaking. Once words were used in sentence structures, most occurrences of pronouns became cross-referential, referring to the same referent as some other word, ultimately some non-demonstrative words, in the flow of speech. The remaining demonstrative occurrences (even perhaps in e.g. *It's a nice day*) could then be artificially reconstructed as cross-referential to a zeroed metalinguistic *We are speaking of something; that something is a nice day*. Starting from the demonstratives as having a unique kind of referent, approximately an unidentified specific referent in the extra-linguistic situation, we have thus reached cross-reference within the sentence-sequence (5.3.4).

In this way, the few essentials of syntactic structure not only constitute a stable system, but also are consolidated by being able to house almost everything else in language, including word meanings. Considering that language also includes its own metalanguage, we see that it is a self-contained system. It is exhibited and recognized by constraints on repetition and combination of parts, and is defined within itself. While this analysis shows something about how language could have come to be, it leaves open the question of single or multiple origin. The relation of word meanings and constraint meanings to the phenomena of the world, and the communication-success criterion for institutionalization, suggest that language could have had various primary–parallel origins, no less than a single origin, or possibly a mix of the two.

12.4. *The nature of language*

12.4.0. *Introduction*

The theory of language structure and development presented here makes it possible to draw some conclusions about the nature of language. The conclusions are the more relevant since the theory finds language to be a self-sufficient system which includes its metalinguistic apparatus and its informational interpretations (all of which have to be accounted for, because any account of them would all have to be stated in language itself). We have seen in 12.3 how language could develop out of a situation in which there was as yet no language. We now consider what the whole theory shows about the properties and functions of language.

12.4.1. *The language system; language and thought*

It should first of all be clear that language is not directly a more developed form of animal communication. Not to enter into the discussions of animal communication and ethology, we may nevertheless judge that the difference between human and animal 'language' is one of kind, not of degree. Animal communication contains sounds or behaviors which one might consider the equivalent of words; but, in present knowledge, it has nothing of the dependence relation which creates sentences and predication. A less obvious but fundamental difference is that there is little present evidence of the respondent animal exercising any choice in type of response (to stimulus or 'symbol'), other than whether (and perhaps how strongly) to respond. Existing evidence suggests that animals have developed innate propensities to particular behavior in response to stimuli; the stimuli include behavior (sounds, postures, etc.) of others in their species. This can be called rudimentary communication; but we cannot say that the behaviors are subjectively signals or symbols, and there is no systemic signalling structure with potential for variation and expansion. The apparent behavioral sequences, as in some aggression and mating behavior, are not innate sequences within an animal, but are a co-evolved interaction produced by each given participant responding to the stimulus offered by the other in response to the animal's own behavior or presence. The whole is more like the co-evolution of flowering plants and pollinators such as bees than like the functioning of human language.[9]

[9] K. von Frisch, *The Dance Language and Orientation of Bees*, Harvard Univ. Press, Cambridge, Mass., 1967; G. Baerends, C. Beer, A. Manning, eds., *Function and Evolution in Behaviour*, Oxford Univ. Press, Oxford, 1975; D. J. McFarland, *Feedback Mechanisms in Animal Behaviour*, Academic, London, 1971; M. Lindauer, *Communication among Social Bees*, Harvard Univ. Press, Cambridge, Mass., 1971.

As to whether animals can learn language, such work as that with chimpanzees has not differed in essentials from other animal training, except for the greater complexities of response made possible by the chimpanzees' general intelligence. The impression that there is language comes from choosing appropriate words as names for the buttons the chimpanzees are trained to push.

A very different distinction is to be made between languages and thought. Again without entering into this vast problem, several considerations are obvious. There is certainly some type of thinking that takes place without language and is unrelated to its structure.[10] And once language is in existence, there are occurrences of structured language without comparably organized thought, as perhaps in illusions, in alliteration, in poems of mood, in nonsense and nonsense-verse, not to mention small-talk. The decisive factor is that language has additional conditions over and above any representation or facilitating of thought: being public (and pre-setting a public stock of words and co-occurrence constraints); efficacy for transmission; and realizability in speech with its linearity, with the limitations which are imposed on speech by articulation and auditory distinguishability, and those which speech imposes on structural representation of partially ordered and complex relations among words. Some structural features of language are of little or no relevance to any language-independent thought: for example, phonemic distinctions and phonemic composition of words, reduction and other transformations, and the loss of specific meaning for words with broad selection.

Evidence that thought and information can be independent of certain grammatical features is seen in the occasional circumventing of grammar to express meaning (as in expressing time without tense in nominalized sentences), and more dramatically in the fact that writing in a given science can be described in the same science-language grammar no matter in which natural language it is written (10.5). Nevertheless, most orderly thought and most language go together: the structure developed by language makes precise, and to an unknown extent may shape, and even direct, thought and the perception of the world, even while it limits the character and direction of these.[11] This situation of something functioning both for service and for

---

[10] L. Weiskrantz, ed., *Thought without Language*, Clarendon Press, Oxford, 1988. In the large and uneven literature on this subject, special reference should be made to two of the early modern contributions: the writings of Jean Piaget, especially *Le Langage et la pensée chez l'enfant*, Delachaux et Niestle, Paris 1924, and Jean Piaget and Bärbel Inhelder, *Mémoire et intelligence*, Presses universitaires de France, Paris, 1968; and also L. S. Vygotsky, *Thought and Language*, MIT Press, Cambridge, Mass., 1962. For the status of language in respect to thought, it may be relevant that almost all writing systems, though they start with pictures as direct representations, develop into signs representing words and their sounds.

[11] This vague relation of thought to the basic structure of language is quite different from the unfounded suggestion that the detailed grammar of a particular language affects or determines the thinking of its speakers, as in the claim that languages which have no grammatical tenses might block the development of physicists who need *t* as an essential variable.

control is well known: aside from those pervasive social institutions which serve solely for control over the people, the bulk of social, cultural, and personal conventions both facilitate as well as limit—often simultaneously— the behavior of the people who function within them.

What has been seen in 11.1, 11.6, about the meaning of the structural entities of language suggests that the system developed in language is consonant with important features of advanced thinking. In overview, the relation of language structure to thought may be summarized as folows. (1) For humans to use words (singly), i.e. sounds that are associated with ('mean') the user's states or received stimuli, is consonant with animal behavior (even if the animals are not thereby subjectively 'sending messages'); hence, this in itself may not constitute thought as a substantively new phenomenon peculiar to humans. (2) Using word combinations in the presence of combinations of the stimuli referred to need not be considered as essentially different from using single words, even if in the objective world certain of the words in the combination 'mean', roughly, objects, while the other words mean states or properties of those objects. (3) However, the fixing of an 'operator–argument' relation between the object words and the property words in each combination, even though it presumably was only a standardization of a naturally-occurring constraint on the combinations, created predication as a semantic relation between the words in a combination, whatever they be. This had two semantic effects. (4) One was that any new combination had meaning by virtue of the predication relation between known words, not only by virtue of the association of the combination with particular experience; thus, the meaning of words shifted from being extensional (i.e. the specific associated referents) to being intensional (i.e. anything that could fit the predicational relations in which the word was used). (5) The further effect of (3) was that predication could be said independently of actual experience, since a predication would have the meaning of one word predicating about another even if there was no experience with which that word combination was associated. It is reasonable to see (4) and (5), both of them due to (3), as important features— perhaps facilitators—of thought going clearly beyond any pre-human level.

Before stating this explicitly, we survey the structural meanings. As to words: first, words (or morphemes) are discrete entities, since they are sequences of discrete phonemes (or of phonemic distinctions): this makes the meaning-bearing symbols explicit. Second, when discrete entities are used to represent in a regular way the objects or events (and relations) with which they are associated (end of 1.4), they necessarily come to mean selected perceived aspects of the objects or events, rather than the less-selective representation contained in, say, a picture. Third, since by 11.1 all words can be used not just for a unique object or event but also for various ones (any ones that fit the meaning the word already has), each word is in effect defined not by a specified set of referents (objects or events), but by the properties that are common either to all its referents or to various subsets of them. As to sentences: the operator–argument relation over a finite vocabulary creates sentences that suffice to indicate or represent not the real world so much as perceptions related to the phonemena of the world.

In addition, the informational structure of language provides structural possibilities that go beyond information, expressing and perhaps even fuelling the imaginative capacities of thought. This potentiality in language arises from the fact that words whose meaning comes from association with perceptions of the world are naturally restricted in their co-occurrences to the way perceptions in nature co-occur (11.6.1). As a result, the restriction of combination (selection) among words reflects the meaning of those words. But this restriction leaves open the possibility of using the unused combinations. The unused combinations present new meanings, which may be nonsensical, interestingly imaginative, etc., in respect to the perceptions that had restricted those combinations. And old words, or newly made phoneme-sequences, can have new meanings which are determined by their new selection of word combinations rather than by direct association with observed phonemena (e.g. *unicorn*).

A crucial contribution of language to thought is, then, the separation of predication (grammaticality) from word-choice (meaning), due here to distinguishing operator–argument dependence (non-zero co-occurrence likelihood) from all the specific likelihoods. Due to this, when we want to make a new combination of meanings we find at hand a structure that puts it into a publicly-recognizable predication, without our having to develop a new bit of grammar for the new word-combination—a new consensus for what the new juxtaposition of meanings should mean. This permits one important type of thought—predication, with the interrogative, optative, and other modes derivable therefrom—to go beyond the direct meaning-combinations of our experience. (Compare the lack of useful new meanings in totally ungrammatical sequences of words.)

A second crucial contribution is the recursive extension of predication to create predication on predication (more generally, all the second-level operators, such as O_o). Selectional restrictions on these in respect to their arguments (especially in the case of connectives) create the machinery for structured extended thought, discourse, and proof.

We see that the projection of phenomena onto perceptions, the ability to define by properties rather than by denotational extension, the ability to imagine beyond what has been observed, all arise in the structuring of language. But these are also part of the apparatus of advanced thinking. It is not a question of whether these capacities came from language to thought, or the converse. One can reasonably suppose that these developments in language capacity facilitated developments in thinking, and advances in thinking facilitated the utilization of language-structure potentialities.

It should perhaps be noted here that because of the structural differences between language and logic (primarily due to the different selections that individual words have, 3.2.1), the attempts to represent the meaningful structures of language in the structures of mathematical logic have not been successful. Logic goes beyond language in the capacity to structure proof, and language goes beyond logic in the capacity to differentiate meaning.

It may also be mentioned that the way complex thinking is expressed in language structure casts doubt on the claimed potentiality of computers to carry out an equivalent to thinking. Computers can indeed be programmed to analyze sentences in respect to the known structure of their language (ch. 4 n. 1). But, first, we know that this structure is based on the recognition of phonemic distinctions, the phoneme-sequences that constitute words, and the operator–argument relation among the words of a sentence. We have seen (12.3.5) how these could develop in the course of the intricate and public behavior of conscious living organisms interacting with their environment, but we do not know how computers could develop such structures without being given them in a program. Second, nothing known today would suggest that computers can carry out, sufficiently for any serious purpose, the selective perception and the subjective response to the outside world, let alone the awareness, that determine the ongoing use of human language— the word-choice of a sentence and the sentence-sequence of a discourse. The fact that this response and awareness must have material grounds within the processes of the human organism does not mean that the particular man-made mechanisms we know today should be expected to simulate those processes.[12]

It follows that language cannot be understood as a further development of some existing system, but has to be explained *de novo*. To do so we have to go by the properties which seem to be necessary and universal for language. Before doing so, in 12.4.2–4, we note here which universal properties are not systematically necessary. Some, such as linearity and the limits on distinguishability of sounds, are, though universal, not in principle necessary to language but are due to the fact that language developed in speech rather than, for example, in gesture or markings. Indeed, language has had to circumvent some properties of speech: to circumvent the continuous sounds by means of segmentation, the continuity in differences among sounds by means of distinctions, the paucity of sound distinctions by using phoneme-sequences for words, and the linearity of word-sequence by taking it as a projection of a partial order on words (which is retained in writing even though there a lattice representation would be possible).

Other universal properties are by-products of the linguistic system rather than being themselves historically formative factors: such are the capacity of a language to be its own metalanguage, the capacity for cross-reference, the

---

[12] Against the claims that computers can be 'taught to think', publicized under the name of Artificial Intelligence, see Roger Penrose, *The Emperor's New Mind*, Oxford Univ. Press, Oxford, 1989.

possibility of classifying words structurally, and the emergence of a mathematical object (6.3).

Another relevant universal property of languages is their intertranslatability. The meaning-ranges of words (their operator–argument selection), and the allusion based on phonemic similarity, and some reductions, and also any paradigmatic requirements, are all not directly translatable unless the two languages in question happen to be similar in the given respect. However, the inter-language similarities of the selections and meanings of many words, the modifiability of a word's meaning by attaching modifiers, and the virtual identity of the meaning of the partial order (predication, and the meanings of constraints defined from it), together make approximate translation possible between any two languages, within reasonable limits of length for any needed strings of modifiers.

As to content, it has been seen that language developed as a vehicle of public, transmissible information. But not all information is or can be carried by the structure of language. More precisely, language is seen to contain a deposit of publicly recognizable aspects of experience. Furthermore, the speaker's intent in a sentence may be to refer to a unique object or situation, a token, but the sentence as said refers to a class of referents, a type. To go beyond this public reference one has to go beyond single sentences into an additional structural dimension—the double array of discourse or the word subclasses of sublanguages, all of which capture the meaning of subject matter as going beyond separate observations. Or else one has to go to the sentence-sequence structure of argumentation, of reasoning and drawing conclusions. Sentence structure and word-choice alone cannot give these additional informational meanings, even as they cannot distinguish truth, subjective 'truth', plausibility, or nonsense. But it is uniquely language structure that can admit of these major further developments.

12.4.2. Systematizing factors

To understand the nature of language we may ask what created the order—the structure—that constitutes that system, and how this structure came to be. Syntactic order does not come from phonemes or meanings. The vocabulary by itself has no intrinsic structure that can be stated: the phonemic structure of words is not relevant to their syntactic behavior, and their meanings do not fall into a sufficiently sharp and orderly system.

Thereafter, the formation of sentences consists in imposing a meaning-bearing ordering (dependence) of word-occurrences. This dependence is generalized by aggregating words into classes in such a way that all sentences have the same dependence formation. The formation of elementary sentences is then repeated recursively on the already-formed sentences, to make the remaining sentences of the language.

Meaning-bearing orderings beyond sentence-formation are created by the similarities among sentences in sequences or in sublanguages. There are also miscellaneous structurings in particular languages, which are meaning-bearing only in a secondary way if at all.

These are in broadest terms the contributions to making the order, or structure, which we see in language. We now ask how these contributions came about.

In 12.4.2–4 we ask what are the formative factors in language structure and how they determine its nature. In the first place, the fact that language is a public activity entails its being much the same both for speaker and for hearer. In this it differs somewhat from, for example, music, which, while it is often a public activity, can also occur privately, and even when public does not have to mean the same to producer and recipient. Within this public status, language developed not so much for face-to-face communication as for error-free transmission. This can be seen, as noted, in the development of discrete elements out of the continuous phenomena of speech, in the establishment of phonemic distinctions among the complexly differing sounds, in institutionalizing repetition as replacement for imitation, and in the public pre-setting of a vocabulary of phoneme-sequences—ones which are arbitrary (not onomatopoetic) in respect to the word meaning.

Transmissibility without error-compounding requires not only that the elements of the material to be transmitted be discrete, but also that they be pre-set in sender and receiver. When the speaker and the hearer are referring to a set of elements known to both, the hearer need receive only enough of a signal to distinguish a particular element—phoneme or word—in contrast to all other elements that could occur there. When the hearer then transmits (repeats) the utterance, he pronounces his own rendition of the pre-set (i.e. known) elements which he has distinguished. This means that both must learn to recognize a set of grammatical elements, primarily particular phonemes (or phonetic distinctions) and secondarily vocabulary (morphemes, words), in respect to which they speak and perceive utterances. It is this public institutionalization that makes the transmission of an utterance a repetition, whereas an attempt to redo or transmit something whose elements are continuous or not pre-set is an imitation. It is well known that all languages have to be learned; and that when a speaker says something which he has heard he is repeating it, not imitating it (whereas he would be imitating in the case of a yell or other non-linguistic sound). Those linguistic phenomena which are not represented by discrete pre-set elements may indeed carry information (e.g. intonations of hesitation or of exaggerated matter-of-factness), but they do not enter into grammatical relations to other elements.

The restriction to discrete elements is uniquely appropriate for error-free transmission of utterances. There has been much discussion as to whether language is an instrument of expression or of communication. Rather than either of these, its structure is suited for an instrument of transmission. Indeed, expression and face-to-face communication are served quite well by such continuous phenomena as intonations, phonetic characteristics of individual speakers, gestures, etc.; and the hearer may garner a great deal of information from such continuous data, over and above what is grammatically structured in the speaker's utterance. But when the spoken material is transmitted, i.e. spoken by the hearer to a new hearer, and so on down the line, there is a possibility of compounding errors when the continuous elements are transmitted. In contrast, when the utterance is characterized by discrete elements only, it can tolerate a considerable amount of 'noise' in transmission from one hearer to the next without producing error.

The non-grammatical status of the non-discrete, non-repetitional elements underscores the fact that the grammatical elements and relations that arose were specially suited to the transmissible information. In these circumstances, what was primarily communicated was not private feelings and desire, or interpersonal comments and requests, but statements of information. The private and interpersonal can in various degrees be said, but only by making the usual kind of transmissible statement about them (*I am happy* or *How are you?* or *Please*).

In addition to these factors in the formation of phonemes and words, much of the structure of syntax is explicable as developments which were appropriate and effective for such public communication. This is especially the case for the main creators of syntax, namely, the massive constraints on word combination. As noted in 12.3.5, these constraints, which state non-equiprobabilities of combination, are public standardizations of the public use of words in combination, whereby first of all the combinations that never had occasion to be used became proscribed, as an institutionalization of their non-use. Going further, the fact that certain words (here termed operators) normally appeared in combination with others was then institutionalized into a requirement for those operator words, as depending on their argument words and so creating a sentence with them.

Standardization of behavior is not surprising in a public context. Already in the pair test that established phonemic distinctions (7.1), what created the distinction was not the meanings of the words, nor differences in meaning, and not even the behavior of the speaker and hearer in recognizing the same phonemic distinction, but rather the institutionalization of these recognitions, so that they were standardized throughout the population. It is the absence of such standardized distinctions in areas other than language that makes the equivalent of phonemes unavailable for culture analysis and for other fields where export of linguistics' phonemic success has been sought. In the far more complex situation of syntax, the driving-force in standardization is the adequacy and effectiveness of common use and of its institutionalization as a stable system of communication. This systematizing process may be not only a codification of custom but also a honing of the behavior as a tool, an increasing of the efficiency of language as a vehicle for information.

Such efficiency-increasing systematizing seems to have taken place not only universally for the basic constraints of language, but also locally for less basic ones. Thus in many languages, very common operators and modifiers (which are operators in attached secondary sentences) are attached as affixes to the words (arguments) to which they apply: e.g. causatives and verbalizers, as the prefix in *ensure* (with its near synonym *make sure*); possessive suffixes on nouns and tense affixes on verbs; also affixed arguments, as in subject and object affixes on verbs in some languages. In some of the most common situations, these are conventionalized into required paradigms for the given language. Another reduction in distance is having a modificatory secondary sentence interrupt a primary sentence at the word being modified (3.4.4). In addition, there are widespread standardized reductions in the phonemic shape of high-likelihood word-occurrences, including the dropping of words which are reconstructible from the environing words, given the hearer's expectations about word combinations.

Beyond the establishment of the dependence relation, which sees a word as requiring one or another member of the set with which it in fact occurs, there was a further systematization which increased the generality of the system and its adequacy for new and changing word combinations. This was to admit as possible arguments of a given operator A all arguments of operators whose argument sets intersected the argument set of A. This procedure creates a partition of the vocabulary into a few operator and argument classes. We thus move from fuzzy argument sets of each operator to a fixed set of arguments from which each operator of a given level selects its more likely arguments. This generalization makes possible the move from dependence to the dependence-on-dependence of 3.1.2, creating the mathematical object of Chapter 6. It gives the linguistic system stability and flexibility in the face of change and innovation.

One other stabilizing flexibility is given to language by the close relation between the meaning of a word and its higher-than-average co-occurrence (selection) with particular operators and arguments: the meanings of a word can be extended or modified simply by extending or modifying their selection (i.e. what they combine with) without having to exhibit to the hearers the new aspects of the perceived world with which the speaker is now associating the word.

At this point it may be useful to attempt an orderly survey of the motive sources for the shape and change of language.

1. There are initial representational factors: in associating short sound-sequences with perceived objects and events; in associating a dependence (of operator on argument) with the co-occurrence of events and relations on

the one hand and objects on the other; in the different but relatively stable likelihoods of particular operators in respect to particular arguments.

2. There are ongoing communicational factors. These are primarily in the service of effectiveness: in instituting distinctions (first of all, yielding phonemes); in serving public consensus, so that speaker and hearer understand language approximately in the same way; to facilitate transmission beyond face-to-face; to reduce the length (and possibly, in some cases, the complexity) of messages without loss in their information. Some communicational changes may also be in the service of social convenience: e.g. the introduction of more attention-catching forms, as in the periphrastic tenses; or the avoidance of socially undesired forms, as in euphemisms. These motive sources may come into conflict with each other, as when certain reductions create ambiguities (primarily degeneracies) that complicate the hearer's decoding of what the speaker meant, i.e. complicate their consensus.

3. There are ongoing influences of one set of forms on another, within the forms of language and their ways of combining. These include the development of patterns within grammar, and also the many kinds of analogical formation, including the cases where word-frequency and word combinations are extended on the basis of how other words are used.

The third type above may be considered 'internal', the first two 'external' and in various senses 'adaptive'.

12.4.3. *A self-organizing system*

That language is self-contained we know from its containing its own metalanguage, and from the properties seen in the preceding chapters and in 12.4.1–2. Here we will survey briefly how this condition, though roughly stable, nevertheless made possible the growth of the complicated systems which are to be found in language. It will be seen in 12.4.4 that language structure is accretional, and it has been seen in 12.3.4–5 that both language structure and language information stay within certain limits. In this sense we can say that language makes itself, as indeed everything that is not made or planned by an outside agent must do. The overall self-organizing condition in language is that on the one hand the existing forms limit what is available for use, and also to some extent affect (direct) this use, while on the other hand use favors which forms are preserved or dropped, and in part also how they change; this includes the ability to reinterpret borrowed and other external material as though structured by the existing system of the language.

Throughout this study, we have seen that the non-random combinations of words in sentences are best described by constraints, and that each constraint is defined on previous constraints, down to the unstructured sounds. Unless—surprisingly—all constraints arose simultaneously, we have to judge that they came to be as successive stages of structure. When later structures develop out of earlier ones without being due to external intervention, we may consider that self-organizing features are displayed. As most notably in Darwinian, Marxist, and Freudian theories, we find, in describing what coexists in an organism or a structured aggregate, that further and explanatory regularities are reached if we ask what processes form the observed interrelating phenomena.

(1) The self-organizing process begins after the existence of some (arbitrary) stock of words. In phonemics we see a structure that developed around its existing elements. Phonemes could not have existed before words; hence words must have originally existed as sound-sequences without phonemic distinctions. The development of phonemic distinctions among these sound-sequences has taken place in all languages, apparently, and has made specific distinctions among words. In so doing, it has created fundamental features of structure: discreteness of the ultimate elements of language (since a difference is a discrete entity), fixedness of word-form, and repetition as against imitation in transmission of speech (and writing, in contrast to pictorial messages).

(2) The chief process (in 12.3.5) begins after the existence of word combinations, which were presumably at first casual and unstructured. The institutionalization of the actual non-occurrence of combinations into a restriction on combining creates a structural requirement (dependence) out of a meaning-related customary behavior.

(3) The reduction and zeroing of words whose occurrence is predictable, originally also a meaning-related customary behavior, became an item of structure as particular domains of words came to be reducible in particular environments.

(4) In various languages, particular restrictions on word combination created new word classes and structure, again largely by developing meaning-related customary combinations into requirements. This is seen, for example, in particular domains of transformations, as in the verb–adjective distinction, and the English auxiliaries. It is also seen in the requirement of certain small subclasses of words (often in affix form), such as the time, person, and number affixes of conjugations and declensions, and in requiring the subject-repetition under the auxiliaries.

(5) At some point language came to predicating about sentences much as in the existing (elementary) sentences the operators predicate about their word (N) arguments. For this, the same dependence that characterized the structure of the elementary sentences was extended to the further operators in respect to the elementary sentences. This created second-level operators and sentence expansion.

(6) This extended dependence invited generalization from dependence on specified word lists to dependence on the dependence properties of words, creating a mathematical object at the core of language structure, as a stable criterion for that structure and for further self-organization within it.

(7) Although the dependence structure would appear to be just an abstract construct, which constitutes a description of regularities in the speech events of the real world, it had sufficient reality in people's word-choice behavior so that new combinations, which were not born out of this dependence, were formed on its analogy. Analogic construction, which is not only apparently universal but also very prominent in language, thus extends the domain of the dependence relation, even where it was not the sentence-forming process on the underlying elementary words. In addition, there is evidence that sentences and sentence-parts which were not formed by the dependence relation (borrowed, reduced beyond recognition as in *goodbye*, proverbs carried over from past grammar, etc.) can be reinterpreted as resultants (or exhibitors) of dependence.

(8) The later regularities are not within the sentence, which has already reached a flexible stability, but within subsets of sentences, such as sentence-sequences (ch. 9) and referent-restrictions (or, equivalently, restrictions in communicational situations, ch. 10). The well-formedness of sentences and the structural conditions for sentence-sequences, as in proofs in logic and mathematics, are special examples of requirements that institutionalize, and in this instance codify, the common behavior found to be effective for discourse in those fields.

This brief survey has shown a few self-organizing features: the development from differences to explicit distinctions (1); the institutionalization of customary behavior into requirement and structure (2, 3, 4, 8); and the extending and generalizing of relations (5, 6, 7).

In all, we have here various examples of how structure (and, in some cases, the impression of purpose) arises out of non-structured (or non-purposive) events—as was also seen for the alphabet (7.1.2), and as is a known issue in the theory of evolution. In some cases, the events create not directly the structured forms, but the conditions for structured forms: thus the establishment of phonemic distinctions created the discreteness of phonemes and words which made it easy for the relations among these to form classes of them.[13] As to the phonemes, the roles and physiologic forms of the vocal organs (vocal chords, mouth, etc.) allowed particular variations and combinations of sounds which yield particular cross-classifications ('phonetic patterning') of the phonemes. In syntax, parallel events on similar domains can create subclassification, as in the bundle of reductions

---

[13] Events can also disturb or modify structure, as in phonemicization, when a phonetic change in an allophone (a sound which is, in a particular environment, the phonetic value of a phoneme) can make it become a different phoneme.

that creates the English auxiliary verbs, where some of the reductions acted on *can*, *may*, *need*, etc., while others acted on *can*, *may* but not on *need*, and so on. And the utility of repeating a grammatical relation between particular words in successive sentences of a discourse can lead, over the whole discourse, to a double array of its sentences.

Furthermore, the accretional property, whereby events act on the resultants of prior events, has in many situations the effect of preserving the structure created by the prior events, and even of strengthening its importance by providing substructures within it—in some cases creating the effect of a direction of development. For example, it has been seen that second-level operators act on first-level operators (hence on the elementary sentences created by the latter) in the same way as the first-level operators had acted on their zero-level arguments in forming those elementary sentences (and comparably when the second-level operators act on second-level operators). This recursivity establishes a single structural process for all sentences. Thereafter, transformations are able to act in the same way on all sentences, in terms of this common structure, thus strengthening the importance of the structure. A different kind of example is the extension of prepositions from their original argument-requirement status to many other syntactic statuses. Each such extension was probably fostered by all the previous extensions of the prepositions.

We thus have a positive feedback, in which the formation of structure favors accretions—including subsidiaries—of various kinds to that structure, all of which further establishes the structure.

The further development of language constraints—in discourse structure (9.3), argumentation (10.5.4, 10.6), and sublanguage (ch. 10)—has a property relevant to this survey. For while all these further structures are clearly fostered by what is being talked about—by the nature of discourse, argument, and subject matter—it can be seen that the apparatus which is here developed is not something new unrelated to the language structure so far, but is rather a set of further constraints in the spirit of language structure and within openings for further meaningful constraints that had been available in language structure.

12.4.4. *An evolving system*

We now consider the motive sources which, given the existence of words and structures as in 12.4.3, directed the many steps of structural development, and affected their survival. It will be seen that these steps have features of evolutionary processes.

(1) **Toward efficacy of communication**. First, many of the developmental steps increased the effectiveness of communication and transmission of words and then of sentences. When words developed from being sound-

sequences to being phoneme-sequences, they became more distinguished one from the other, and their transmission (as repetition rather than imitation) became more error-free. The use of words in various combinations permitted more meanings to be expressed with a given stock of words.

When word-combination likelihood is institutionalized into word-class dependence, with the unused being now excluded, we have a facilitating (and limiting) contribution toward expressive and communicative effectiveness.[14] This is seen also in various later developments: in the fixed (word-subset) domains for many reductions; in the replacing of custom by requirement (including the case when the absence of an affix means its opposite or complement, as when absence of plural means singular); in the creation of word subclasses for science sublanguages by excluding other words from the subclass's environment; in the requiring of specified sentence-structures and sentence sequences for proof in the language of mathematics.

Communicative and transmissional efficacy is also gained from abbreviation. The grammatical reductions abbreviate certain words when no information loss results (3.2.6, 3.3). They thus reduce the redundancy in a sentence, but at the cost of complexity in decoding and at the cost of adding a constraint— if one first constructs the source sentence and then reduces it.

There is also a question of efficacy in the encoding of a sentence by the speaker when he makes the sentence, and in its decoding by the hearer in recognizing it as a sentence. Encoding advantages may perhaps be seen in the survival of exceptionally frequent forms (e.g. the complementary *am*, *are*, *is*, *was*, *were*, of *be*), and also of common idiomatic phrases (e.g. *For God's sake!*) which are gestural in their semantic character and in their inertness in respect to the grammar. (Such possibilities of survival do not apply to how words are affected by sound-change; the latter affects the pronunciation of phonemes rather than the use of words.)

Decoding advantages are seen in the reduction of distance between an operator and its argument, as when the operator becomes an affix on its argument (*child-like* from *like a child*, *remorseful* from *full of remorse*). So also when secondary sentences which modify a word are inserted after the host word in the primary sentence, as an interruption, rather than being said after the primary sentence (where they would need an address-laden pronoun to identify their host, 3.4.4).

There are also encoding and decoding advantages, in the long run, when the grammatical apparatus becomes as simple as it can be. This is seen in the possibility of a least grammar (one or more, 2.2–3), and in the non-development of a complete addressing system (5.3.2). It is also seen in the fact that the dependence relation (3.1) is characterized not by complicated differences among probabilities but by the simple contrast of zero versus all positive probabilities. And it is seen in the fact that in the base form of many

---

[14] Presumably the immediate advantage of facilitating the communication outweighs, in the ongoing evolution of language use, the long-range disadvantage of limiting the communication.

languages, words have fixed argument-requirements (syntactic classification, G65) rather than having one classification in one occurrence and another in another. (The many English words, such as *walk*, which occur both as nouns and as verbs are not a counter-example: for each word, one status is derivable in a regular way from the other, G66.)

There is also an encoding and decoding simplification when grammatical processes take place as soon as the conditions for them are satisfied, and not after later additions or processes take place. Reductions can be shown to take place, in the course of making a sentence, as soon as the two words between which the high-likelihood condition is satisfied are both present (G207). Also, the selectional limitation of a word, whereby it has higher-than-average frequency only with particular co-occurrents, holds mostly between words and their immediate operators or arguments (G5).

Major systemic simplifications were gained at relatively small structural cost when the (first-level) operator–argument relation in elementary sentences was extended to further (second-level) operators acting on the sentence (or on its operator) as argument, yielding expanded sentences; and when the domain of referents of words was extended to include words and word-occurrences, yielding cross-reference (pronouns) and metalanguage.

The generalization of dependence, once both first-level and second-level operators existed, to being a dependence on dependence, meant that operators were not defined by specific lists of argument words. This left room for words to be dropped or added, or to change in meaning (by changing their selection, or by adding modifiers), and thus made the basic dependence system flexible and stable in the face of language change and development.

Finally, there is the use of analogic construction and the reinterpretation of non-dependence utterances as pseudo operator–argument sentences. These extend the dependence structure over virtually all utterances, and thus simplify the carrying out of grammatical variations on them.

(2) **The stopping of developments**. A telling feature of language is that the universal structure-making processes can stop when further development would add little to communicational effectiveness or to information capacity. For example, the number of positionally successive arguments which operators have stops generally at two, although a few three-argument operators may be found (e.g. in English *put*, *attribute*). Here, given enough two-argument operators acting under further operators, one can get the semantic effect of n-argument operators, to any n: e.g. *He placed the book so that the book was on the table* for *He put the book on the table*. Similarly, there are only two levels of operators: those that act (depend) on words that depend on nothing, and those that act on words that depend on something. There seems to be no clear case in natural language of dependence on dependence on dependence (e.g. a verb that can act only on such verbs as act only on

verbs). Without the second level, language would have only elementary sentences. But by having second-level operators act on second-level operators, we can get the effect of higher-level operators. The reasons for the length and depth of arguments stopping at two may thus be not very different from the evolutionary advantages in biology that let sex differentiation stop at two.

In respect to the stopping of developments, it may be relevant to note a similarity and difference between the evolving of language and biological evolution. In language, communicational ease and distinguishability can be considered to correspond to the success-in-the-environment in the 'survival of the fittest' model; and developments toward such efficacy may be expected to go no further than the environmentally imposed needs. In contrast, the old question of why some evolutionary developments seem to go on beyond what is needed for survival in the environment can be met in biological evolution by taking into account a second 'cutting-edge' which differs somewhat from the demands imposed by the external environment: namely, the effect of intra-species competition, e.g. for territory or in mating selection, and the effect of co-evolution between species or between individual behaviors within a species (e.g. in animal communication, 12.4.1). In this latter cutting-edge, the favoring of particular ones among individuals already changed by favoring up to that point can continue in one direction without obvious brakes. This situation does not arise in language, except possibly in such favorings as for constructions or word uses that are more attention-getting than the current usages. Language therefore does not seem to have runaway directions of development.

(3) **Lack of plan in detail**. Although the development in (1) above may give a global direction to language, many detailed developments are less consistently directional. This limits the regularity and simplicity of the system. For example, many formations are not fully carried out. The differences among the sounds ('allophones') which are complementary alternants of a phoneme are in part regular over various similar phonemes and in part unique; similarly for the morphophonemic alternants of a word. Reductions are also in some cases similar to others, and in some cases unique (e.g. in zeroing *to come* under *expect*).

At any one time, various structural developments are incomplete, and some remain so. Special constructions such as the auxiliary verbs can be the product of cumulative centuries-long changes in form and in restrictions of use; but not all of the words have all the characteristic properties, so that the subclass remains a fuzzy one. Regularization is a tendency, not a plan or a law, and it can stop short at certain points (e.g. at some of the most frequently used suppletions such as *is–am–are*) or at half-way stations (as happened in the development of *do* toward being a pro-verb). Most generally, there is the conjecture of Henry Hiż that every word may have a unique

choice of transformations in whose domain it belongs,[15] supported now by
the discoveries of Maurice Gross on the unique constellations of grammatical
and transformational properties for individual French verbs.[16]

There is also, in various languages, a scatter of required reductions, going
against the general property of optionality of reductions. And various
languages and language families have grammatical complexities at various
points, in ways that are obviously not essential for language as a whole and
that have no measurable communicative import for the particular language.

In the light of all these irregularities in the grammar of any given language,
it is not surprising that in describing the grammar of a language, or in
analyzing particular constructions, we may find things that have non-unique
analyses—that can be fitted into the grammar, or into the general theory of
language structure, in different alternative ways. This does not mean that we
should not analyze language in terms of specific processes, or that there is
unclarity about the effects of each type of constraint.

Another kind of non-plannedness is the under-utilization of structural
potentialities due to cost or inconvenience of use. Thus, as has been noted, a
complete addressing system for sentences is available both in the partial
order of dependence and in the linear order of speech, but only rudimentary
bits of addressing are used (as in *former–latter* referring to parts of an
utterance). Instead, the needs of cross-reference are met, for example, by
the roundabout but simpler addressing of interrupting one sentence (after a
given word in it) with a secondary sentence that begins with the same word
(the basis of modifiers, 3.4.4), and by leaving certain pronouns, such as *he*,
to be ambiguous for the hearer as to the unaddressed antecedent. Also,
language can provide grammatical grounds for factoring certain parts of its
vocabulary into selectionally more independent word-components (e.g. *He
made the recruits walk ten miles* for the 'causative' *He walked the recruits ten
miles*), but again such possibilities do not carry through regularly. Many
languages also place a two-argument operator between its two arguments,
although various ambiguities would be avoided if the operator were either
before or after all its arguments (as in Polish notation).

In the same spirit, languages are not fully adequate for all informational
purposes. Typically, each language has some features of information, e.g.
certain time-relations, which are difficult to express in it, especially in a way
that would distinguish them from other possible interpretations. In English,
aside from various difficult distinctions in tense, there is no tense in nominalized sentences (i.e. when a sentence is the operand for a further operator):
one says *His being there yesterday sufficed*, with no past tense, though

[15] H. Hiż, 'Congrammaticality, Batteries of Transformations, and Grammatical Categories',
Proceedings, Symposium in Applied Mathematics 12, American Mathematical Society, 1961,
pp. 43–50.
[16] M. Gross, *Méthodes en Syntaxe*, Hermann, Paris, 1975, and *Grammaire transformationelle
du français: Syntaxe du verbe*, Larousse, Paris, 1968.

periphrastic tenses are used to circumvent this in English (12.2, 12.3.4). Also, English (differently from some other languages) lacks a third-person imperative: one circumvents this by saying *Let him go there*, which puts it into a second-person imperative with an informationally irrelevant *let*. (Also a first-person imperative in *Let's go*.) Many languages lack the form and meaning of a three-stage comparative: no *John is older than Mary is older than Tom*, though one can say (not using the comparative form) *John's age exceeds the excess of Mary's age over Tom's*. In many languages, too, the speaker's knowledge of the antecedent of a pronoun cannot be communicated to the hearer: in *John phoned Tom yesterday; I know because I bumped into him*, the speaker knows to whom *him* refers, but he cannot tell the hearer without cumbersome additions.

(4) **Accretions**. The structure of language is descriptively accretional: phonemic distinctions, words, word combinations, the dependence relation in word combinations (which makes elementary sentences), the extension from dependence on words to dependence on dependence-properties (which makes a system for complex sentences). The whole system is a resultant of successive contributory departures from equiprobability. The system is constructive in that virtually everything is reached by constraints acting on previous constraints, i.e. built on the resultants of previous constructions. But it is also accretional in that the previous constructions are virtually never lost; they retain some presence in the later resultant. Thus the second-level operators re-enact the process of the first-level operators, and act on the first-level operators, leaving the constructions of the latter (on zero-level arguments) intact.

The reductions are shape-changes within the underlying sentence: the changed—even zeroed—words are grammatically still present, and their dependence structure is retained (or paraphrastically transformed) in the reduced form. (In addition, the optionality of most reductions leaves the original sentences still in the language.) Hence, when we say that a sentence is derived from another, we could better say that it contains that other. Thus the expanded sentence, *I hope he will go*, contains not just the words of *He will go* but also the grammatical relations and meanings of the independent *He will go*. And so in reduced sentences, *I expect John momentarily*, from *I expect John to come momentarily*, still contains the *to come* of *John is to come momentarily*, since *momentarily* is selectionally inexplicable unless the *to come* is still present even if in zero form. The accretional character is also seen in the fact that one can so define the later operators that they never have to alter or correct the information in their operand sentence. In particular, as noted above, *John will not go* can be derived not from *not* on an asserted *John will go* (which it then denies) but from *I deny*, parallel to *I say*, on a composed but not yet asserted *John will go*; *I deny* acts here in place of *I say*. In this way the source is not nullified or corrected but is added to, grammatically and informationally, by every added word.

More generally, the dependence-on-dependence relation is generalized from the direct dependence relation, but without eliminating the latter: within each language in each period, the argument-requirement of each operator can in principle be given as a list of words. Somewhat similarly, transformations act on expanded sentences in the same way they act on elementary ones—and this without affecting how they act on the elementary ones themselves. Also, the later constraints on sentence-sequences, in conjoined sentences (9.1), in the double array of discourse (9.3), and in proof structure (10.5.4), leave the internal structure of the component sentences unaffected.

It should be noted that languages have ongoing changes now too: consider, in English, such local ongoing developments as the tense-like used to (*I used to go there*) and *going to* (*I'm going to go there*), or the more systemic trend to make the 'perfect' (e.g. *has gone*) into a past tense, to free the passive from its completive aspect, and to put tense-indication into nominalized sentences. Changes of this kind can in the long run have an effect on the general structure of the language.

(5) **Form and function**. We have seen that when the whole structure of language is analyzed into its least constraints, we obtain a step-wise connection between each contributory constraint and each contribution to the information in the utterance (11.4). In consequence, the development of form and the development of informational capacity go hand in hand. There are some similarities here to the general, phenomenological, relation of structure and function, especially in living organisms but even elsewhere in the physical world. In biological evolution, structural changes arise independently of any function, by mutations, but their 'survival' depends largely on the contribution of their engendered physiological activity or organismic behavior to the survival and procreation of the organism in which they are located or of individuals with similar genetic equipment. In this way, structure and 'function' are evolutionarily related.

More generally, one might raise the well-known question as to how a direction of development or evolution gets built up in a theory that seems to speak of arbitrary changes. In the case of language the attempted changes themselves may not be arbitrary, but may be due to any of several, perhaps conflicting, tendencies (simplification, colorfulness, precision, etc.), although no such general sources of change had been definitely established. The key supplier of direction, however, may be that each change operates on something in the language as previously changed, so that for example once dependence has been institutionalized any change in word combination would have the effect of creating further detail (likelihood change, transformational shift of word use, etc.) rather than cutting across the previously established structure.

In language, the types and sources of change are studied in historical linguistics and in some sociolinguistic investigations; some changes may arise from the regularizing pressure of the existing system (12.4.2–3). In any case, the survival of the changes depends in part on their contribution to the informational and communicational efficacy of the linguistic system. In this way, the relation of structure and function, which originates in the form–content correspondence of 11.4 and 11.6, is maintained through the history of language change.

In (1)–(5) above, we see processes characteristic of evolving systems. The whole is of course very different from biological evolution, the more so since language lacks essential properties of organisms, such as necessary death, descendants, and speciation. The source of variation is not mutational or 'random' in the case of language; in language we have to explain not only the survival-value of a form, but also its source—how it came to be. (The sources of variation would be considered 'random' if they are unrelated to what affects survival.) Also the driving-force or 'cutting-edge' of evolutionary change is different, though with suggestive similarities. In both cases, a variant could take hold primarily when it satisfied certain external conditions. In biological evolution, as noted above, one can distinguish on the one hand natural selection's species-survival which includes increased opportunity of having descendants for the individual or for others in his group (some of whom may be expected to have much the same as his genetic equipment), and on the other hand mating-favoring properties which also increase the chances of having descendants, but which are less limited to what is favored by the environment, and indeed may be in general more unlimited.

In language, a slightly comparable distinction exists between informational efficacy and competition within an informational niche. The efficacy criterion holds that survival of a linguistic form (phonemic distinction, word, word class, construction, reduction, etc.) is favored if it is more efficacious in communicating information—including here the cost to speaker (especially in encoding) and to hearer (in decoding). This efficacy can be understood as success in eliciting a response appropriate to the speaker's message; this formulation skirts the subjectivity issue (in animal communication, is the 'message'-issuing 'speaker's' cry or posture subjectively communicational, or is it just an expressive behavior which constitutes a stimulus eliciting a particular response?). In the competition for niche, sentences containing a particular form may be favored in a particular language community due to a variety of properties: they may be clearer, more forceful or contrastive, easier to say or to understand, more comparable systematically to related sentences, etc.—all in a never-ending series or escalation of innovation.

The changes due to efficacy are for the most part long-range, universal and basic for language, more regular with very few exceptions, and may present the long-range picture of a progression. The changes due to niche-competition are relatively short-range, and particular to individual languages, language families, or periods. They are less closely bound to—and limited by—informational efficacy, and if some have a directional character, it is rather like cultural trends, cumulative to the point of constituting the 'genius' of a language, or its 'drift'.

To all this must be added that the definability of a metalanguage and of cross-reference within a language and the double array of discourse structure, the specializing of science sublanguages, and the defining of logic and mathematics as linguistic-type systems, all show that major informational capacities can still be evolved out of the sentence-making structure.

Throughout, we can see that the structure of language relates in part to its history, not only in the specific relation of language change to language structure but also in the fundamental relation of the structure to the evolution of language.

More generally, while language is a subset of general cultural behavior, it develops under the shaping force of having it measure up against what would be effective in communicating information about the perceived world. Both of these conditions give the impetus and direction for the origin and development of language. Different factors thus shape the development of language than that of other behaviors, and occurrences of language are so readily distinguishable that they can be studied separately from other events, to see their systemic character.

12.4.5. What is needed for language to exist?

The properties of the self-organizing and evolving systems discussed above show language to be a self-contained stable system, resting on no prior framework, not even on a prior space in which its events are located, and creating in itself the metalanguage that can describe it. However, whereas a metalanguage that describes a system can be created after the system exists, a metalanguage that forms or generates a system would have to exist before the system does. We have seen that the latter is impossible in the case of language because the metalanguage of a language is itself a linguistic system (2.1). In accord with this, we have seen that the non-equiprobability of word combination exhibits the elements and their combinabilities even in the absence of a metalinguistic generator. Then in the absence of a metalinguistic generator we understand the various departures from equiprobability, which together comprise the total non-equiprobability, as being each formed by a particular constraint. We thus see language not as a single object or whole system, but as the cumulative product of constraints on word-occurrence, which could arise out of a language-less situation, and which developed in the further constraints of Chapters 9 and 10 above, and in the course of historical changes (such as are studied in historical and comparative linguistics, and in linguistic change).[17]

[17] In considering the developing use of language both in the species and in the individual, it may be noted that the descriptive stages of sentence construction, which provide the apparatus

This picture is of some relevance if one considers the early development of language, about which virtually no evidence exists. There could of course have been a single source, developing with the development of early man from some point onward, with the secondary (and often less regular) constraints of particular language families arising later in separated communities. However, the self-organizing and evolutionary character of almost all the structural developments suggests that the universal constraints in language could possibly have arisen independently, or quasi-independently, in separated communities.[18] In either case, it is plausible to think that the development of sentencehood—the dependence relation—accompanied a major social or productional development in early man.

In any case, it was the arising of constraints that constituted the development from words and unstructured co-occurrences of words to sentences and to the existing language structure, in a situation where there was as yet no language having that structure. These constraints are more regular than the word combinations they produce, because complexities and non-regularities can be relegated to their particular domains.

Furthermore, while language structure is certainly unique, the constraints that create it are not unique except in detail. They are not so different from some constraints in other aspects of human life. For example, grammatical relations, such as being the subject or object of a verb, exist nowhere but in language. But to have the occurrence of certain things depend on classes of other things is a kind of dependence that may not be unique to language—yet such is the constraint that makes particular word-occurrences into subject or object. Also, language is a structure of demands: some combinations are in it and therefore correct, while others are not in it and therefore wrong. But these demands, as was seen in 12.3.5, can be seen as institutionalizations of less demanding and more naturally-occurring customary uses in word combining. There are demands that are unique to language; but they can be reached by institutionalization of custom—of customary combining of particular bits of behavior. That bits of behavior should be combined, and in customary ways, is not unique. And for custom

---

of O_o (operator on sentence) and O_{oo} (operator on two sentences), can be considered to parallel the grammatical apparatus needed to describe neural processing and recall of language material. In neural processing and recall, the need to recognize, classify, and locate each sentence parallels the O_o operator class, and any processing of associations between two sentences parallels the O_{oo}. In the spirit of Harry Stack Sullivan's view that what is unconscious is what has not had to be raised to consciousness, the apparatus of bringing memories into awareness and recall involves the ability to process the memories with an apparatus paralleling O_o and O_{oo}; and it could be facilitated by initial sentential deposit of the memories. This is in contrast to involuntary recall after brain injury, where no specific association (O_{oo}) seems to be involved.

[18] This picture of how language could have developed does not mean that a different system could not have arisen instead to fill the same functional niche. Language is not like the integers, which have properties independent of people knowing them, and for which we cannot imagine alternative structures to have been able to develop in any human community.

to be institutionalized, whether for convenience of living or for social control, is no novelty in culture and in social organization. The effect of this view of language is to avoid both the non-explanation of saying that language is just something *sui generis*, and also the reductionism of saying that language is no more than some simpler communicational system.

The lack of uniqueness is seen not only in the constraints that create language but also in the knowledge or capacities needed to speak and understand language, in the absence of any metalinguistic direction. The constraints of Chapters 3–5 indicate all the kinds of departures from equiprobability which carry the grammatical structure and the information of a language. They are ordered in respect to each other, so that we know which has a more pervasive effect, and so that we can avoid counting any constraint or its effect twice.

We can therefore count all the kinds of constraint in a language; and in any sentence we can count all the constraints that go into making it. We have to count, for a particular language: (a) its phonemes (its phonemic distinctions and their allophones); (b) a finite stock (up to some limit) of those phoneme-sequences that constitute words or morphemes at a given time, with the main meanings associated with each; (c) the argument-requirement for each word (in most cases fixed for all base-occurrences of the word); (d) roughly the selection for each (i.e. with what operator or argument words the given word has higher-than average co-occurrence, yielding its environment-specific meanings; this is primarily a set of word-pairs); (e) the main reductions in phonemic shape, with their conditions for application and their domains (which may be individual words, or all words in a given position); (f) also the linearization of the operator–argument pairs or triples, and the alternative linearizations.[19] While no one will count all this, it is clear that the amount of knowledge, and—what is more important—the mental capacity required for using and sequencing the particular kinds of knowledge, can be within the ordinary mental capacities of human beings.[20] Indeed, all of these contributions to the sentences of a language are specifically learned by each speaker and hearer (over a given vocabulary), first simply by long

[19] This catalog is not entirely the same as the least-grammar form of characterization required in the present theory. And neither of these is entirely the same as the way an infant or an adult learns a language, or the way language came to be historically, or the way language information is processed in the brain.

[20] How the brain processes both the speaking and the recognizing of speech, and what specific capacities of the brain are involved, is still little known. Clinical research in aphasia and neurological research have both shed some light on the problems. For relevant evidence from aphasia, see in particular: A. R. Luria, *Traumatic Aphasia*, Mouton, The Hague, 1970; M. Critchley, *Aphasiology and Other Aspects of Language*, Arnold, London, 1970; K. Goldstein, *Language and Language Disturbances*, Grune and Stratton, New York, 1948; H. Head, *Aphasia and Kindred Disorders of Speech*, Macmillan, New York, 1926. In any case, what has been seen in the present book about the relevant structure of language suggests that it is not a direct reflection of the structure of neural action in the brain. Rather, language may be understood as a particular relation of its own—constraints, above all dependence—among neural events and mechanisms.

exposure to the constrained word combinations in the language, and perhaps later by metalinguistic (grammatical) description.[21]

This Gedankenexperiment shows that there is no single 'language capacity' involved in the ability to use a language, but rather that language use is a learned activity drawing on various otherwise-existing mental capacities. It also shows that perhaps none of the mental capacities involved are solely linguistic. If one of them is, it is the association of particular words (or sound-sequences) with certain perceptions of the world; but this is the one language-type capacity that animals have too (though to a far more limited extent, and normally as a direct response to a stimulus).[22]

In spite of the mass of detail and of complex resultant forms, what characterizes language is a small set of relatively simple and regular constraints which created phonemes and sentences and further constructions on the basis of words, and which constitute the set of behavior patterns which a person has to acquire in order to use a language freely. No special structuralism or capacity is required; in any case, it would be inexplicable how such could come into existence before language itself existed.

[21] This survey indicates what is needed in order to use a whole language. To understand an individual sentence, all that is essential is a recognition of the phonemes of the language and the argument-requirements of the words present, hence (a) and (c) above. This is seen, for example, in the fact that any finite syllable-satisfying phoneme-sequence of a language is acceptable as a sentence if we can take an initial sub-sequence in it to be some name or newly created technical object and the remainder to be some action or state of that person or object; preferably either the initial segment should end in the phonemes of the plural suffix or the second segment should end in the phonemes of a singular tense suffix.

[22] The distinction made in 12.4.1 between human and animal language does not mean that human language requires a totally different *sui generis* principle, nor does it deny the possibility of evolutionary continuity from animal communication (cf. D. R. Griffin, *The Question of Animal Awareness*, Rockefeller Univ. Press, New York, 1981). The fundamental new relation (dependence) among 'symbols' (words), and the far greater stock of symbols (with their varying phonemic shapes), can presumably all be learned within the general structure of the so much more massive and complex human brain. It should be noted in this connection that there is no present evidence as to how old is human language as a syntactic system. Since a large stock of words may have existed for a long time before the syntax-creating dependence relation developed, such evidence as there is for the existence of Broca's area in the brain of early man (R. E. Leakey and R. Lewin, *Origins*, Dutton, New York, 1977, p. 190) may mean only that a large stock of words was then growing. Syntactic language, with its facilitation of complex sustained thought, might have developed only much later, perhaps at the time of the rapid growth of material culture and co-operative production.

Index